Optical Components, Systems, and Measurement Techniques

OPTICAL ENGINEERING

Series Editor

Brian J. Thompson

Provost
University of Rochester
Rochester, New York

Optical Components, Systems, and Measurement Techniques

RAJPAL S. SIROHI
MAHENDRA P. KOTHIYAL

Department of Physics
Indian Institute of Technology
Madras, India

Marcel Dekker, Inc.   New York • Basel • Hong Kong

Library of Congress Cataloging-in-Publication Data

Sirohi, Rajpal S.
   Optical components, measurement techniques, and systems / Rajpal S.
Sirohi and Mahendra P. Kothiyal
      p.   cm. -- (Optical engineering ; v. 28)
   Includes bibliographical references and index.
   ISBN 0-8247-8395-6
   1. Optical measurements.   2. Lasers.   3. Optical instruments.
I. Kothiyal, Mahendra P.   II. Title.   III. Series: Optical
engineering (Marcel Dekker, Inc.) ; v. 28.
TA1522.S57   1990
621.36–dc20   90-20940
                                                               CIP

This book is printed on acid-free paper.

MARCEL DEKKER, INC.
270 Madison Avenue, New York, New York 10016

Current printing (last digit):
10 9 8 7 6 5 4 3 2 1

PRINTED IN THE UNITED STATES OF AMERICA

About the Series

The series came of age with the publication of our twenty-first volume in 1989. The twenty-first volume was entitled *Laser-Induced Plasmas and Applications* and was a multi-authored work involving some twenty contributors and two editors: as such it represents one end of the spectrum of books that range from single-authored texts to multi-authored volumes. However, the philosophy of the series has remained the same: to discuss topics in optical engineering at the level that will be useful to those working in the field or attempting to design subsystems that are based on optical techniques or that have significant optical subsystems. The concept is not to provide detailed monographs on narrow subject areas but to deal with the material at a level that makes it immediately useful to the practicing scientist and engineer. These are not research monographs, although we expect that workers in optical research will find them extremely valuable.

There is no doubt that optical engineering is now established as an important discipline in its own right. The range of topics that can and should be included continues to grow. In the "About the Series" that I wrote for earlier volumes, I noted that the series covers "the topics that have been part of the rapid expansion of optical engineering." I then followed this with a list of such topics which we have already outgrown. I will not repeat that mistake this time! Since the series now exists, the topics that are appropriate are best exemplified by the titles of the volumes listed in the front of this book. More topics and volume are forthcoming.

Brian J. Thompson
University of Rochester
Rochester, New York

Preface

Optical techniques, because of their noncontact and noninvasive nature, have been employed for the measurement of a variety of parameters. The advent of lasers has extended the range and accuracy of the measurements. Some new measurement procedures have also been developed. The material on these techniques is largely available but scattered in journals and books. This book aims at consolidating such material in one place. For the sake of completeness and to provide a basic background to the reader unfamiliar with optical principles, some discussion on light sources, including lasers, detectors, optical components, and basic optical systems, has been included.

The contents of this book evolved as a result of lectures offered to engineering graduate students and students at short-term schools over a period of many years. Active participation by the students has led to a significant improvement in the quality of the contents. The book will appeal to both students and practicing engineers.

The first three chapters provide basic background in optics and lasers. The first chapter deals with thermal and laser sources, detectors, and recording materials. A few laser sources are treated in detail in view of their use in metrology. Semi-conductor laser diodes are discussed for their applications in fiber optic (FO) communication and FO sensors. Photo-detectors are also discussed due to their wider applications. The materials required for holography have been summarized. Various optical components are discussed in Chapter 2. These include refracting, reflecting, diffracting, thin-film, and polarization elements. Some simple optical systems are described in Chapter 3.

Optical techniques play an important role in the measurement of length, angle, and alignment. Chapter 4 and 5 are devoted to these techniques. Heterodyne and phase-shifting techniques have emerged as powerful measurement tools in interferometry. The basic principle and some applications are covered in Chapter 6.

For a variety of NDT applications and stress analysis, holography and speckle techniques have proved invaluable and are being routinely employed. Chapter 7 is devoted to these useful optical techniques. Optical processors perform better and faster in many applications. Chapter 8 describes the FT processor and its capabilities, image subtraction techniques, and other operations.

For stress analysis, photoelasticity is a well-established technique. Holography has supplemented this technique. Chapter 9 contains a discussion of various aspects of photoelasticity. Freedom from RF, EMI, and EMP disturbances and other useful characteristics have made fiber optic sensors extremely attractive. Use of fiber optic components for sensor applications is the subject matter of Chapter 10.

Finally, a brief discussion on a few other important optical techniques is given in Chapter 11. These techniques include optical surface sensing, surface evaluation, ellipsometry, and laser anemometry.

Cooperation from the members of the Applied Optics Laboratory of IIT, Madras, in the preparation of the manuscript is gratefully acknowledged. We have also recieved help and cooperation from many of our friends and teachers, and we would like to express our gratitude to them. Comments and criticism from the readers are welcomed.

Rajpal S. Sirohi
Mahendra P. Kothiyal

Contents

Contents

Optical Components, Systems, and
Measurement Techniques

CHAPTER 1

Light Sources, Detectors, and Recording Media

1.1 LIGHT SOURCES

Light is that portion of electromagnetic spectrum that is responsible for the sense of seeing. Light can be produced by suitable conversion of other forms of energy. Nuclear reactions at the core of the sun and the other stars produce an almost inexhaustible supply of light energy. Artificially, light is produced by heating, electrical discharge, electrical current injection, etc. Since our visual experiences are due to light, and since also many physical processes can be understood by the light generated, the measurement or quantification of light is essential [1-3].

It may be mentioned that radiation sources as standards of radiance and of wavelength are used for calibration. Some applications require pointlike sources of high irradiance, while others may require high irradiance sources regardless of size. We shall discuss only a few different types of sources.

1.1.1 Black Body Sources

The energy density $\rho(v)$ per unit frequency range from a blackbody is given by

$$\rho(v) = \frac{8\pi v^2}{c^3} \frac{hv}{e^{hv/kT}-1} \tag{1.1}$$

The energy density does not depend on the size or material of the body but only on its temperature. The total energy over the whole frequency range is

$$W = \int_0^\infty \rho(v)dv = \sigma T^4 \frac{W}{m^2} \tag{1.2}$$

where σ is the Stefan's constant.

Blackbody radiators are only idealizations. They can be approximated by cavity radiators. A cavity radiator consisting of a graphite rod with a coaxial cylindrical hole having a large length-to-diameter ratio and walls having an isothermal temperature profile is a suitable radiation source up to temperatures of 3000K. The graphite wall is resistance-heated and temperature uniformity is achieved by a variable heat source distribution obtained by varying the cavity wall thickness. The rod is radially surrounded by radiation shields. Insulation by radiation shield is very effective. The graphite radiator is to be operated either in a vacuum or in an inert gas atmosphere. In the temperature range of 1200 to 2200 K, it can be heated in vacuum. For higher temperatures (>2200K), the graphite is heated in argon gas atmosphere.

Calibration of radiation detectors, particularly radiation pyrometers, is done with blackbody radiation. Calibration of halogen lamps and of ultraviolet UV lamps used for cosmetic surgery and pharmaceutical research is also carried out with blackbody radiation. Moreover, blackbody radiation can be used as a standard for the transmittance and reflectance measurements of optical elements such as windows, filters, prisms, mirrors, etc.

1.1.2 Filament Lamps

Filament lamps come in a wide variety of shapes, sizes, and power ratings. They are both vacuum and gas filled and are used in a large number of applications. The lamps used as intensity standards are carefully manufactured; the filament plane is well defined, and the base is accordingly fitted. Further, the filament is aged. Other applications may also require planar filaments, but the requirements are not so stringent. For example, the projector lamps usually have flat filaments; normally, halogen lamps are used. Halogen lamps can be operated at higher color temperatures, because the halogen inside the envelope prevents oxidation of tungsten filament.

1.1.3 Arc Lamps

The earliest arc lamp was the carbon arc, an open arc; it was and is widely used for its high radiation and color temperatures ranging from 3600 K to 6500 K. It contains two electrodes in which the arc is maintained. The electrodes are moved toward each other to compensate for the rate of consumption of the material. The rate at which material is consumed (5 to 30 cm/hr) depends on the intensity of the arc. The anode forms a crater of decomposing material that provides a center of very high luminosity. Some electrodes are hollowed out and filled with a softer carbon material to help keep the arc fixed in the anode. The carbon arc is used in three forms: the low intensity arc, the flame arc, and the high intensity arc. The low and high intensity arcs are generally operated on DC; the flame type adapts to either DC or AC. In all cases a ballast must be used.

1.1.4 Enclosed Arcs

The carbon arc is not very efficient and has a short life; and moreover undesirable combustion products during operation are produced. Therefore enclosed arcs, particularly mercury arcs, are preferred. A coiled tungsten cathode is usually coated with thorium, and an auxiliary electrode is used for starting. A high resistance limits the starting current. Once the arc is started, the operating current is limited by the ballast supplied by the high reactance of the power transformer.

Multivapor arcs are also used; they contain other materials besides argon and mercury. Also available are compact arc lamps. Extreme electrical loading of the arc gap results in very high luminance. The lamps have internal pressures of a few atmospheres, and bulb temperatures go as high as 900°C.

1.1.5 Concentrated Arc Lamps

Zirconium arc lamps use cathodes made of a hollow refractory material containing zirconium oxide. The anode, a disk of metal with an aperture, resides directly above the cathode with the normal to the aperture coincident with the longitudinal axis of the cathode. Argon gas fills the tube. The arc discharge causes the zirconium to heat to about 3000K and produce an intense, very small source of light.

The tungsten arc lamp is also used as an intense and small source of light. It contains a ring electrode and a pellet electrode, both made of tungsten. The arc forms between these electrodes and causes the heating of the pellet to incandescence. The ring also incandesces but to a lesser extent.

1.1.6 Discharge Lamps

^{86}Kr lamp: One discharge lamp that remained as a standard of wavelength until 1983 is the ^{86}Kr lamp. It consists of a capillary with an internal diameter of 2 to 4 mm and a wall thickness of about 1 mm; it is filled with ^{86}Kr and run at the triple point of nitrogen. The recommended current density is 0.3 ± 0.1 A/cm^2. The wavelength of the radiation arises from the transitions between the $2p_{10}$ and $5d_6$ levels of the isotope of krypton. The vacuum wavelength is given as $\lambda_{vac} = 6057.80210$ Å. The meter is then defined as 1,650,763.73 wavelengths.

Low-pressure lamps containing cadmium, mercury, thallium, sodium, etc. are used as spectral lamps in many spectroscopic and measuring instruments.

There are other kinds of sources that are used for certain applications. For example, UV lamps are used for cosmetic surgery and photodynamic or photodissociative work in many industries. Similarly, infrared lamps may be used for other curative applications.

1.2 LASER SOURCES

The laser [4,5,6,7] is a source of coherent radiation. It consists of an active medium in a cavity. The active medium may be solid, liquid or gaseous. The cavity is an open resonator bounded by two mirrors, one of which is partially transmitting. When the active medium is excited, the atoms or molecules are raised to higher energy levels. They relax to the ground state by spontaneous emission. It was shown in 1917 by Einstein that there is another process of relaxation called the stimulated emission. If the population of higher energy level is raised to the extent that it is more than that of the lower level, a population inversion is created between these levels. A spontaneously emitted photon will cause the excited atoms to relax and release photons in phase. These photons bounce back and forth in the resonator, and the wave grows; a part of the wave is transmitted through the output mirror forming the useful laser beam. In order to create the population inversion the atoms/molecules are to be raised to the higher energy level. This is achieved by the process called pumping. Solid lasers in which the active centers are embedded in a dielectric matrix are optically pumped by intense light pulses or sometimes by intense continuous sources. Almost all gaseous lasers are electrically pumped. A CO_2 laser can also be pumped by adiabatic expansion, resulting in a gas-dynamic laser. All semiconductor lasers are pumped by current injection. Some solid lasers are pumped by semiconductor lasers. Other forms of energy, for example solar, chemical, etc. have been used for pumping. Figure 1.1 shows a schematic of a laser. The laser may also contain some other elements like a Q-switch, a prism, an etalon, etc. A classical treatment of the physics of lasers can be found in a number of books [4,5,6,8].

A passive resonator can support infinite modes. All those modes that satisfy the resonance condition $2d = m\lambda$ are supported, where d is the separation between the mirrors. These are called longitudinal modes and are separated by $c/2d$ in frequency. When the active medium is present in the resonator, then only those modes that lie within the gain curve are supported, as shown in Fig. 1.2. The gain curve is the fluorescent line profile above the loss line. For a Ne line at 633 nm, the width of the fluorescent line is about 1500 MHz. If the gain curve has large width, more of these modes are supported.

These modes have random phases, and hence energy in each mode is less. The modes can be made to behave by the process of the "survival of the fittest" if we introduce periodic losses in the cavity. The phases of all these modes will then be constant and hence the output will be a high energy pulse. This procedure is known as mode locking. Extremely narrow pulses can be produced by this method [9].

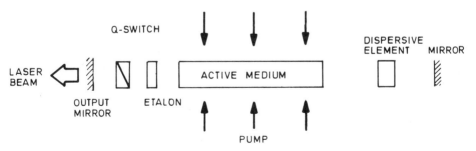

Figure 1.1   A schematic of a laser.

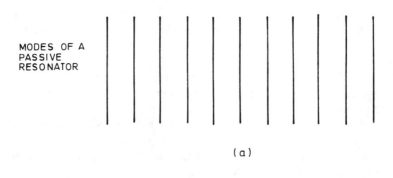

MODES OF A
PASSIVE
RESONATOR

(a)

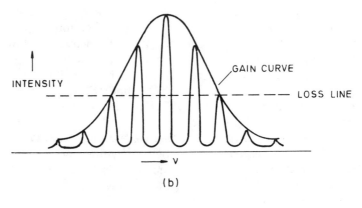

INTENSITY

GAIN CURVE

LOSS LINE

ν

(b)

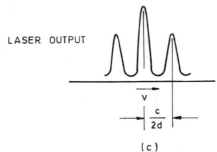

LASER OUTPUT

ν

$\dfrac{c}{2d}$

(c)

Figure 1.2 (a) Modes of a passive resonator, (b) gain curve, and (c) multimode output from a laser.

The He-Ne laser is the most frequently used source both in industry and in the laboratory [6,7,10]. It radiates usually at 633 nm but can be made to lase at a number of wavelengths in the visible region. The output of the laser varies from a few tens of μW to 100 mW or so. The laser usually operates in fundamental mode. It is a very high radiance source. The radiance of the source is defined as the power radiated per unit area per unit solid angle. A power of 1 mW in a beam approximately 1 mm in diameter and of 1 mrad of divergence corresponds to a radiance of approximately 10^8 W/(cm$^2 \cdot$ steradian) in a single transverse mode. The laser can be operated in a single longitudinal mode giving a coherence of thousands of meters.

An argon ion laser can be operated at a number of wavelengths in the visible region and delivers a power output in the range of a few watts. Sufficiently large coherence length is obtained using thermostatically controlled etalons for use in hologram interferometry and the generation of optical elements on photoresist and on gelatin.

Ruby, because of its extremely narrow spectral line, is used for hologram interferometry. Equal intensity but variable time delay pulses can be easily achieved for studying transient phenomena holographically.

Metal/material processing is done with high power lasers. CO_2 operates at 10.6 μm and is used for metal working such as cutting, etc. Nd:YAG and Nd:glass, operating at 1.06 μm, are used for cutting, drilling, and other applications.

The details with reference to the technology of some of the common lasers are given here.

1.2.1 The He-Ne Laser

The He-Ne [7] laser is the most commonly used laser; it has a continuous power output from a fraction of a mW to tens of mW. It is relatively easy to construct and is reliable in operation. The laser transitions occur in the neutral Ne atom; the He plays the role of selectively populating the pertinent energy states of Ne so as to produce population inversion. There are three principal laser transitions at: 0.6328 μm, at 1.15 μm, and at 3.39 μm.

This laser has, however, been operated to lase at discrete wavelengths in the whole visible region. A laser named Ge-Ne and operating at 543.5 nm is commercially available.

The energy level diagrams of He and Ne are given in Fig. 1.3. That of Ne is represented by Paschen notation and that of He by Russel-Saunders (R-S) notation. The Ne states shown in Fig. 1.3 correspond to the following electron configuration:

States	Electron configuration	No. of levels
ground	$2p^6$	1
$^1s_{2-5}$	$2p^53s$	4
$^2p_{1-10}$	$2p^53p$	10
$^2s_{2-5}$	$2p^54s$	4
$^3p_{1-10}$	$2p^54p$	10
$^3s_{2-5}$	$2p^55s$	4

Mechanism of Population Inversion

The He-Ne laser is excited by a stationary glow discharge fired by direct current or by radio-frequency current. The mixture is at a pressure of about 1 torr; the partial pressure of the He is 5 to 10 times that of the Ne. Helium is used to populate selectively the Ne levels. Free electrons of the gas discharge collide with the helium and neon atoms to excite them by impact energy transfer. Many of these helium atoms collect in metastable 2^1S and 2^3S levels. Since the only lower lying state is a singlet ground state, 1^1S, no

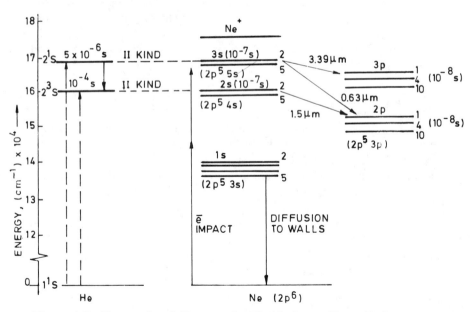

Figure 1.3   Energy level diagram for He-Ne laser. (From Ref. 16.)

transitions from 2^3S to 1^1S are optically allowed. These metastable states of He nearly coincide in energy with the 3s and 2s states of neon. The 2^3S state of He lies only 300 cm^{-1} (0.039 eV) above the $2s_2$ level of Ne, and the 2^1S state of He lies about 375 cm^{-1} (0.048 eV) below the $3s_2$ level of Ne. The He atoms in the mestastable levels can thus tranfer their energies to the Ne atoms at ground state through collisions of the II kind thus raising them to 2s and 3s states. The excitation process can be represented as

He + e $\rightarrow$ He*

He* + Ne $\rightarrow$ He + Ne*

The small differences in energy are taken up by the kinetic energy of the atoms after collision. This is the main pumping mechanism in a He-Ne laser. The upper laser levels 3s and 2s have lifetimes of the order of 10^{-7}s and lower laser levels of the order of 10^{-8} s. Thus the conditions sutiable for laser action exist.

Oscillation at 0.6328 μm

The 0.6328 μm transition occurs between the $3s_2$ level and the $2p_4$ level of Ne and lies in the red region of the spectrum. The terminal 2p group of levels decays radiatively with a lifetime of about 10^{-8}s to the long-lived 1s state. Therefore the atoms tend to collect in the 1s state. These atoms are excited by collision with electrons to the lower laser level $2p_4$ resulting in reduction of inversion. To avoid this, the atoms in the 1s state are brought to the ground state by collisional deexcitation with the walls of the tube. For this reason the gain in the 0.6328 μm transition is found to increase with a decrease in tube diameter.

Oscillation at 1.15 μm

The 1.15 μm transition occurs between the $2s_2$ level and $2p_4$ level of neon. The $2s_2$ level is populated by the resonant transfer of energy by He atoms in the 2^3S state to the neon atoms in the ground state. The lower laser level for this transition is the same as that of the 0.6328 μm transition. Thus it tends to quench the visible laser transition. Further, the gain at this transition also depends on the tube diameter.

Oscillation at 3.39 μm

The 3.39 μm transition occurs between the $3s_2$ level and the $3p_4$ level of neon. It shares the upper level with the 0.6328 μm laser transition. It has a very high gain of about 50 dB/m, which is attributed to two factors, gain that increases quadratically with wavelength and the short lifetime of the $3p_4$ level. Because of the high gain in this transition, oscillations normally will set in at 3.39 μm rather than at 0.6328 μm. The laser oscillation at 0.6328 μm is obtained by suppressing the gain for 3.39 μm by introducing elements like Brewstor windows of quartz or glass that absorb at this wavelength. Because of the high gain of the 3.39 μm transition, the laser at this wavelength can operate in a mirrorless configuration, or in superradiant mode if the discharge tube is sufficiently long. For this reason, He-Ne lasers oscillating at 0.6328 μm seldom use tubes exceeding 1.5 to 2.0m in length.

The gain for these transitions are listed here.

Wavelength of the transition (μm)	Gain (dB/m)
0.6328	0.4
1.15	0.4
3.39	25–50

Frequency Stabilization

The frequency stabilization of a laser is defined [11] as

$$S_v(\tau) = \frac{v_L}{\Delta v_L(\tau)}$$

where v_L is the average frequency of the laser and $\Delta v_L(\tau)$ is some measure of frequency fluctuation during the period of observation τ.

The long term wavelength stability or frequency stability of the laser is mainly determined by the stability of the resonator. Because of temperature changes, the cavity length varies. This is primarily responsible for the long-term drift in the laser frequency. For applications in metrology, and in particularly long-path interferometry, long-term wavelength stability is required and lasers have been developed that provide stabilized wavelength output. Usually the gain curve of lasers and the absorption line of molecules are used for stabilizing the oscillation frequency of the lasers.

Gain Curve Frequency Stabilization

Stabilization schemes based on gain curve [12] fall under two groups, modulation and nonmodulation, depending on the method of generating the error signal for the control of cavity length. In the modulation scheme, one of the mirrors of the resonator is mounted on a piezoelectric transducer on which the sinusodial wave voltage is applied. This modulates the laser cavity, and as a result the laser intensity varies. This intensity modulation can thus be used to servocontrol the frequency of the laser at the peak of the gain curve.

In the nonmodulation scheme, an internal mirror laser that supports two longitudinal modes is used. These two modes are orthogonally polarized and hence can be separated using a polarizing beam splitter (Wollaston prism). The intensities of these modes are compared and used to generate the error signal. When the cavity length changes, these modes drift in the gain curve, the result being unequal intensity. The error signal is used to change the tube current to bring the cavity length to the original value. In other lasers, an external heating element is wrapped around the tube and the laser is operated at a temperature higher than the ambient temperature. The error signal changes the current in the heating element and consequently the temperature to keep the cavity length constant. Long-cavity-length lasers are frequently stabilized by using either an internal or an external mode selector [13]

In another scheme, a single transverse and single longitudinal mode He-Ne laser is stabilized using Zeeman splitting [12]. The single-longitudinal-mode operation is achieved by using a mirror spacing of approximately 13 cm. An axial field is applied to the laser. The output consists of right and left circularly polarized waves of slightly different frequencies as shown in Fig. 1.4. First order theory predicts that the frequency splitting will be proportional to the magnetic field strength and the ratio of line Q to cavity Q. The two waves are of equal amplitude when the mode splitting is symmetrical about the line center. The proper cavity tuning can be obtained by maintaining equal intensities of the two waves. The waves are separated by a polarization beam splitter. A typical example is a Hewlett-Packard stabilized laser that operates in two modes by Zeeman effect. The magnitude of the axial magnetic field is such that the frequency difference it produces is about 2 MHz. This laser shows a good stability of 10^{-9} and operates without modulation.

Lasers stabilized using the gain curve are inferior in frequency stability by at least one order of magnitude to those stabilized by absorption line. They are used as secondary standards of length, and meet fully the accuracy requirements of industrial measurements.

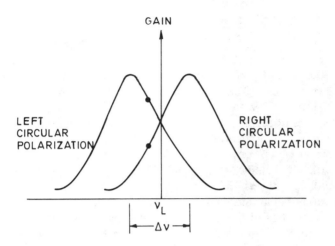

Figure 1.4   Zeeman splitting of Ne line.

Stabilization by Saturated Absorption

Stabilization by saturated absorption involves placing inside the laser a tube containing a gas at low pressure. The gas has an absorption line within the gain curve of the laser. The strong standing-wave electric field inside the cavity gives rise to a saturation of the absorption, and a Lorentzian-shaped dip forms at the center of the line. The laser is stabilized to the center of the dip in absorption. Frequency stability better than 1 part of 10^{10} may be expected using suitable molecular lines.

The 3.39 μm line of the He-Ne laser is stabilized with the methane absorption line. In the visible region, the He-Ne laser is stabilized on $^{127}I_2$ absorption lines [14].

The lasers stabilized using absorption lines have better stability because (a) absorption takes place from the ground state of the molecule, (b) the molecular lifetimes involved are much longer than those of Ne-laser-level life-times so that there is a much smaller natural width, and (c) the pressure broadening of the dip is small since absorption gas at low pressure is used. Because of better frequency stability, these lasers have been adopted as primary standards of length. The accuracy with which the laser wavelength is determined is far better than that of ^{86}Kr.

Table 1.1 gives a summary of stabilized He-Ne lasers. The frequency and wavelength of iodine- and methane-stabilized He-Ne lasers are given in Table 1.2. The wave length calibration is easily performed by using the

Table 1.1   He-Ne Laser Stabilized by two Methods

By gain curve	By absorption curve
Nonmodulation	Iodine($^{127}I_2$) stabilized at 0.63 μm and 0.61 μm
Two mode intensity comparison at 0.63 μm and 1.15 μm	Methane (CH_4) stabilized at 3.39 μm
Modulation	
External mirror at 0.63 μm and 1.15 μm	

beat-frequency method with these lasers. Their complexity, however, does not permit their use in the industrial environment.

1.2.2 Ion Lasers: The Argon-Ion Laser

The first laser to work on ionic transitions was the mercury-vapor laser, with the oscillations occurring in the transitions of a singly ionized mercury. Since then a number of ionized atom lasers [16] have been invented. Of these, the most useful and powerful are those employing argon, krypton, xenon, and cadmium. We shall discuss the argon-ion laser invented by Bridges at Hughes in 1964 [17]. The Ar^+ laser emits at a number of discrete wavelengths between 350 and 520 nm, the most intense transitions being at 488, 496.5, and 514.5 nm.

Figure 1.5 shows the energy level diagram of the Ar^+ laser. The ground state of the Ar^+ is $3s^2\,3p^5 2p_{3/2}$, which lies about 15.75 eV above the ground state of the neutral atom. The three wavelengths mentioned above correspond to the following transitions:

488 nm	$4p^4\,D_{1/2} \rightarrow 4s^2 p_{3/2}$
496.5 nm	$4p^2\,D_{3/2} \rightarrow 4s^2 p_{1/2}$
514.5 nm	$4p^4\,D_{5/2} \rightarrow 4s^2 p_{3/2}$

The Ar^+ laser can be operated in a pure argon discharge that contains no other gases. The argon gas is ionized by using a high current of 15 to 50 A through the discharge tube; the diameter of the tube is small, so that current densities of 10^3 A/cm^2 result. Further, an axial magnetic field is applied, which results in an increase in the current density and decrease in the number of collisions of ions and electrons with the wall. The ions are

Table 1.2   Frequency and Wavelength of Iodine- and
Methane-Stabilized Lasers

Absorbing molecule	Frequency (MHz)	Vacuum wavelength (fm)	Laser
$^{127}I_2$	473612214.8	632991398.1 ($\pm 1 \times 10^{-9}$)	He-Ne
	489880355.1	61970769.8 ($\pm 1 \times 10^{-9}$)	He-Ne
	520206808.51	576294760.27 ($\pm 7 \times 10^{-10}$)	Frequency doubled He-Ne
CH_4	88376181.608	3392231397.0 ($\pm 2 \times 10^{-10}$)	He-Ne

Source: From Ref. 15, used with permission.

further excited by electron collisions, and an inverted population of the ion energy levels is achieved. Inversion is possibly brought about by the two-step excitation

$$Ar + \bar{e} \rightarrow Ar^+ \text{ (ground state)} + 2\bar{e}$$
$$Ar^+ + \bar{e} \rightarrow Ar^+ \text{ (excited state)} + \bar{e}$$

Very high current densities in the argon-ion laser require specialized construction of the tube. The tube may be made of BeO, which offers many advantages over the other materials. It has staggered off-axis holes in the bore material to provide a gas return path between anode and cathode. Another approach uses segemented metal bores for plasma confinement. Segmented graphite bores are good because of their high thermal emissivity and high sputtering potential.

To dissipate a large amount of heat, almost all CW argon ion lasers are water cooled. Low power CW lasers, in the range of 20 mW, are air cooled. Some Ar$^+$ lasers that can be operated in pulse mode are convection cooled. The average power output is only a few mW, but the peak power could be several watts. The pulse widths are approximately 5 to 50 μs and repetition rates go up to 50 Hz.

The Ar$^+$ laser can be operated in broad band where all line output is available. Use of a dispersive element in the cavity leads to single line operation. Addition of an inter-cavity etalon will yield single-frequency operation; the line width is of the order of 3 MHz.

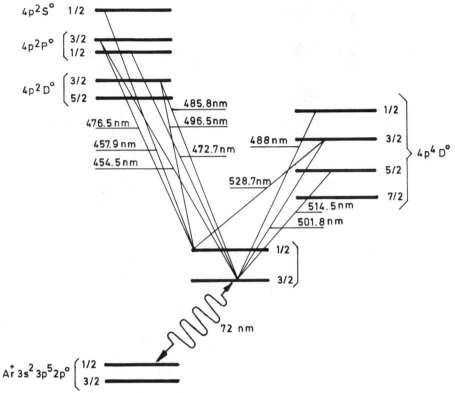

Figure 1.5   Energy level diagram of argon-ion laser. (From Ref. 17.)

1.2.3 The Carbon Dioxide Laser

The most useful and best understood molecular laser is the CO_2 laser [18]. It is capable of continuously generating as much as 10 KW of power at a relatively high efficiency (up to 40%). The active medium of the laser is a gas mixture consisting of carbon dioxide, molecular nitrogen, and diverse additives such as helium and water vapor. The active centers are CO_2 molecules lasing on the transitions between the vibrational levels of the electronic ground state. CO_2 lasers usually use a glow discharge for their excitation.

The CO_2 molecule consists of three atoms, and it can execute three basic internal vibrations of which the bending mode is degenerate. They are shown in Fig. 1.6 (a) [16,19]. These three modes of vibration are in-

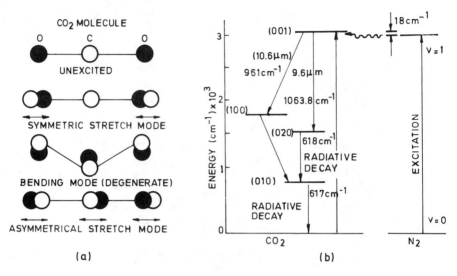

Figure 1.6 (a) Modes of vibration of a CO_2 molecule and (b) pertinent vibrational levels of CO_2 and N_2 molecules. (From Ref. 16.)

dependent of each other. The total energy of the molecules is expressed as

$$E = h\upsilon_1(\tfrac{1}{2} + v_1) + h\upsilon_2(\tfrac{1}{2} + v_2) + h\upsilon_3(\tfrac{1}{2} + v_3)$$

where $\upsilon_1, \upsilon_2, \upsilon_3$ are the frequencies of symmetric-stretch, bending, and asymmetric-stretch modes, respectively. v_1, v_2, and v_3 are integers corresponding to the degree of excitation of the three modes, respectively. For example, the designation (001) indicates that only the asymmetric-stretch mode is excited and that it contains only a single quantum of energy $h\upsilon_3$.

Since N_2 is added to CO_2 for enhancing the power output, we consider the N_2 molecule also. It is a homonuclear diatomic molecule with no permanent dipole moment. It cannot decay radiatively from the v = 1 to the v = 0 level of electronic ground state.

The pertinent vibrational energy levels of the CO_2 and N_2 molecules are shown in Fig. 1.6 (b) suppressing the rotational structure. The laser operates at 10.6 µm, but replacement of the rear mirror by a grating gives a single rotational line output. The laser can be tuned between 8 and 11 µm. At higher pressures, lines overlap and the tuning is continuous. The laser can also be operated in Q-switch mode.

The excitation mechanism of the CO_2 laser is through electron collisions of the CO_2 molecules and the resonant transfer of energy from the

N_2 molecules.

Electron Collision

$$CO_2(000) + \bar{e} \rightarrow CO_2(000) + \bar{e}$$

Electron collisions preferentially populate the (001) level, as the collision cross-section is very large.

Resonant Energy Transfer from the N_2 Molecule

$$N_2(v = 0) + \bar{e} \rightarrow N_2(v=0)+\bar{e}$$

N_2 (v = 1) is a metastable level, and the transition to the ground state (v = 0) is optically forbidden. It relaxes to the ground state by resonant transfer of energy to the CO_2 molecule [16]:

$$N_2(v = 1) + CO_2(000) \rightarrow N_2(v = 0) + CO_2(001) - 18 \text{ cm}^{-1}$$

The small difference of 18 cm^{-1} comes from the thermal energy of the N_2 molecule, which becomes slow.

The N_2 molecule, because of electronic collisions, may be raised to higher vibrational levels. Fortunately, higher vibrational levels of N_2 molecules are closely resonant with the corresponding CO_2 levels up to (004), and the transitions between (00n) and (001) are fast according to

$$CO_2(00n) + CO_2(000) \rightarrow CO_2(00n - 1) + CO_2(001)$$

Thus the (001) level of CO_2 is preferentially filled.

The population inversion is created between (001) and lower-lying levels (100) and (020). These levels have rapid deexcitation rates; deexcitation by collision with the CO_2 molecule at the ground state is according to

$$CO_2(100) + CO_2(000) \rightarrow CO_2(020) + CO_2(000) + \Delta E$$

The energy ΔE (~103 cm^{-1}) is less than the thermal energy, and the process is near resonant. Further,

$$CO_2(020) + CO_2(000) \rightarrow CO_2(010) + CO_2(010) + \Delta E$$

where $\Delta E \approx 1$ cm^{-1}, again $\Delta E < kT$ and the process is near resonant.

The (010) level is depopulated by coupling its energy nonlinearly into the kinetic energy by the process

$$CO_2(010) + CO_2(000) \rightarrow CO_2(000) + CO_2(000) + \Delta E$$

Addition of He in CO_2 helps bring about the relaxation of the (010) state by collisional transfer of energy:

$$CO_2(010) + He \rightarrow CO_2(000) + He + \Delta E$$

This process is found to be at least 20 times faster than the process mentioned earlier. Further, He also reduces the kinetic and rotational temperatures of CO_2 in the plasma. This also increases the gain, since the sum total population is distributed in fewer levels.

The presence of traces of water vapor, H_2, or H_2 and 0_2 overcomes the problem of the dissociation of CO_2 molecules into CO and oxygen, and sometimes into carbon itself. The process is

$$CO^* + OH \rightarrow CO_2^* + H$$

where the asterisk indicates excited states. Self-contained CO_2 lasers of low power output are available for scientific work at 10.6 μm.

1.2.4 The Ruby Laser

Ruby is a crystal of Al_2O_3 in which some of the Al^{3+} ions are replaced by Cr^{3+}. Chromium ions are the active centers in the ruby laser; the laser levels are those of the Cr^{3+} ion in the Al_2O_3 lattice. The absorption spectrum of a typical ruby with orthogonal polarizations relative to the optic axis (the c axis) is shown in Fig. 1.7(a).

The pumping of the ruby is performed usually by flash lamps, say a xenon flash lamp. A portion of the light that corresponds in frequency to the two absoprtion bands, $^4F_1$ and $^4F_2$, is absorbed, thereby raising the Cr^{3+}ions to these bands. The decay time from the so-called pump bands to the upper laser level is sufficiently fast ($\approx 10^{-7}$s). We may therefore assume safely that the pump bands are empty.

The energy-level diagram of ruby is given in Fig. 1.7(b) [20]. Laser action usually occurs on the $\bar{E} \rightarrow 4A_2$ transition (R_1 line, $\lambda_1 = 694.3$ nm). The laser action can also be obtained on the $2\bar{A} \rightarrow 4A_2$ transition (R_2 line, $\lambda_2 = 692.8$ nm) by using a suitable dispersive cavity tuned to suppress oscillations at $\lambda = \lambda_1$ and allow them at $\lambda = \lambda_2$. The lifetimes of levels $2\bar{A}$ and $\bar{E}$ are 3×10^{-3}s each at T = 300K. The frequency separation between $2\bar{A}$ and $\bar{E}$ levels is 29 cm^{-1} (i.e., 870 GHz). These two levels are connected by a very fast nonradiative relaxation, so that even during laser action the corresponding relative populations are still given by Boltzmann statistics. This explains why the laser usually oscillates at the R_1 line.

The laser levels are degenerate; the ground level $4A_2$ is four fold degenerate, while $2\bar{A}$ and $\bar{E}$ are each twofold degenerate. We can however

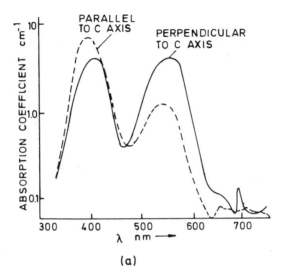

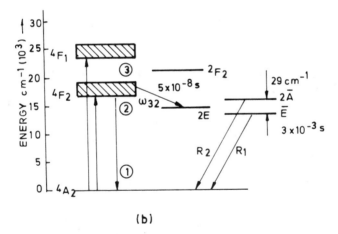

Figure 1.7 (a) Absorption spectrum of a typical ruby for orthogonal polarizations and (b) energy level diagram of ruby. (From Ref. 20.)

treat the ruby as a three-level laser, provided we make the assumption that the populations of $2\bar{A}$ and $\bar{E}$ levels are equal, which is true within 15%. The R_1 line, to a good approximation, is Lorentzian in shape with a width (FWHH) = 11 cm^{-1} ($\approx$330 GHz) at T = 300K.

A free-running ruby laser exhibits spiking. It can be Q-switched to obtain giant pulses. The ruby laser is used for hologram interferometry. Using etalons in the cavity, coherence length can be increased to more than 5m. The Q-switching can be performed in such a way that two pulses with a time delay can be produced. The double-pulse ruby laser is used for studying transient phenomena and also the dynamical behavior of objects to external force fields using hologram interferometry.

1.2.5 The Nd:YAG Laser [16]

The most popular solid-state laser system is yttrium aluminium garnet ($Y_3Al_5O_{12}$) in which a few percent of Al^{3+} ions have been replaced by Nd^{3+}ions. It is a four-level laser and hence has a low excitation threshold. The YAG material has high thermal conductivity; so that the laser can be operated at a high pulse rate and also as a continuous-wave laser. The efficiency of this laser is comparatively high, running to a few percent. The Nd:YAG lasers are optically pumped with krypton flash lamps. The YAG lattice is doped with chromium ions in addition to Nd ions, so that a xenon flash lamp could be used for pumping. The excited Cr^{3+} ions transfer their energy to the Nd^{3+} ions. On the other hand, semiconductor-diode-array-pumped Nd:YAG lasers are very compact and efficient.

Fig.1.8 (a) shows the energy-level diagram of Nd^{3+} ions in the YAG lattice. The optical pumping raises the ground state Nd atom (the atomic term $^4I_{9/2}$) to a few states identified by the terms $^4G_{7/2}$, $^2G_{7/2}$, $^4S_{3/2} + {}^4F_{7/2}$, $^4F_{5/2} + {}^2H_{9/2}$, and $^4F_{3/2}$. These five groups of states give rise to five bands in the Nd:YAG absorption spectrum as shown in Fig. 1.8(b).

The upper laser level is the state $^4F_{3/2}$, which is metastable and has a fluorescent lifetime of from 100μs to 1 ms. The Nd^{3+} ions can be pumped on all the transitions A to F. The laser action can take place on about twenty transitions, but the most common is $^4F_{3/2} \rightarrow {}^4I_{11/2}$, which gives out about 60% of the energy. Therefore, $^4I_{11/2}$ is taken as the lower laser level. This energy state lies at $E_2 \approx 2111$ cm^{-1} from the ground state. Therefore, at room temperature, this level is practically empty. The most intense lines are infrared at 1.0615 and 1.0642 μm.

The cross-section at the center of laser transition at T = 300K is σ = 9 $\times 10^{-19}$ cm^2 as compared to σ = 1.22 $\times 10^{-20}$ cm^2 for the ruby laser. Therefore, for a given inversion the gain constant in the Nd:YAG laser is approximately 75 times that of the ruby laser. This results in a low oscillation

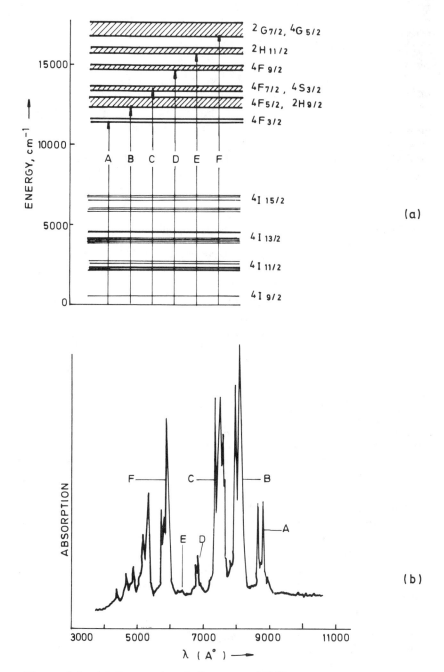

Figure 1.8 (a) Energy level diagram of Nd^{3+} ions in YAG lattice and (b) absorption spectrum of Nd:YAG.

threshold; the CW threshold of a few hundred watts and a CW output at $\lambda = 1.06$ μm of a few watts. The width of the line at room temperature is $\Delta v = 6$ cm^{-1}.

The main disadvantage of the Nd:YAG laser is that good quality Nd:YAG crystals are difficult to grow in sizes bigger than 1 cm in diameter and 10 cm in length. Glass, which can be produced in large rods of optical quality, is used as a host. The line width is broad, of the order of $\Delta v = 200$ cm^{-1}, and hence a pulse width as small as 5 ps can be obtained in a mode-locked operation. The thermal conductivity of glass is at least an order of magnitude smaller than that of YAG, and hence Nd: glass lasers are operated in pulse mode.

1.3 SEMICONDUCTOR LIGHT SOURCES

Semiconductor optical sources are basically of two types, light-emitting diodes (LEDs) and lasers. It is known that p-n junctions made from many materials from groups III and V of the periodic table emit external spontaneous radiation in the visible, near infrared, and far infrared regions of spectrum under appropriate conditions. Such devices are known as LEDs. Their typical spectral width is between 30 and 50 nm. The laser works on the principle of stimulated emission and gives bandwidth in the range of 0.5 to 3 nm.

1.3.1 LEDs as Incoherent Emitters

LEDs [2,16] are available both as homostructure (single-material) and heterostructure diodes. The latter generate more optical power. LED structures have been developed that can be classified broadly as: surface emitters and edge emitters. Figure 1.9 shows the schematic details of these sources.

The GaAs diffused junction (a homostructure) was developed by Burrus et al in 1970 and implemented at a number of places. The Burrus structure is shown in Fig. 1.9(a). These devices are pigtailed to a fiber, and are excellent sources for multimode optical fiber systems. They are capable of generating about 2 mW of optical power from a surface area of 50 μm in diameter in a solid angle of 2π steradian, corresponding to a radiance of 25 W/(cm^2 · steradian).

Edge emitters are offshoots of the resonant-cavity semiconductor lasers in which the resonant cavity is spoiled so that incoherent radiation occurs as in Fig. 1.9(b). The device operates in a superradiant mode, below the lasing threshold. The output exhibits good spatial coherence. The total power generated by the edge emitter is smaller than that from a surface

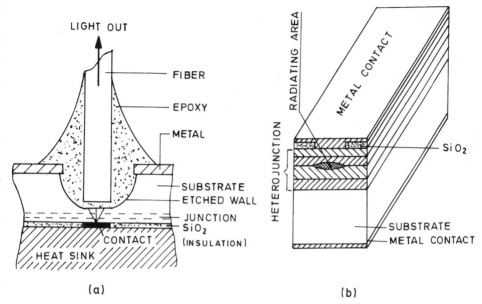

Figure 1.9 (a) Schematic of a Burrus diode structure and (b) schematic representation of an edge emitting diode.

emitter, but the radiance is significantly higher. At 1 mW, an edge emitter may have a radiance of over $10^3 W/(cm^2 \cdot steradian)$.

The LED may be considered as band-limited Gaussian noise source, i.e., the radiation is effectively shot noise limited unless spurious noise is transmitted from the bias supply network. The effect of the electrical bias supply on the optical output is of the first order: both frequency and output intensity will be functions of electrical bias. The LEDs may be directly amplitude modulated simply by varying the bias current.

1.3.2 Semiconductor Lasers

A primitive form of semiconductor laser is shown in Fig. 1.10(a). It is made by cutting a chip of GaAs that already has a p-n junction. It is desirable to have the plane of the junction parallel to the (100) face of the GaAs crystal so that the natural cleavage planes (110) are perpendicular to the junction. The cleaved ends provide the necessary feedback as the reflectivity of the semiconductor-to-air interface is about 30%, which is sufficient for lasing to occur. Metal contacts provide a supply of electrons and holes to maintain the population inversion. The output is obtained from both faces, and

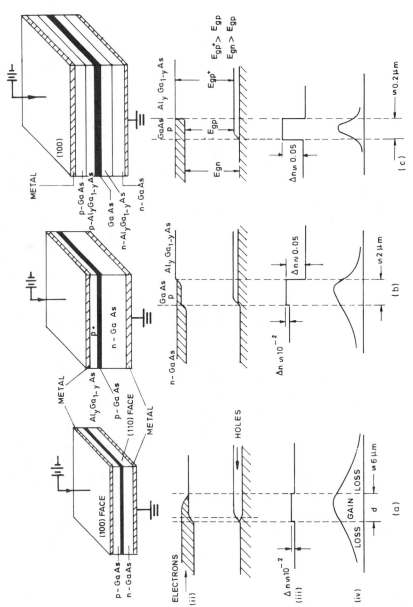

Figure 1.10 (a) Schematic of a homojunction semiconductor laser, (b) schematic of a heterojunction semiconductor laser, and (c) schematic of a double heterojunction semiconductor laser: (ii) energy band diagram at forward bias, (iii) refractive index variation, and (iv) intensity profile of the emitted radiation.

it is highly astigmatic. The gain occurs in the active region, which is approximately 3-8 μm. The difference between the refractive indices of p-type region and that of the n-type region is very small, and hence there is almost no optical confinement to the active region and the light spreads out into the surrounding lossy regions. Figure 1.10(a) shows (ii) the electron and hole diffusion distances, (iii) the index of refraction, and (iv) the intensity profile of the emitted radiation.

The homojunction structure has several disadvantages. Optical and carrier confinements are poor; there is almost no optical confinement because the refractive indices of the n region, p region, and the diffusion layer are almost equal: and there is no carrier confinement, because of the absence of a potential barrier across the junction. These devices require very high threshold currents and are operated at the temperature of liquid nitrogen. The disadvantages are removed in heterojunction devices. Here a layer of semiconductor of relatively narrow energy gap is sandwiched between two layers of wider-gap semiconductor. The narrower-gap semiconductor constitutes the active region. Further, the two different semiconductors should have different refractive indices and they should have the same lattice constants. Figure 1.10(b) shows a heterojunction laser structure. Such a junction provides an effective energy barrier to electron diffusion, and the resulting carrier confinement leads to a reduced threshold current. The active region is the p-type GaAs region in the middle. A layer of $Al_xGa_{1-x}As$ is grown on an n-type GaAs. This layer is heavily doped with Zn to form a p^+ region. During growth Zn diffuses into GaAs(n) substrate a distance ≈2μm, forming a p-type GaAs layer that acts as an active layer. The band-gap energy E_{gp}^+ of the $Al_xGa_{1-x}As(p^+)$ region is higher than that of the GaAs(p) region. This forms an energy barrier for confining the electrons to the active p region. The Refractive index step at the $Al_xGa_{1-x}As$ (p^+) and GaAs (p) interface is quite large, providing a partial optical confinement. Figure 1.10(b) shows (ii) the energy-band diagram at forward bias, (iii) the refractive-index variation, and (iv) the intensity profile of the emitted radiation.

Threshold current density for the single heterostructure laser is ≈9,000 A/cm² as compared to ≈ 26,000 A/cm² for the homostructure laser. Further improvement is possible by changing the structure. A double heterostructure laser consists of a sandwich of p-type GaAs between n-type and p-type $Al_xGa_{1-x}As$ layers. In this structure, the active region (GaAs p-type) has a very small width, as low as 0.1 μm. The p-n interface has a large refractive index step that improves optical confinement. The barrier confinement is achieved by the large difference between the energy gaps of the materials of the p-n interface. Figure 1.10(c) shows a schematic of a double-heterojunction semiconductor laser with (ii) an energy-band diagram at forward bias, (iii) a refractive-index step, and (iv) the intensity profile for the emitted radiation.

The semiconductor lasers generally employed for mode-locking purposes are double-heterostructure (DH) stripe-geometry structures that, in addition to optical and barrier confinement, also confine the injection current to a small region along the injection plane. A schematic of a DH stripe-geometry laser is shown in Fig. 1.11. The insulator SiO_2 allows the injection current to pass only through the narrow stripe region. Stripe geometry lasers have several advantages over other lasers; a reduction in threshold current; a small emitting area, facilitating the coupling to optical fibers; and operation in fundamental mode.

Semiconductor lasers are generally used in fiber-optic systems for optical communication, fiber-optic sensors, etc. Lasers that maintain high powers under continuous operation are now available. They are thermally stable devices that provide single-wavelength operation and can be pulsed at a high rate. These are now being used for hologram interferometry and in digital-speckle-pattern interferometers leading to compact systems. These lasers are also being used as pump sources for Nd:YAG lasers.

1.4 DETECTORS

A detector converts the incident optical energy into electrical energy, which usually consists of a signal and random fluctuations known as noise. Many physical effects have been used to detect optical radiation. Detectors based on either photoexcitation or photoemission of electrons provide the highest performance. There are detectors that convert photons into thermal effects within the medium. These are pyroelectric detectors.

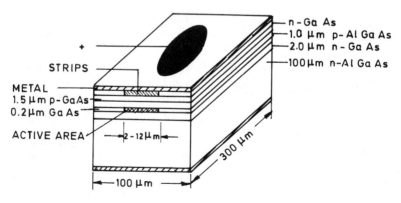

Figure 1.11   Schematic of a stripe-geometry DH laser.

Detectors are available that span the wavelength range from deep ultra-violet to far infrared.

The response of a detector can be described in terms of responsivity R_v, which is defined as the ratio of the rms value of the signal voltage V to the rms power P incident on the detector [21]:

$$R_v = \frac{V}{P} = \frac{V}{IA} \quad \frac{volt}{watt}$$

where I is the intensity in watt/cm^2 and A in cm^2 is the area of the detector. Another parameter to characterize a detector is its noise-equivalent power (NEP). It is specified under a set of operating conditions. The NEP is defined as the power required to generate an output equal to the noise. For example, the NEP for detection under the set of conditions including a 600K blackbody radiation source modulated at 1 KHz with a bandwidth of 1 Hz is written as

$$NEP(600K, 1\ KHz, 1\ Hz) = \frac{P}{S/N} = \frac{IA}{S/N} = \frac{IA}{V_s/V_n} \quad watt$$

where P is the detected power, I the intensity, A the detector area and S and N the measured signal and noise, respectively. V_s and V_N are the voltages proportional to signal and noise. Often the reciprocal of the NEP is designated as the detectivity and used to characterize some detectors.

We are limiting our discussion to semiconductor detectors and photo-multiplier tubes, as these are normally used in most systems; semiconductor detectors are small and hence compatible with fibers and integrated optics.

All these photodetectors rely on the energy of incident photon either to create an electron-hole pair in the semiconductors or to cause the release of an electron from the cathode of a PMT. Because of the photon nature of light, the shot noise is inherent in all detectors. There is an optical wavelength above which the detector does not respond to the optical radiation. This is governed by the equation $h\nu \geq E_{threshold}$ where h is the Planck's constant and ν is the frequency of optical radiation. The detector may be characterized by its quantum efficiency. This is defined as the fraction of useful electrons to incident photons. A typical value of quantum efficiency is 70%.

Four types of semiconductor detectors are in use in optical systems; the PIN diode, the avalanche photodiode (APD), the PIN-FET hybrid module, and the photoconductor. The basic detection process is identical in all these devices, it is the creation of electron-hole pairs by incident photons. Therefore the wavelength response of the detector is determined primarily by

the band gap of the semiconductor materials. We describe here only the
PIN diode and the APD.

1.4.1 The PIN Diode

A PIN diode [22] consists of three regions: a p-type region, an intrinsic (i)
region, and an n-type region. The free electrons from the n-type region
diffuse to the intrinsic region. Similarly, the holes from the p-type region
diffuse to the intrinsic region. This diffusion process, however, does not
continue indefinitely. For every free electron leaving an n-type region, an
immobile positive charge is left behind in the n-type region. The amount
of positive charge increases as the number of departing electrons increases.
Similarly, as the holes depart from the p-region, immobile negative charges build up in the p-region. Therefore a potential difference exists between the two regions. This potential difference is observed externally as
a contact potential. Figure 1.12 (a) shows the microscopic picture in the
three regions, while Figure 1.12(b) shows the variation of the electric field
in the structure. It is this electric field that stops the diffusion of free
carriers to the intrinsic region. The density of mobile carriers drops rapidly
as one moves away from the p-i and n-i boundaries toward the middle of
the intrinsic region. That portion of the intrinsic region that has a scarcity
of mobile charges is called the depletion region. The PIN diode is usually
operated with a reverse bias voltage. With this bias, an additional electric
field is created that reinforces the internal electric field. This field further
exerts force on the free carriers; consequently, the depletion region ex-
pands. Further, the conductivity of the intrinsic region is low, while that
of the p and n regions is high because of the abundance of free carriers.
When an external potential is applied, the conductivity distribution causes
the electric field to build up primarily in the intrinsic region and not in p-
and n-type regions. When a photon whose energy hv exceeds the threshold
value enters the intrinsic region, an electron-hole pair is created. Under the
action of an applied field, the photogenerated electrons and holes are
swiftly swept towards n- and p-type regions respectively and create a
signal current.

   Figure 1.2 (c) shows a cross-section of a typical PIN diode. The width
of the intrinsic region is designed to absorb the maximum amount of light
at a wavelength of interest. A typical width of an intrinsic region ranges
from 10 to 20 μm for the Si photodiodes, and the maximum sensitivity is
in the 0.8 to 0.9 μm spectral range. The light reaches the absorption region
via an antireflection coating to avoid reflection losses and a thin p+ region
of high conductivity.

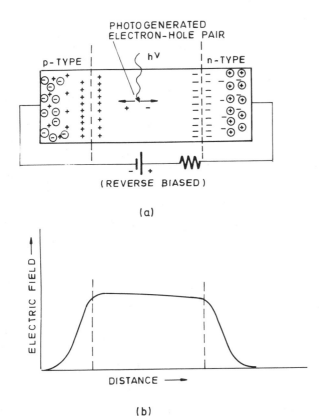

(a)

(b)

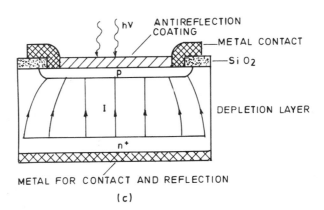

(c)

Figure 1.12   (a) Microscopic picture in a PIN Diode, (b) variation of electric field with distance, and (c) cross-section of a typical PIN Diode.

1.4.2 The Avalanche Photodetector

Figure 1.13 (a) gives the cross-section of an avalanche photodetector (APD) [22,23]. Its structure is similar to that of the PIN diode except for the insertion of an additional p-type multiplying region. The intrinsic region is slightly doped to reduce its resistivity to around 300 Ω cm. The n-type region is made thinner but is more doped than the multiplying p-type region. The incident photons are absorbed in the intrinsic region, and photogenerated electrons and holes drift towards n- and p-type regions respectively. It may be noted that the region of highest resistivity is concentrated near the junction between the n-type region and the multiplying p-type region when reverse bias is applied. The variation of the electric field as a function of distance is shown in Figure 1.13(b). The highest field exists in the region of highest resistivity. The photogenerated electrons, before reaching the n-type region, pass through the multiplying p-type region. If the electrons have acquired sufficient acceleration on reaching the multiplying region, new electron-hole pairs are generated by the collision process. The newly created electron-hole pairs will collide with the crystal lattice and produce yet further electron-hole pairs and thus initiate the process known as avalanche multiplication. Indeed, the APD utilizes this process to achieve higher electrical output. A typical gain factor from an APD may range between 10 and 100.

Two major sources of noise in the PIN diode and the APD are thermally generated noise and shot noise. Thermal noise is independent of signal current whereas shot noise depends on it. Therefore in practice the APD is used in situations where the shot noise before multiplication is well below the thermal noise. An optimum signal-to-noise ratio is obtained when the multiplication process brings the shot noise up to the level of thermal noise.

1.4.3 The Photomultiplier Tube

The photomultiplier tube (PMT) [21] is the most sensitive of optical detectors: photon flux as low as one photon per second can be detected. Figure 1.14(a) shows a cross-section of a PMT. It consists of a photocathode, dynodes, and an anode in an evacuated tube. The beam of light incident on the cathode releases photoelectrons, which are attracted to positively charged electrodes called dynodes where more electrons are splashed out as secondary emission. The secondary electrons are accelerated toward the more positive dynode, and the process is repeated. The gain at each dynode is typically by a factor of 4. After a multiplication over say ten dynodes, the original photocurrent is increased roughly a million-

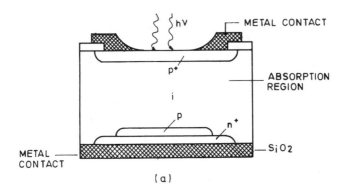

(a)

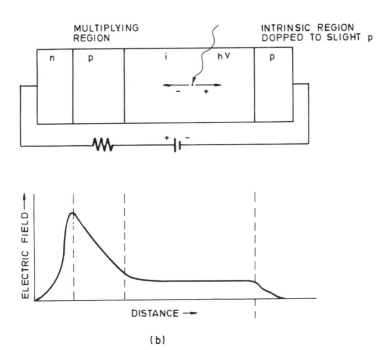

(b)

Figure 1.13   (a) A cross-section of a Si-APD and (b) various regions in an APD and variation of electric field with distance when reverse biased.

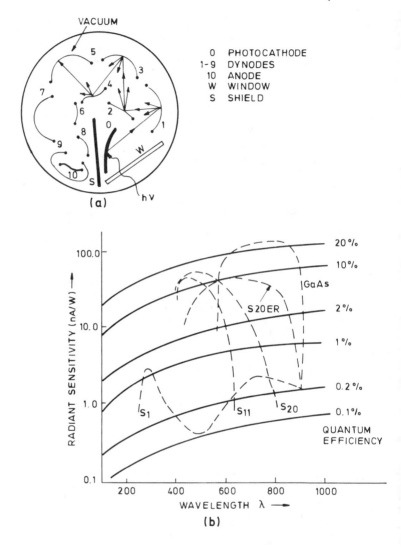

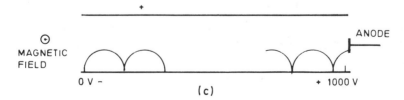

Figure 1.14 (a) A cross-section of a squirrel-caged PMT, (b) spectral response of some common cathodes, and (c) a schematic of a cross-field PMT.

fold and is easily measured. Even an individual event due to a single photoelectron is easily detected as a pulse on the anode. Since the number of photons in a beam of monochromatic light is proportional to the intensity, and the probability of photoemission is the same for each photon, the average number of photoelectrons emitted per second should be exactly proportional to the light intensity over a large dynamic range. Of course this holds good for wavelengths less than the threshold wavelength.

It may be remarked that metals are not good candidates for photocathode materials. The first practical photoemitters were hydrides of alkali metals. Figure 1.14(b) shows the spectral response of some commonly used photocathodes. S-1 or Ag-O-Cs cathode contains metallic silver dispersed as small single crystals and a nearly equal amount of cesium oxide (Cs_2O). The S-1 cathode resembles the cesium-rich cesium telluride (Cs_2Te) cathode. Cs_3Sb is used both as an opaque (metal-backed) cathode and as a semi transparent (glass-backed) cathode. Opaque cesium antimonide cathodes are designated S-4 and semitransparent ones as S-11. The S-20 cathode is obtained by multilayer alkali deposition whose infrared optimized version (S20ER,S25) gives a quantum efficiency of 2 to 10 between 400 and 850 nm. The GaAs photocathode, characterized by a monocrystalline semiconductor layer activated by cesium or oxygen, gives a quantum efficiency of 20% between 600 to 900 nm.

The secondary emission is similar to the photoemission; the main difference is that the secondary electrons are produced by energetic electrons, while the photoelectrons are produced by energetic photons. Therefore the material used for photocathodes can be used for dynodes as well. Since the electron can be made more energetic and the optical absorption is irrelevant to secondary emission, various wide-band materials such as MgO_2 and KCl are also used as secondary emitters.

A good electron multiplier should have a configuration that accelerates secondary electrons and collects them efficiently. The PMTs are available in a number of configurations, namely linear focused, squirrel cage, venetian blind, and box and grid. The configuration shown in Figure 1.14(a) is the squirrel cage; it is the most compact. The fast tubes use the focused configuration. In all the configurations the anode is nearly caged by the last dynode.

The PMTs so far described depend solely on an electric field, which puts the electrons into their proper places. These tubes are adversely affected by magnetic fields of even a few gauss. Another kind of tube that requires both electric and magnetic fields for its operation has become popular in some applications. This is a resistance-strip or cross-field multiplier as shown in Figure 1.14(c). The magnetic field is directed into the paper. Two parallel strips of resistive material, one coated with secondary emitter, are subjected to a large potential gradient. The upper strip is made more positive than the lower one. A photoelectron produced at the lower strip is subjected to the cross electric and magnetic fields and hence takes the path shown in Figure 1.14(c). Since the place where the electron restrikes the lower strip is 50 to 100 volts more positive than the place it left, it acquires enough energy to produce secondaries. These secondary electrons hop along the strip and multiply at each strike. The number of hops is usually around 20. These tubes are simple in construction.

1.5 RECORDING MATERIALS

A large number of recording materials [24] are available. We will concentrate only on those materials that are used for holography. The high-resolution demands on the recording materials result in a very slow response. There are, however, materials that can be used almost in real time.

The response of the recording materials to exposure and development, if any, may be described in terms of absorption, refractive index change, and thickness variations. Usually thickness variation is present in both the cases where absorption and refractive index changes take place, but its influence is very small. On the other hand, there are materials wherein the recording is only in terms of surface relief (thickness variation). We usually characterize materials as amplitude modulating or absorption materials and phase modulating or phase materials. Surface relief recording also modulates the phase of the wave. One of the most important performance parameters of holographic materials is MTF, since it appears as a square in the calculation of diffraction efficiency. The hologram recording materials have been classified in a number of groups and are listed in Table 1.3 [25,26]

1.5.1 Photographic Materials

Photographic materials [24] are by far the most sensitive holographic materials. They can be sensitized over the widest spectral range, and even at the emission wavelength of Nd:YAG or Nd:glass lasers. An emulsion about 6 μm thick is put on the plate or film having an Estar base (Kodak), an acetate base (Agfa) or a polyster base (Ilford). Exposure to the emulsion generates a latent image. The latent image is made visible by development; the silver halide crystals containing latent image specks of at least critical size are completely reduced to metallic silver grains. The development yields a magnification of about a million. The halide crystals, which do not contain a latent image, are dissolved and removed by fixing and rinsing

Table 1.3

S. No.	Class of material	Spectral range (nm)	Recording process	Spatial frequencies cycles/mm	Types of grating	Processing	Readout process	Maximum diffraction efficiency %
1.	Photographic materials	400–700 (<1300)	Reduction to Ag metal grain	>7000	Plane/vol. ampl.	Wet chemical	Density change	5
			Bleached to silver salts		Plane/vol. phase	Wet chemical	Refractive index change	20–50
2.	Dichromated gelatin	250–520 and 633	Photocross linking	>3000	Plane phase, volume phase	Wet chemical followed by heat	Refractive index change	30 >90
3.	Photo resists	UV–500	Photocross linking or photopoly-merization	<3000	Surface relief/phase blazed reft.	Wet chemical	Surface relief	30 70–90
4.	Photo polymers	UV–500	Photo polymer-ization	≈200–1500 band pass	Volume phase	None or post-exposure and post-heating	Refractive index change or surface relief	10–85

Table 1.3 (*Continued*)

S. No.	Class of material	Spectral range (nm)	Recording process	Spatial frequencies cycles/mm	Types of grating	Processing	Readout process	Maximum diffraction efficiency %
5.	Photoplastics/ photoconductor thermoplastics	Nearly panchromatic for PVK TNK photoconductor	Formation of an electrostatic latent image with electric-field-produced deformation of heated plastic	400–1000 band pass	Plane phase	Corona charge and heat	Surface relief	6–15
6.	Photochromics	300–450	Generally photo induced new absorption bands	>2000	Volume absorption	None	Density change	1–2
7.	Ferroelectric crystals	488	Electrooptic effect	>1000	Volume phase	None	Volume phase	60

baths. This procedure yields amplitude holograms. Metallic silver grains can be converted to dielectric silver salts by bleaching, which then forms a phase hologram. The recording can be either two-dimensional or in volume depending on the interbeam angle.

1.5.2 Dichromated Gelatin

Dichromated gelatin [24] is among the oldest of photographic materials. Gelatin can be put on glass plates and sensitized. Alternatively, the photographic plates can be fixed, rinsed, and sensitized. Ammonium, sodium and potassium chromates have been used as sensitizers. Gelatin sensitized by ammonium dichromate is the most often used. It exhibits sensitivity in the wavelength range of 250 to 520 nm. It is sensitized at 633 nm by the addition of methyl blue dye. The exposure causes cross-linking between gelatin chains and changes swelling properties and solubility. Treatment with warm water dissolves the unexposed gelatin and thereby forms a surface relief pattern. On the other hand, rapid dehydration in alcohol creates cracks or tears in the softer gelatin. These strain-induced cracks are very efficient in redirecting the incident light into a first order diffracted beam; the volume holograms in gelatin yield diffraction efficiency of 90% or more.

1.5.3 Photoresists

Photoresists [24] are organic materials that are sensitive to ultraviolet radiation. A thin layer (≈ 1 μm) of photoresist is obtained on the substrate by either spin coating or spray coating. On exposure, recording involves one of the following three processes: formation of an organic acid, photo crosslinking, or photopolymerization. There are two types of photoresists, negative and positive. In the negative photoresist, the unexposed regions are removed on development, while in the positive photoresist, the exposed areas are removed on development. Therefore a surface relief pattern is formed. A grating recorded on photoresist can be blazed either by ion bombardment or by optical means. Holograms recorded on photoresist can be replicated by mechanical means.

In most negative photoresists the emulsion next to the substrate is the last to photolyze when UV light enters from the emulsion-air interface side. Until photolysis occurs at the emulsion-substrate interface, the material that has not been photolyzed will simply dissolve in the developer even though it was in an exposed area. This nonadhesion of negative photoresist in holographic recording is a serious problem. To overcome this problem, the photoresist should be exposed from the substrate-emulsion interface side so as to photolyze the resist better.

1.5.4 Photopolymers

Photopolymers [27] layers for use in holography can be prepared by enclosing an aqueous solution between glass plates; or they can be dry layers. Exposure causes polymerization of the monomer, which results in modulation of the refractive index, which may or may not be accompanied by surface relief. The sensitivities of photopolymers are greater than those of photoresists and photochromic materials but less than those of silver halide emulsions. They also possess the advantage of dry and rapid processing. Thick polymer materials like PMMA and CAB can change their refractive index by 10^{-3} on exposure, thereby making highly efficient volume holograms. They have band-limited responses primarily because of the limitations imposed by the diffusion length of the monomer at the lower end and the length of the polymer at the higher end of the response curve.

1.5.5 Thermoplastics

A thermoplastic [27] is a multilayer structure having a substrate coated with a conducting layer, followed by a photoconductor layer, and on top of which is a thermoplastic layer. A photoconductor that works well is the polymer poly-N-vinyl carbazole (PVK) to which is added a small amount of electron donor 2,4,7-trinitro-9-fluorenone (TNF). The thermoplastic is a natural tree-resin, Staybelite.

The recording process involves a number of steps. First a uniform electrostatic charge is established on the surface of the thermoplastic with a corona discharge assembly. The charge is capacitively divided between the photoconductor and the thermoplastic layers. Then the thermoplastic is subjected to exposure; the exposure causes the photoconductor to discharge its voltage at illuminated areas. This does not cause any variation in the charge distribution on the thermoplastic; the electric field in the thermoplstic remains unchanged. The charge variation and consequently the electric field variation is accomplished by recharging the thermoplastic. In this process the charge is added at the exposed areas. Thus an electric field distribution is established in the thermoplastic which on heating deforms it under the electrostatic forces until these forces are balanced by surface tension. Cooling quickly to room temperature freezes the deformation; the recording is thus in the form of surface relief. The recording is stable at room temperature but can be erased by heating the thermoplastic to a temperature higher than that used for development. At the elevated temperature the surface tension evens out the thickness variation and hence erases the recording. The thermoplastic can be reused several hundred times. The response of these devices is band limited dependent on the thickness of the thermoplastic and other factors; but the almost instant development and reusability make thermoplastic a very attractive recording medium for hologram interferometry. As for storage applications, the information can be updated only on an entire page.

1.5.6 Photochromics

Materials that undergo reversible color change on exposure are called photochromic material [24]. Photochromism occurs in a variety of materials organic and inorganic. Colorless materials exposed to ultraviolet radiation become dark and may absorb at different wavelengths. These wavelengths may be used for bleaching. The recording can be performed in either darkening or bleaching mode. Organic films of spiropyrane derivative have been used for recording in darkening mode at 633 nm, in silver halide photochromic glasses either in darkening mode at 488 nm or in a bleaching mode at 633 nm and in doped crystals of CaF_2 and SrO_2 in the bleaching mode at 633 nm. Erasure can also be accomplished by the application of heat.

The sensitivity of photochromics is very low because the chemical reaction occurs at the molecular level. For the same reason they are essentially grain free and resolution in excess of 3000 1/mm has been achieved. These materials are attractive, as no postexposure processing is required, since only energy is needed for in situ recording and erasure. These materials, however, have not found favor because of the fading out of recording and the low diffraction efficiency. Further, the readout degrades the recording, particularly when the readout wavelength is the same as the recording wavelength, a primary requirement for volume holography.

1.5.7 Ferroelectric Crystals

Certain ferroelectric [FE] crystals [21,24] like lithium niobate ($LiNbO_3$), lithium tantalate ($LiTaO_3$), barium titanate ($BaTiO_3$), strontium barium niobate (SBN), etc. exhibit small changes in refractive index after an intense exposure to light. This photoinduced refractive index change can be reversed with the application of heat and light. The reversibility of refractive index changes in these crystals is similar to that of color in photochromics. The mechanism of recording in FE materials is as follows: The electrons freed by exposure migrate and get trapped at regions of low intensity. This produces a net space charge pattern and corresponding electric field pattern. The electric field modulates the refractive index through the electrooptic effect and creates a volume phase hologram. These are real-time recording materials; the records are stable, as the charges are bound to the localized traps. The recording can, however, be erased by illuminating it with a light beam of a wavelength that can release the trapped electrons. The sensitivity of $LiNbO_3$ is very low, but it has been improved considerably by Fe doping. In fact, Fe-doped $LiNbO_3$ is one of the best ferroelectric materials for optical storage, and the mechanisms of writing and erasure are well understood [24].

This list of recording materials is not exhaustive. There are a number of other materials that have been used for holographic recording. Magnetooptic materials in thin layers record the information through the Faraday effect or the magnetooptic Kerr effect. Transparent electrophotographic films have been developed that hold promise for photographic recording

REFERENCES

1. Driscoll, Walter G., and Vaughan, William, eds. (1978). *Handbook of Optics,* McGraw-Hill, New York.
2. Pressley R. J. (1971). *Handbook of Lasers,* CRC Press, Cleveland.
3. *RCA Electro-Optics Handbook* (1974). Technical series EOH-11 RCA Corporation, Harrison, New Jersey.
4. Steele, Earl L. (1972). *Optische Laser in der Electrotechnik,* Verlag Berliner Union, Stuttgart.
5. Young, M. (1977). *Optics and Lasers,* Springer-Verlag, Berlin.
6. O'Shea, D. C., Callen, R., and Rhodes, W. T. (1978). *Introduction to Lasers and Their Applications,* Addison-Wesley, Reading.
7. Marshall, S. L., ed. (1968). *Laser Technology and Applications,* McGraw-Hill, New York.
8. Shimoda, K. (1986). *Introduction to Laser Physics,* Springer-Verlag, Berlin.
9. Demokan, M. S. (1982). *Mode Locking in Solid-State and Semi-Conductor Lasers.* Research Studies Press, Chichester.
10. Ross, M. ed. (1971). *Laser Applications,* Vol. I, Academic Press, New York.
11. Birnbaum, G. (1967). Frequency stabilization of gas lasers, *Proc. IEEE, 55:* 1015–1026.
12. Wallard, A. J. (1973). The frequency stabilization of gas lasers, *J.Phys. E: Sci. Instrum., 6:* 793–807.
13. Smith, P. W. (1965). Stabilized, single-frequency output from a long laser cavity, *IEEE J. Quant. Electron. QE-1:*343–348.
14. Wallard, A. J. (1972). Frequency stabilization of the helium neon laser by

saturated absorption in iodine vapor,*J.Phys. E: Sci. Instrum.*, 5: 926–930.

15. Matsumoto, H. (1984). Recent interferomertic measurements using stabilized lasers, *Precision Engineering, 6:*, 87–93.
16. Yariv, A. (1985).*Optical Electronics,* Holt, Rinehart and Winston, New York.
17. Bridges, W. B. (1964). Laser oscillations in singly ionized argon in the visible spectrum, *Appl. Phys. Letts. 4,* 128.
18. Witteman, W. J. (1987). The CO_2 Laser, Vol. 53, Springer-Verlag, Berlin.
19. Patel, C. K. N. (1968). High Power Carbon Dioxide Lasers, *Sci Amer., 219:* 22–23.
20. Maiman, T. H. (1960). Optical and micro-optical experiments in ruby, *Phys. Rev. Letts., 4:* 564–566.
21. Poehler, T. O. (1988). Detectors, *Methods of Experimental Physics,* Vol. 26, Academic Press, New York.
22. Sharma, A. B., Halme, S. J., and Butusov, M. M. (1981).*Optical Fiber Systems and Their Components,* Springer-Verlag, Berlin.
23. Kressel, H. ed. (1982). *Semiconductor Devices,* Springer-Verlag, Berlin.
24. Smith, H. M. (1977). *Holographic Recording Materials,* Springer-Verlag, Berlin.
25. Gladden, J. W., and Leighty, R. D. (1979). Recording media *Handbook of Optical Holography,* Academic Press, New York.
26. Biedermann, K. (1975). Information storage materials for holography and optical data storage, *Opt. Acta, 22:* 103–124.
27. Collier, R. J., Burckhardt, O. B., and Lin, L. H. (1971).*Optical Holography.* Academic Press, New York.

CHAPTER 2

Optical Components

2.1 INTRODUCTION

The manipulation of light beams is performed by a variety of optical components. Some change the direction of the beam; others introduce power to the beam. The functioning of these elements is based on optical phenomena such as reflection, refraction, interference, diffraction, etc. Detailed treatment of these can be found in any textbook on optics, but we briefly describe them again here.

2.1.1 Reflection

A beam incident on a smooth boundary between two media is partially reflected and partially transmitted. The direction of the reflected beam is governed by the law of reflection, which is usually stated in two parts, namely (a) the angle of incidence is equal to the angle of reflection, and (b) the incident ray, the reflected ray, and the normal to the surface at the point of incidence lie in the same plane. The intensity of the reflected beam depends on the state of polarization of the incident beam, the optical constants of the media, the angle of incidence and the wavelength. It can be calculated using Fresnel formulae. If the boundary is not smooth, the light is scattered over a solid angle; one may speak of diffuse reflection.

2.1.2 Refraction

Both reflection and refraction take place simultaneously at the smooth boundary between two media. The direction of the refracted ray is governed by Snell's law, namely (a) $n_i \sin \theta_i = n_r \sin \theta_r$, and (b) the incident ray, the refracted ray, and the normal at the point of incidence lie in the same plane. Here n_i and n_r are the refractive indices in the first and second media, and θ_i and θ_r are the angles of incidence and refraction, respectively (Fig. 2.1). It is seen that for $n_i < n_r$, $\theta_i > \theta_r$, the refracted ray bends towards the normal, while for $n_1 > n_r$, the refracted ray bends away from the normal. When the angle of refraction is $\pi/2$, the corresponding incidence angle is called the critical angle. For all angles of incidence greater than the critical angle, there is no refracted ray: the ray is totally reflected. According to em theory the field decays exponentially in the second medium. Therefore if the thickness of the second medium is of the order of a wavelength, the field can be coupled out, thereby frustrating the total internal reflection.

The intensities of reflected and refracted beams are obtained using Fresnel formulae. It is seen that the reflectivity of a metallic surface approaches unity at higher wavelength.

When the refraction takes place at a boundary between an isotropic and an anisotropic medium, in general two rays exist in the anisotropic medium. The intensities of the two beams are governed by the state of polarization of the incident beam and the orientation of the optic axis of the

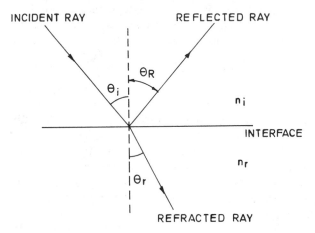

Figure 2.1   Reflection and refraction at an interface between two dielectric media.

anisotropic medium. Many optical components of anisotropic materials are available for specific tasks.

There are optical components that are inherently based on the interference of light. Thin-film devices like filters, high- and low-reflectivity mirrors, dichroic filters, etc. fall under this category. Then there are devices that exploit both interference and diffraction of light: gratings are an example. We will briefly describe some of the components used extensively in optical instrumentation and measurement.

2.2 MIRRORS

2.2.1 The Plane Mirror

A plane mirror is a flat surface that is high-reflection coated. A plane mirror always produces a virtual image of ideal, i.e., diffraction-limited quality. The image is left-right reversed Fig. 2.2 (a). When the mirror is rotated by an angle θ, the reflected ray is rotated by 2θ. If two plane mirrors enclose an angle θ, then $360°/\theta° - 1$ images are formed of an object placed inside the mirrors. The mirror is used for path folding, path switching, in optical levers for measurement, in kaleidoscopes, etc. Semitransparent mirrors are used as beam splitters. The reflectivity of a mirror can be

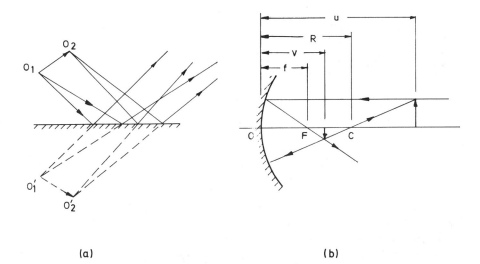

(a)   (b)

Figure 2.2   (a) Image formation by a plane mirror and (b) image formation by a concave spherical mirror.

enhanced by depositing a dielectric coating on top of a metallic coating. Such enhanced reflectivity mirrors are also commercially available.

2.2.2 The Spherical Mirror

Spherical surfaces, when metallized, become spherical mirrors. If the reflection is taken from a concave surface, we have a concave mirror. The power of the mirror is $2/R$, where R is the radius of curvature of the surface. The image formation, under paraxial approximation, is governed by

$$\frac{1}{u} + \frac{1}{v} = \frac{1}{f} = \frac{2}{R} \tag{2.1}$$

where u and v are the object and image distances and f is the focal length. The usual sign convention is employed here. A concave mirror can produce magnified, demagnified, erect or inverted images of an object depending on its location. Figure 2.2 (b) shows the image formation by a concave mirror. A convex mirror always produces an erect, demagnified image.

The spherical mirror suffers from monochromatic aberrations, namely spherical aberration, coma, and astigmatism (see Section 3.1.1). The magnitudes of various abberrations for objects at infinite distance are (Figure 2.3) [1]

$$\text{Spherical aberration} \quad \frac{y^2}{4R} \tag{2.2a}$$

$$\text{Sagittal coma} \quad \frac{y^2(l_p - R)u_p}{2R^2} \tag{2.2b}$$

$$\text{Astigmatism (separation} \quad \frac{(l_p - R)^2 u_p^2}{2R} \tag{2.2c}$$
$$\text{between S and T)}$$

$$\text{Petzval curvature} \quad \frac{u_p^2 R}{4} \tag{2.2d}$$

where y is the semiaperture, R is the radius of curvature of the mirror, u_p is the half field angle, and l_p is the mirror to stop distance. When $l_p = R$. i.e., the stop is located at the center of curvature, coma and astigmatism are zero, and the image surface is a sphere of radius approximately equal to the focal length, centered about the center of curvature. Only spherical aberration is present.

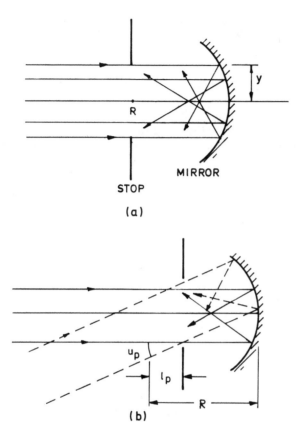

Figure 2.3 (a) Spherical aberration of a concave spherical mirror and (b) coma and astigmatism in a concave spherical mirror.

The minimum diameter of the blur spot B is

$$B = \frac{y^3}{4R^2}$$ (2.3)

The angular blur β, expressed in radians, is

$$\beta = \frac{y^3}{2R^3}$$

$$= \frac{1}{128(f\#)^3}$$ (2.4)

where $f\# = 2y/f = 4y/R$ is the f-number of the mirror. This expression for angular blur size is exact only for third-order spherical aberrations, but it gives good results up to f number 2.

2.3 THE WEDGE PLATE

A wedge plate deviates a ray by an angle $\delta = (n-1) A$, where n is the refractive index of the material and A is the wedge angle; see Figure 2.4 (a). When the deflection angle is 0.01 rad; a power of 1 prism diopter is assigned to it. A variable power wedge system can be realized by a pair of wedges that may be oppositely rotated; see Figure 2.4(b). The total deviation when two wedges enclose an angle β is

$$\delta = \sqrt{\delta_1^2 + \delta_2^2 + 2\delta_1\delta_2 \cos \beta} \quad \text{and} \quad \tan \Phi = \frac{\delta_2 \sin \beta}{\delta_1 + \delta_2 \cos \beta} \qquad (2.5)$$

where δ_1 and δ_2 are the deviations due to individual wedges. A variable power system can also be realized by laterally shifting a matching pair of planoconvex and planoconcave lenses. For use with white light, an achro-

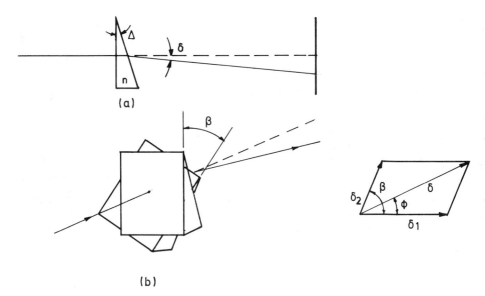

(a)

(b)

Figure 2.4   (a) Deviation produced by a wedge plate and (b) variable deviation produced by a pair of wedge plates. The emergent ray in general is not on the plane of the paper.

matic wedge plate is used. This is realized by a pair of wedge plates of different glasses; this gives deviation without dispersion.

When placed in a collimated beam of monochromatic light, the wedge introduces a variable path difference across the beam. The path difference Δ is given by

$$\Delta = (n - 1)A\,x \qquad\qquad (2.6)$$

where x is the distance from the apex of the wedge.

Wedge plates are used as beam splitters in interferometers using lasers; in survey instruments; as Fabry-Perot etalon plates (air/gas or liquid filled); for producing holographic gratings; in cameras for focusing; for measuring binocular accomodation, checking collimation of a laser beam [2], etc.

2.4 THE PLANE PARALLEL PLATE

A plane parallel plate (PPP) placed obliquely in a collimated beam displaces it by Δs, see Fig. 2.5(a), where

$$\Delta s = t\,\frac{\sin (i - r)}{\cos r} \qquad\qquad (2.7)$$

$$= t\,\frac{(n - 1)}{n}\,i \qquad \text{for small i}$$

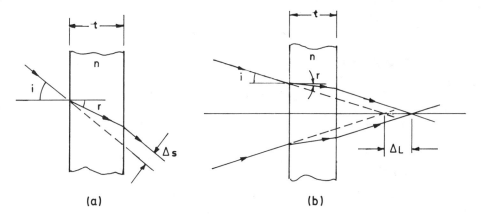

(a)   (b)

Figure 2.5   (a) Lateral displacement of a collimated beam by a PPP in oblique incidence and (b) longitudinal shift of focus by a PPP.

where t is the thickness of the plate and i and r are angles of incidence and refraction. It also introduces a path difference Δ which is given by

$$\Delta = t \,(\sqrt{n^2 - \sin^2 i} - \cos i) \qquad\qquad\qquad (2.8)$$

$$= (n - 1)t \qquad \text{when } i = 0$$

The beam may be multiply reflected inside the plate, however; so that a number of beams occur both in reflection and in transmission. The path difference between any two consecutive beams is $2nt \cos r$.

The PPP when used in convergent (or divergent) beams results in longitudinal shift of focus and also introduces aberrations. The longitudinal shift of focus (see Fig. 2.5 (b)) is

$$\Delta L = \frac{t}{n} \left(n - \frac{\cos i}{\cos r} \right) \qquad\qquad\qquad (2.9)$$

Under paraxial approximation this is given by

$$\Delta L = \frac{(n - 1)}{n} \, t \qquad\qquad\qquad (2.10)$$

The equivalent air thickness of the plate is equal to its actual thickness, less the axial shift, i.e.,

$$t_{eq} = t - \Delta L = \frac{t}{n} \qquad\qquad\qquad (2.11)$$

The equivalent air thickness is often used in the design of prisms.

The PPP introduces aberrations when used in noncollimated beam. The expressions for the various aberrations introduced are given in various books [1,4].

PPPs are used as compensating plates in interferometers; as shear plates for testing and collimation checks [3]; in optical micrometers; in photogrammetric cameras; as Brewster windows in lasers; in pile-of-plates polarizers; in the Savart polariscope; in solid Fabry-Perot resonators; for mode structuring; as bases for scales, graticules, gratings, etc.

2.5 PRISMS

A prism is a piece of a denser medium bounded by an entrance face and an exit face. Prisms come in a variety of shapes and sizes. They can, however, be grouped functionally into two categories, namely (a) dispersion prisms used in spectroscopic instruments, and (b) reflection prisms used in measuring or observation instruments.

   Ideally a collimated beam incident on a prism emerges as a collimated beam. Some prisms can be used in noncollimated illumination.

2.5.1 Dispersion Prisms

Dispersion prisms disperse the beam into its spectral components. Incidence at the entrance face where the dispersion takes place is nonnormal. Common examples are 60° prisms, constant deviation prisms, Littrow prisms, etc. and are shown in Figure 2.6. These prisms can also be used in tandem. The Littrow prism is used both as a mirror and as a dispersive element in some laser cavities to provide wavelength selection. The angular disperson $d\theta/d\lambda$ of a prism is given by

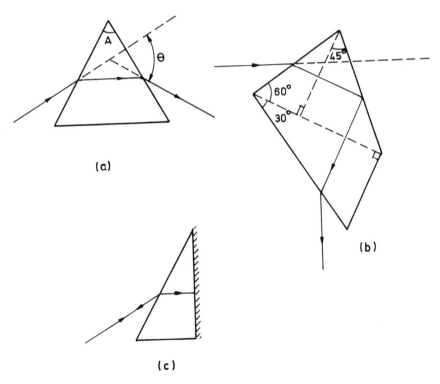

Figure 2.6 (a) 60° prism at minimum deviation, (b) constant deviation prism, and (c) Littrow prism.

$$\frac{d\theta}{d\lambda} = \frac{d\theta}{dn} \frac{dn}{d\lambda} \tag{2.12}$$

where θ and n are the angle of deviation and refractive index of the prism material. For a prism of angle A at minimum deviation condition, $d\theta/dn$ is expressed as[*]

$$d\theta/dn = \frac{2 \sin (A/2)}{[1 - n^2 \sin^2 (A/2)]^{\frac{1}{2}}} \tag{2.13}$$

Further, $dn/d\lambda$ may be obtained from Cauchy's relation:

$$n = A + \frac{B}{\lambda^2} + \frac{C}{\lambda^4} + \cdots \tag{2.14}$$

where A, B, and C are empirically determined material constants. Usually C is very small compared to B, and hence two constants only are used in the visible region. It may therefore be seen that the angular dispersion of the prism depends on the prism angle and $dn/d\lambda$.

The resolving power R of a prism is given by

$$R = \frac{\lambda}{\Delta\lambda} = t \frac{dn}{d\lambda} \tag{2.15}$$

where t is the base thickness of the prism. This relation is derived using Rayleigh's resolution criterion. Therefore the ability of the prism to resolve two closely separated wavelengths λ and $\lambda + \Delta\lambda$ depends solely on its size.

2.5.2 Reflection Prisms

Reflection prisms are used to bend the direction of the beam. They can also compress or expand the beam in one direction and hence they are associated with a magnification M given by

$$M = \frac{\cos i_1}{\cos i_2} \frac{\cos r_2}{\cos r_1} \tag{2.16}$$

---

[*] At minimum deviation, the refractive index n of the prism material is given by

$n = [\sin (A + \theta)/2][\sin (A/2)]^1$

where i_1 and i_2 are the angles of incidence and emergence, and r_1 and r_2 are the corresponding angles of refraction. These prisms are used for beam expansion in dye laser cavities. Also the titled plane processor for synthetic aperture radar data processing is based on prism magnification. The stigmatic output from diode lasers may be modified using a prismatic expander.

A simplest form of prism is a right angle prism, whose angles are 45°, 90°, 45°. It can be used to turn the beam by 90° or by 180°. It can also be used to invert the image in one meridian without deviating the beam.

One of the main uses of prisms is to provide the proper orientation of the image in a system. A right-angle prism with a roof on the hypotenuse surface is called an Amici prism. It is used in a telescope to provide a right-handed and erect image. One must, however, look in a direction normal to the objective axis. A combination of an Amici and a pentaprism removes this limitation. A Dove prism with a roof on the hypotenuse but used before the objective (used in parallel beam) accomplishes the same result. A Porro-system consisting of two right angle prisms with their hypotenuse surfaces in orthogonal planes is often used in binoculars. It not only gives an erect image but also provides the possibility of adjusting the interocular distance between the eyepieces.

The Dove prism is used for derotating the image, but its operation is limited to parallel beam. The Pechan prism, on the other hand, can be used in convergent beam. For giving an erect image and bending the line of sight by 45°, the Schmidt prism is used in measuring microscopes. The design details of some of these prisms are given in Table 2.1.

In most of the reflection prisms, the ray deviation is through total internal reflection; some prisms employ metallic reflection in addition to total internal reflection. Therefore these prisms can only accept rays within a cone called the acceptance cone.

It is convenient in laying out a prism to unfold it about the reflecting surfaces. This generates the so-called tunnel diagram, in which the ray paths can be accurately constructed through the prism by drawing straight lines on the diagram. The rays must be refracted at the entrance and exit faces unless the diagram is reduced to t/n, the equivalent air thickness of the prism.

The tunnel diagram for a right-angle prism (half of a Porro prism) is shown in Figure 2.7. The reflecting surfaces are AB and BC; the corner C is imaged in AB at C_1 and the corner A is imaged in BC at A_1. The ray R_2 is drawn as a straight line in the tunnel diagram, while the rays R_1 and R_3 exhibit refraction at the entrance face. The tunnel diagram is particularly useful in detecting the presence of unwanted reflections. The ray R_3 in Figure 2.7. is obliquely incident near A. It passes through the diagram as

Table 2.1 Design Details of Various Prisms

RIGHT-ANGLE PRISMS

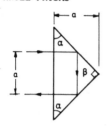 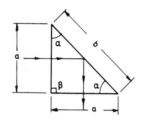

$$b = \sqrt{2}\, a$$
$$\alpha = 45° \quad \beta = 90°$$

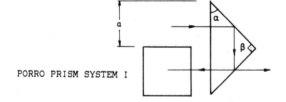

ABBE PRISM

BC = a√2	α = 135°
CD = 1.3094 a	β = 135°
DE = 0.5774 a	γ = 60°
EF = a	δ = 30
HG = 2a	
AE = 3a	

PORRO PRISM SYSTEM I

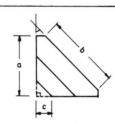

$$\alpha = 45°;$$
$$\beta = 90°$$

AMICI PRISM

$$b = \sqrt{2}a \quad c = a/\sqrt{8}$$
$$\alpha = 45° \quad \beta = 90°$$

Table 2.1 *(Continued)*

PENTAPRISM

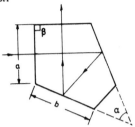

$b = 1.0824a$
$\alpha = 45°$ $\beta = 90°$

DOVE PRISM

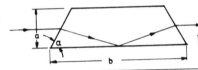

$$b = a \left[1 + \frac{(n^2 - \sin^2\alpha)1/2 + \sin \alpha}{(n^2 - \sin^2\alpha)1/2 - \sin \alpha} \right]$$

PECHAN PRISM

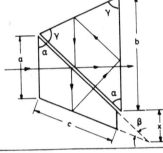

$b = 1.8284a$ $c = 1.0824a$
$x = 0.2071a$
$\alpha = 45°$ $\beta = 22.5°$ $\gamma = 67.5°$

SCHMIDT PRISM

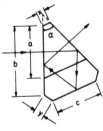

$b = 1.4142a$ $c = 1.0824a$
$\alpha = 45°$
$x = 0.1a$
$y = 1.8478x$

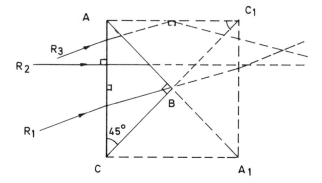

Figure 2.7   Tunnel diagram for a right-angle prism.

reflected from the hypotenuse surfaces, thereby undergoing three reflections. This produces a left-handed image. Since the prism is to be used with two reflections, this ray generates a ghost image. A small notch in the hypotenuse surface can eliminate this reflection and hence the ghost image.

Deviation prisms are classified in four groups according to the way they direct the optical axis in an optical system. (a) The optical axis remains unshifted in the entrance and exit spaces. Examples are the Dove prism and the two-components Amici prism. (b) The optical axes are parallel but shifted. The Fresnel rhomb and the porro-prism are examples. (c) The optical axis is deflected by 90°. Examples are the right-angle prism with hypotenuse as reflecting surface and its roof family and the pentaprism and its roof family. (d) The optical axis deflected by 45°. The Schmidt prism and its roof family are examples in this category. There are some prisms that deviate the optical axis by 30° and other angles.

2.6 LENSES

A lens is formed when a transparent material is bounded by at least one nonplanar surface. Most lenses have spherical surfaces, because these are easy to produce. The elementary lens types are biconvex, planoconvex, positive meniscus, biconcave, planoconcave and negative meniscus. A ray incident on the lens will suffer refraction at the first surface and then at the second surface and will be bent toward or away from the axis. If the ray heights at both surfaces are the same, the lens is called a thin lens. A lens may be considered thin if its thickness is small compared to other distances associated with its optical properties. For further discussion we will work under paraxial approximation, i.e., we consider rays very close to the optical axis. These rays strike the surface nearly normal, and hence angles of incidence and refraction are small. Thus the sines of these angles can be approximated by the angles themselves.

Under paraxial approximation, image formation by a lens in air is governed by the Gaussian form (Fig. 2.8) of the lens formula, i.e.

$$\frac{1}{f} = \frac{1}{v} - \frac{1}{u} \qquad\qquad (2.17)$$

where u and v are the object and image distances as measured from the lens and f is the focal length of the lens. The Newtonian form of the lens formula is

$$x_1 x_2 = - f^2 \qquad\qquad (2.18)$$

where x_1 and x_2 are measured from the corresponding focal points. The focal length of the lens is given by the Lens Maker's formula,

$$\frac{1}{f} = (n - 1)(\frac{1}{R_1} - \frac{1}{R_2}) \qquad\qquad (2.19)$$

The inverse of focal length measured in meters gives the power of the lens in diopters. Therefore the power P of a thin lens is

$$P = \frac{1}{f} = (n - 1)(\frac{1}{R_1} - \frac{1}{R_2}) \qquad\qquad (2.20)$$

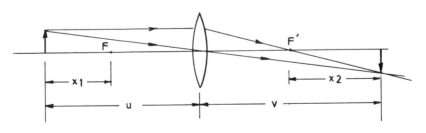

Figure 2.8   Image formation by a convex lens.

The power P of a combination of two lenses of powers P_1 and P_2 that are separated by a distance d is

$$P = P_1 + P_2 - dP_1P_2 \qquad (2.21)$$

If the transfer of a ray from the first surface to the second is carried out, the lens is termed thick. The power of a thick lens is given by [5]

$$P = (n - 1)\left(\frac{1}{R_1} - \frac{1}{R_2} + \frac{t(n - 1)}{nR_1R_2}\right) \qquad (2.22)$$

where t is the thickness of the lens. The properties of a thick lens can be best explained with the help of cardinal planes or points. They are (Figure 2.9) focal planes or points, principal planes or points, and nodal planes or points.

The rays emanating from a focal point (the intersection of a focal plane with the optical axis) will run parallel to the axis after the passage through the lens. Further, the ray bundle parallel to the axis will be brought to a focal point by the lens. The principal planes are planes of unit lateral magnification. The image and object distances are measured from the principal planes. The nodal planes are planes of unit angular magnification. If the medium of object space and image space is the same, the nodal planes coincide with the principal planes.

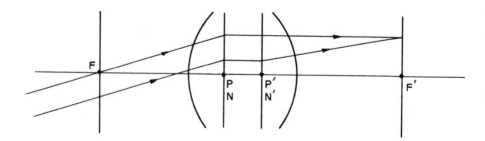

Figure 2.9 Cardinal planes/points of a thick lens. P,P', principal points; N,N', nodal points, coincident with principal points when the object and image spaces have the same refractive indices; F,F', focal points.

2.7 GRATINGS

A grating is an arrangement having a large number of equispaced slits. It is realized by drawing lines with a diamond pencil on a glass plate or a metallic surface. Currently, gratings are fabricated by recording an interference pattern between two waves. These are called holographic gratings. A beam incident on a grating is diffracted in many orders depending on the groove profile. We can classify gratings as plane gratings, concave gratings, and holographic gratings.

2.7.1 Plane Gratings

Plane gratings are realized by engraving on an optically flat surface. Replica gratings can be obtained from masters.

The irradiance distribution in the far field of a grating when a collimated beam of light is incident at an angle i see Fig. 2.10 (b) is given by [5]

$$I(i,\theta) = I_0 \left(\frac{\sin \alpha}{\alpha} \right)^2 \left(\frac{\sin N\beta}{\sin \beta} \right)^2 \tag{2.23}$$

where $\alpha = \pi b(\sin i + \sin \theta)/\lambda$, $\beta = \pi d(\sin i + \sin \theta)/\lambda$, d is the grating period, and b is the slit width; the irradiance is obtained in the direction of the diffraction angle θ.

The first term in the expression for irradiance distribution is the diffraction term, which controls the irradiance of various orders. The second term is the interference term which arises because of superposition of waves from N slits. The direction of the principal maxima are given by

$$d(\sin i + \sin \theta) = m\lambda \tag{2.24}$$

where m is the diffraction order. The equation is known as the grating equation. Figure 2.10(a) and (b) show the schematic of diffraction at a plane grating along with the intensity distribution in the far field.

A polychromatic beam incident on a grating is dispersed into its constituents in each order. The angular dispersion $d\theta/d\lambda$ is given by

$$\frac{d\theta}{d\lambda} = \frac{m}{d \cos \theta} = \frac{\sin i + \sin \theta}{\lambda \cos \theta} \tag{2.25}$$

The linear dispersion of a grating is the reciprocal of the product of the angular dispersion by the effective focal length f. It is given in nm/mm by

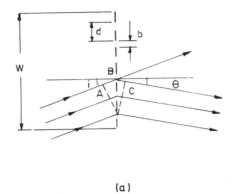

(a)

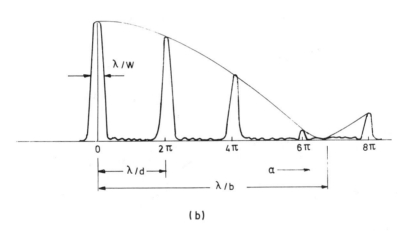

(b)

Figure 2.10   (a) Diffraction at a plane grating and (b) intensity distribution in the far field of a plane grating.

$$\frac{d\lambda}{dx} = \frac{d \cos \theta}{mf} \tag{2.26}$$

The resolving power of the grating is given by

$$\frac{\lambda}{d\lambda} = mN \tag{2.27}$$

The resolving power depends on the product of the order m and the total number of lines in the grating. It can also be expressed as

$$\frac{\lambda}{d\lambda} = \frac{W}{\lambda} (\sin i + \sin \theta) \tag{2.28}$$

where W is the width of the grating.

More spectral energy can be concentrated into a particular order by blazing the grating. A blazed grating has rulings such that the reflecting surfaces are tilted with respect to the grating surface. The irradiance distribution is then given by

$$I(i,\theta) = I_0 \left(\frac{\sin \alpha'}{\alpha'} \right)^2 \left(\frac{\sin N\beta}{\sin \beta} \right)^2 \tag{2.29}$$

where

$$\alpha' = \frac{\pi b \{ \sin(i - \phi) + \sin (\theta - \theta) \}}{\lambda}$$

$$\beta' = \frac{\pi d (\sin i + \sin \theta)}{\lambda}$$

and ϕ is the blaze angle.

2.7.2 Concave Grating

Concave gratings are made by ruling equispaced grooves on a concave surface of radius R. Let (l_A, i) be the coordinates of the point source. The image will be formed, under first order approximation, at (l_B, θ) when the following equations are satisifed:

$$\frac{\cos^2 i}{l_A} - \frac{\cos i}{R} + \frac{\cos^2 \theta}{l_B} - \frac{\cos \theta}{R} = 0 \tag{2.30a}$$

$$\frac{1}{l_A} - \frac{\cos i}{R} + \frac{1}{l_B} - \frac{\cos \theta}{R} = 0 \tag{2.30b}$$

If the point source is on the Rowland circle, i.e., $l_A = R \cos i$, and if it is imaged also on the Rowland circle, then $l_B = R \cos \theta$. Thus we see that (2.30a) and (2.30b) cannot be satisfied simultaneously. This means that on the Rowland circle there is no primary coma but only astigmatism so that a point on the slit is imaged as a line of length z, where

$$z = (\sin^2 \theta + \sin i \tan i \cos \theta)l. \qquad (2.31)$$

where l is the length of the ruled line.

2.7.3 Holographic Gratings

Holographic gratings are realized by recording interference patterns on an optically good surface. The surface can be plane, spherical or aspherical. The fringe spacing may also be variable. Such gratings can be aluminized to give higher irradiance; they can also be blazed to give higher irradiance in the order of interest. These gratings are free from ghost images and have very low scattering.They can also be produced in large sizes. These are available in the following three types:

Type I Ruled Equivalent

Ruled equivalent gratings are recorded on plane or concave surfaces with plane waves. The fringe spacing is same over the surface. These gratings can therefore be used in all applications where ruled plane and concave gratings are employed. Nevertheless, these gratings have the same aberrations as ruled gratings.

Type II Aberration Corrected

Aberration corrected gratings are recorded on concave surfaces by non-parallel beams, and hence correction of certain aberrations is possible; particularly astigmatism is greatly reduced. These gratings are used in spectrographs and Seya-Namioka monochromators [6].

Type III Stigmatic

Stigmatic gratings are also recorded on concave surfaces with nonparallel interfering beams. They are, however, corrected for astigmatism. The focal curve is thus not a part of the Rowland circle but is of a complicated shape.

2.7.4 Gratings for Measurement

Gratings in the frequency range of 10 to 100 lines per mm are used for measurement and stress analysis. A grating is cemented to the object under test. On application of the load, the object and consequently the grating undergoes deformation. The deformed grating is compared with the master grating, and the moiré fringes thus formed are analyzed to give deformation components.

In Talbot interferometry, low frequency gratings are used to study flow fields, thermal fields, etc. [7]. Similarly, in diffractometry low-frequency gratings are used. Talbot interferometry has also been used to measure focal lengths of multifocus lenses and radius of curvature of surfaces and in collimation testing [8,9].

Low variable-frequency gratings are used for measurement of transfer functions of lenses and optical instruments like binoculars.

Length measurement with high accuracy is accomplished by gratings. They are used in angle-coding devices. The grating as an optical element is very interesting both academically and in practical applications.

2.8 POLARIZING ELEMENTS [4, 10]

Light is an electromagnetic wave of wavelength from 0.4 to 0.7 μm in the electromagnetic spectrum and is responsible for the sense of vision. In an isotropic medium the electric and magnetic vectors vibrate perpendicular to the direction of propagation. The electric vector, the magnetic vector and the wave vector constitute the orthogonal triplet set.

It is experimentally established that the electric (E) vector is responsible for photoeffects. Observed from the direction of the source, an unpolarized wave will have its E vector oriented randomly but perpendicularly to the direction of propagation as shown in Figure 2.11(a). This symmetry of the E vector with respect to the propagation direction is characteristic of unpolarized light. The light emitted from incandescent

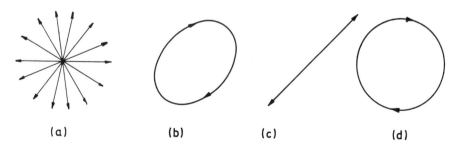

(a)   (b)   (c)   (d)

Figure 2.11   Representation of (a) randomly polarized light, (b) elliptically polarized light (c) linearly polarized light, and (d) circularly polarized light.

lamps, hot bodies, etc. is unpolarized. If the locus of the tip of the E vector is constrained not to change in form and shape, the wave is then polarized. The result is an asymmetry with respect to the direction of propagation. If the tip of the electric vector generates an ellipse that does not change in shape and form, the wave is said to be elliptically polarized; see Fig. 2.11(b) The tip can, however, rotate clockwise or anticlockwise resulting in right-handed or left-handed elliptical polarizations. Two special cases are linearly polarized and circularly polarized light; see Fig. 2.11(c) and (d). In a linearly polarized beam, the tip of the electric vector generates a line. The azimuth of linearly polarized light is defined by the angle that the line makes with the horizontal plane. In circularly polarized light, the tip of the E vector generates a circle. In general we characterize polarized light by four parameters: the magnitude of the E vector (irradiance), the azimuth, the ellipticity and the sense of rotation. These are shown in Fig. 2.12.

There are various ways to describe polarized light, namely,

1. Jones vector and Jones matrix when the light is coherent.
2. Muller matrix and Stokes parameters
3. Poincaré sphere representation

The light emitted from lasers having Brewster windows is linearly polarized. It is possible to obtain polarized light of any state of polarization and to convert elliptical to circularly or linearly polarized light. It is con-

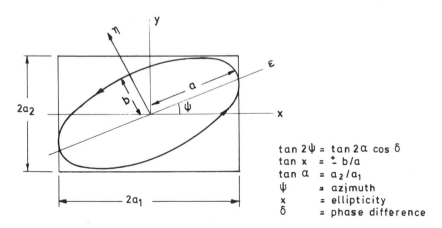

$$\tan 2\psi = \tan 2\alpha \cos \delta$$
$$\tan x = \pm\, b/a$$
$$\tan \alpha = a_2/a_1$$
$$\psi = \text{azimuth}$$
$$x = \text{ellipticity}$$
$$\delta = \text{phase difference}$$

Figure 2.12 Representation of elliptically polarized light showing azimuth, ellipticity and handedness.

venient to work with linearly polarized light.

There are a large number of optical devices that are used to obtain a polarized beam and manipulate the state of polarization. These components are usually made of transparent anisotropic materials. An anisotropic material is one whose physical properties, here refractive indices, are direction dependent. When a beam of light is incident on such a material, two waves, which are linearly polarized with their E vectors orthogonal to each other propagate in the material as shown in Fig. 2.13. Before discussing the polarization components made of anisotropic materials, we shall first consider some definitions:

Plane of incidence: The plane containing the normal to the surface and the incident ray.

Plane of vibration: The plane in which the E vector vibrates.

Plane of polarization: A plane perpendicular to the plane of vibration.

Optic axis: A direction in a crystal (anisotropic material) along which both o and e waves propagate with the same velocity.

Principal section/plane: Contains the incident tray and the optic axis.

o wave: Ordinary wave with the direction of vibration of the E vector perpendicular to the principal section.

e wave: Extra ordinary wave with the E vector lying in the principal plane.

Degree of polarization in reflection: Defined as $P = (R_{\parallel} - R_{\perp})/(R_{\parallel} + R_{\perp})$

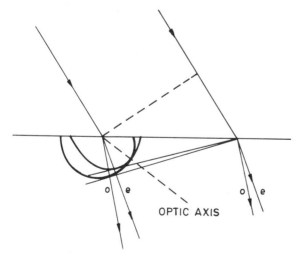

Figure 2.13   Ray propagation in an anisotropic medium.

where $R_{\parallel}$ and $R_{\perp}$ are the reflectivities for parallel and perpendicular components.

Extinction ratio: A ratio of intensities of linearly polarized light transmitted by a polarizer along its transmission and extinction directions.

2.8.1 Polarizers

The elements used for the production of linearly polarized light are called polarizers. Of the many methods available for the production of linearly polarized light, polarizers based either on dichroism or on double refraction are the most commonly used. In dichroic materials, a beam with a certain direction of polarization is transmitted, and the beam with polarization orthogonal to it is strongly absorbed. The large polarizing sheets used in windows and for the fabrication of large polariscopes work on dichroism and are commercially known as polaroid sheets. Their extinction ratio is poor. On the other hand, polarizers made from crystals are of small aperture and high extinction ratio. Fig. 2.14 illustrates the Nicol prism and the Glan-Thompson prism along with the direction of the optic

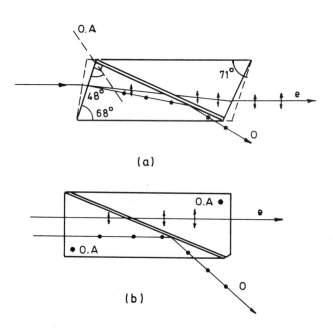

Figure 2.14   Schematic of (a) Nicol prism and (b) Glan-Thompson prism.

axis. A combination of a linear polarizer and a quarter-wave plate is an elliptical polarizer.

2.8.2 Phase Plates/Components

Phase plates are made from uniaxial crystals (or stressed plastic sheets) with optic axis lying in the surface; see Fig. 2.15(a). For a plate of thickness t the path difference introduced by the plate is $|n_o - n_e|t$. The thicknesses for λ plates, $\lambda/2$ plates and $\lambda/4$ plates are given by

$$t = \frac{m\lambda}{|n_0 - n_e|} \quad \text{full-wave plate}$$

$$= \frac{(2m + 1)\lambda}{2|n_0 - n_e|} \quad \text{half-wave plate}$$

$$= \frac{(2m + 1)\lambda}{4|n_0 - n_e|} \quad \text{quarter-wave plate}$$

The compensators most commonly used are the Babinet and Soleil compensators. The Babinet compensator consists of two wedges, one fixed and another that can be shifted parallel to the former as shown in Figure 2.15(b). The path difference Δ introduced by the compensator is given by

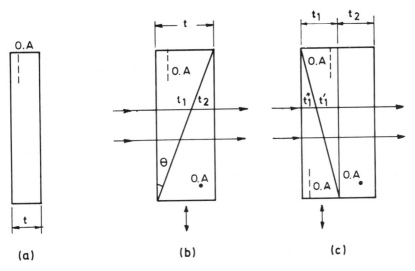

Figure 2.15 (a)A phase plate, (b) Babinet compensator, and (c) Soleil compensator.

$$\Delta = |n_0 - n_e|(t_1 - t_2) \tag{2.32}$$

Since t_1-t_2 varies perpendicularly to the edge of the wedge, the phase difference varies along the length of the compensator.

The Soleil compensator gives only constant phase difference over its surface, as is evident from Figure 2.15(c). The path difference is again given by the above equation, but $t_1 - t_2$ is constant over the whole surface.

2.8.3 The Savart Plate

The Savart plate introduces a linear shear between o and e waves. The magnitude d of the shear is given by

$$d = \frac{t\sqrt{2}(n_o^2 - n_e^2)}{n_o^2 + n_e^2} \tag{2.33}$$

where 2t is the thickness of the plate.

2.8.4 The Wollaston Prism

If the angle of the wedges of a Babinet compensator is large and the wedges are cemented, then the compensator is called a Wollaston prism. It introduces angular shear between o and e waves as shown in Figure 2.16. The angular shear α is given by

$$\alpha = 2 \mid n_o - n_e \mid \tan \theta \tag{2.34}$$

Other prisms that introduce angular shear are the Rochon prism and the

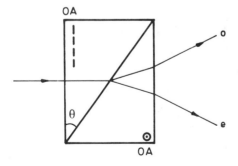

Figure 2.16   Schematic of a Wollaston prism.

Sénarmont prism. These prisms are used as beam splitters in many applications.

2.9 FIBER-OPTIC COMPONENTS

A ray of light is conducted [11,13,14] through a cylindrical fiber of dielectric material by multiple total internal reflections. To avoid cross-talk between fibers when packed, each fiber is coated with a dielectric material of low refractive index. In some cases, another layer of an absorbing material is given. The light-conducting fiber thus consists of a core and a clad. The numerical aperture of a single fiber for meridional rays is $(n^2 - n_o^2)^{1/2}$ where n and n_o are the refractive indices of the core and the clad respectively. For strict analysis, skew rays should be considered. The light transmission efficiency of a fiber depends on its refractive index, the absorption coefficient of its material, and the scattering loss at the boundary between core and clad. If the refractive index of the core is not constant but varies parabolically, the fiber possesses the focusing property and is called by the trade name Selfoc. Selfoc bundles are used for image transfer in table-top Xerox machines and in some scanners [12]. Other refractive index variations have also been studied.

An optical fiber essentially transmits the irradiance distribution at its entrance face to the exit face. If the fiber bundle consists of coherently aligned fibers, a sampled version of the image formed at its entrance face is transmitted. These fiber bundles are thus called coherent bundles. On the other hand, if the fibers are not aligned, the incident distribution is scrambled, and only the light is transmitted. These fiber bundles are called incoherent bundles. Incoherent bundles are mainly used for the conduction of light to inaccessible areas, because of the flexible nature of the fibers. The heat of the source, and vibration influences, on the object are thus eliminated. Coherent bundles are used for the transfer of images, and because of flexibility of the fiber bundle, inaccessible areas can be viewed. Further, coherent bundles are used to manipulate the image, i.e., to magnify and demagnify the image, to invert the image, to correct for image curvature, etc. Some fiber-optic components and their applications are shown in Fig. 2.17. Recently, single-mode fiber has been used to provide a reference beam in electronic speckle pattern interferometery. Fiber optics is becoming an important medium for information transfer. Various elements like couplers, interconnects, etc. are used in these systems. Couplers are available in a variety of forms and designs: the requirement is that they introduce minimum loss [13,14]

Further, the fiber optic sensors are replacing conventional sensors because of their immunity to EMI and EMP, their low power consumption,

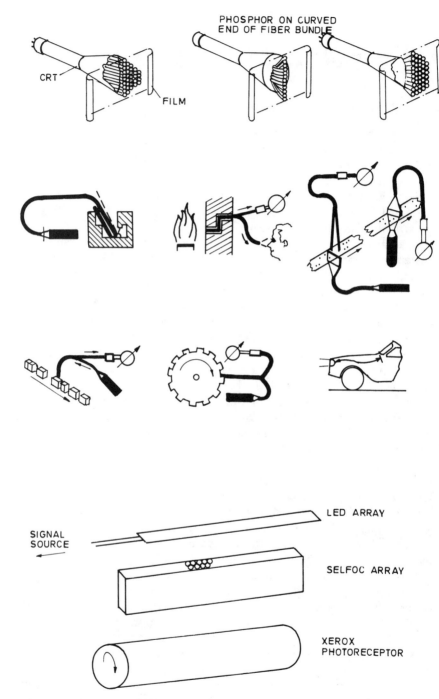

Figure 2.17 Some applications of fiber optics.

their use in hazardous areas, and so on. These sensors are both interfero-metric and noninterferometric and are used for the measurement of process and other variables. Fiber-optic sensors are discussed in detail in Chapter 10.

2.10 THIN FILM COMPONENTS

Deposition of a thin layer on a surface can drastically alter the reflection, transmission, and polarization properties of the surface [15]. Thin films are used to reduce the reflectivity (anti-reflection coatings), to enchance re-flectivity (beam splitters and high-reflectivity mirrors), to change selec-tively the reflectivity or transmissivity (cold mirrors, hot mirrors, dichroic mirrors, filters), and to change the polarization state of the light (polar-izers).

The optical properties of a thin film or a stack of thin films can be obtained by the solution of Maxwell's equations with appropriate bound-ary conditions. We write the reflectivity and transmissivity of a film of thickness h and refractive index n_2 on a substrate of refractive index n_3 as (Fig. 2.18) [5]

$$R = \frac{r_{12}^2 + r_{23}^2 + 2r_{12}\,r_{23}\cos 2\beta}{1 + r_{12}^2\,r_{23}^2 + 2r_{12\,r_{23}}\cos 2\beta} \tag{2.35a}$$

and

$$T = \frac{n_2\cos\theta_3}{n_1\cos\theta_1}\,\frac{t_{12}^2\,t_{23}^2}{1 + r_{12}^2\,r_{23}^2 + 2r_{12}\,r_{23}\cos 2\beta} \tag{2.35b}$$

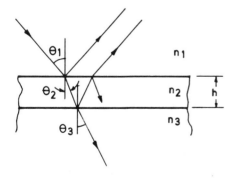

Figure 2.18   Reflection and refraction at a dielectric film on a dielectric substrate.

where r_{ij} and t_{ij} are the reflection and transmission coefficients of the interface between i and j media, n_1 is the refractive index of the first medium, θ_3 and θ_1 are the angles in the first medium and substrate, and β $=(2\pi/\lambda)n_2h\cos\theta_2$. It may be seen from these expressions that the reflectivity and transmissivity remain unchanged when $\beta \rightarrow \beta + \pi$.

It can be shown that the reflectivity will be minimum when a film of quarter-wavelength optical thickness is deposited on the substrate, provided $n_3 > n_2$. The reflectivity for normal incidence is given by

$$R = \frac{n_1n_3 - n_2^2}{n_1n_3 + n_2^2} \tag{2.36}$$

The reflectivity would be zero if $n_1n_3 = n_2^2$. On the other hand, reflectivity is enhanced when a film of quarter-wavelength optical thickness is deposited on a substrate whose refractive index is lower than that of the film.

We can understand the mechanism of decrease and the enhancement of reflectivity of the surface by a film of $\lambda/4$ optical thickness by considering the following simple picture. It is known that a phase change of π occurs when a wave is reflected from a denser medium. Now consider a film of refractive index lower than that of the substrate deposited on it. The wave reflected from the air-film interface suffers a phase change of π. Similarly, the wave reflected from the film-substrate interface also suffers a phase change of π. But this wave traverses the film twice and hence acquires an additional phase delay of π. Therefore the two waves reflected from the upper and lower interfaces are out of phase by π and hence interfere destructively, whereby the reflectivity is reduced. On the other hand, if the index of the film is higher than that of the substrate, the two waves will meet in phase, thereby enhancing the reflectivity.

The optical properties of the surface are further modified using more than one layer of dielectric material of different refractive indices and of suitable thicknesses. Figure 2.19(a) shows the reflectivity of a surface coated with a single layer of low-refractive-index material and a broad band antireflection (AR) coating. Mirrors of reflectivity approaching 100% can be achieved by multilayer stacks of high and low refractive indices. Figure 2.19(b) shows the calculated reflectivity of a multilayer of four periods on a glass substrate. It exhibits strong high-reflectivity zones that can be suppressed by the proper choice of the ratio of thicknesses of high and low refractive index films. Filters are obtained by sandwiching a half-wavelength layer between two high-reflectivity stacks. Dielectric coatings are used for laser mirrors, Fabry-Perot etalons, filters, etc. Antireflection coat-

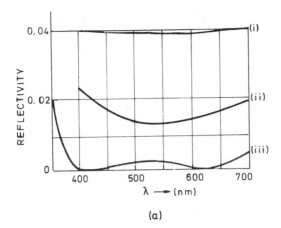

(a)

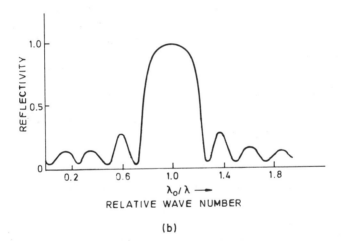

RELATIVE WAVE NUMBER

(b)

Figure 2.19 (a) Reflectivity of a bare substrate (i) n = 1.52; antireflection coating, single layer (ii); broad-band antireflection coating, multilayer (iii). (b) Calculated reflectivity for a multilayer of 4 periods: air $(HL)^4$ glass; $n_H h_H$ =$n_L h_l$ = $\lambda_0/4$; n_H = 2.36, n_L = 1.38, and n_{glass} = 1.52.

ings are used in imaging systems.

The reflectivity of metal-coated substrates can be enhanced by providing dielectric layers.

2.11 GRATICULES

Graticules are employed in instruments used for both observation and measurement. Graticules are of two kinds. *Setting graticules* are used for setting a cross wire of a reference mark before a measurement is made. A number of graticules along with their setting accuracies are shown in Figure 2.20 [16]. It may be noted that graticules with symmetrical settings provide better setting accuracy. *Measurement graticules* are used for comparison, measurement, etc. The magnification of an optical system plays an important role when graticules are used for measurement. The accuracy in the graticule is equally demanding. A good example of both the setting and the measurement graticule is a graticule employed in the optical micrometer described in Chapter 4. Other graticules are those used for measurement of angles, comparison of thread profiles, etc; they are also described in Chapter 4.

2.12 MODULATORS

There are a number of methods for modulating light beams. Of these, amplitude modulation, phase modulation, frequency modulation, and po-

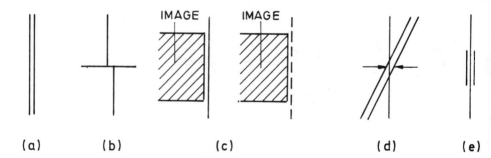

(a)   (b)   (c)   (d)   (e)

Figure 2.20   Various kinds of setting graticules with their setting accuracies.

larization modulation are commonly used.

2.12.1 Amplitude Modulation

Some kind of attenuation of light is obtained in amplitude modulation. One of the simplest ways to modulate the beam is to chop it mechanically using a multiaperture rotating disc. Many spectroscopic instruments make use of this kind of modulation to remove dc or when the detector is followed by a lock-in amplifier.

2.12.2 Electrooptic Modulators

The phase of a light wave can be varied by passing it through a material whose length or refractive index or both can be changed as a function of time. [17] A plate of thickness t and refractive index n introduces a phase change δ when placed in a collimated beam of light of wavelength λ, where

$$\delta = \frac{2\pi}{\lambda} (n - 1)t$$

The phase δ is modulated by varying n by the electrooptic effect. Both amplitude and phase modulations are obtained through electrooptic effects and also through acoustooptic effects.

The Electrooptic Effect

Certain transparent materials exhibit birefringence on application of an electric field. The electrooptic effect, that is, change in $1/n^2$ on application of a field E, is expressed as

$$\Delta(1/n^2) = \gamma E + pE^2 \tag{2.37}$$

where n is the refractive index. γ is the linear electrooptic coefficient and p is the coefficient associated with the quadratic effect. In solids, the linear variation is called the Pockels effect, while the quadratic is called the Kerr electrooptic effect.

The Pockels Effect

Let us assume [17,18] that a wave with E vector vertical is propagating along the z direction in an electrooptic medium with axes along x and y. The axes x and y are oriented at 45° with the vertical. The electric field components at the entrance face of the medium are

$$E_x = \frac{E_0}{\sqrt{2}} \cos wt \quad \text{and} \quad E_y = \frac{E_0}{\sqrt{2}} \cos wt$$

The index changes, $n-n_0$, on application of an electric field axially (along the z-axis) is given by

$$n - n_0 = \pm \frac{1}{2} \gamma\, n_0^3\, E \tag{2.38}$$

If the crystal axes are properly chosen, the indices of refraction along the x and y axes are

$$n_x = n_0 + \frac{1}{2} \gamma\, n_0^3\, E \tag{2.39a}$$

$$n_y = n_0 - \frac{1}{2} \gamma\, n_0^3\, E \tag{2.39b}$$

Therefore the field components of the wave at the exit face of the electrooptic material of thickness L will be given by

$$E_x = \frac{E_0}{\sqrt{2}} \cos(wt + \phi_x) \quad \text{and} \quad E_y = \frac{E_0}{\sqrt{2}} \cos(wt + \phi_y)$$

where ϕ_x and ϕ_y are the phases introduced by the material. These are expressed as

$$\phi_x = \frac{2\pi}{\lambda} n_x L = \phi_0 + \frac{\pi}{\lambda} \gamma n_0^3\, EL \tag{2.40a}$$

$$\phi_y = \frac{2\pi}{\lambda} n_y L = \phi_0 - \frac{\pi}{\lambda} \gamma n_0^3\, EL \tag{2.40b}$$

The field independant phase ϕ_0 is $(2\pi/\lambda)n_0 L$.

An analyzer with transmission axis horizontal is placed after the electrooptic medium. It allows only the horizontal components of the field through. The field transmitted is therefore given by

$$E_H = \frac{E_x}{\sqrt{2}} - \frac{E_y}{\sqrt{2}} = \frac{E_0}{2}[\cos(wt + \phi_x) - \cos(wt + \phi_y)]$$

$$= E_0 \sin(\frac{\Delta\phi}{2}) \sin(wt + \phi_0) \tag{2.41}$$

where $\Delta\phi = \phi_x - \phi_y = (2\pi/\lambda)\gamma\, n_0^3 EL = (2\pi/\lambda)\, n_0^3\, V$ and V is the potential applied across the medium. When the phase difference $\Delta\phi = \pi$, the voltage V_0 applied across the medium is known as the half-retardation voltage. It is given by

$$V_0 = \frac{\lambda}{2\gamma n_0^3} \tag{2.42}$$

The amplitude E_H (Eq. 2.41) can be expressed as

$$E_H = E_0\left(\sin \frac{\pi}{2} \frac{V}{V_0} \right) \sin(wt + \phi_0) \tag{2.43}$$

The intensity of light transmitted by the analyzer is given by

$$I = |E_H|^2 = I_0 \sin^2\left(\frac{\pi}{2} \frac{V}{V_0}\right) \tag{2.44}$$

The transmitted intensity varies sinusoidally with V. When V=0, the transmitted intensity is zero. When $V = V_0$, however, the transmitted intensity is the maximum. If we consider an ammonium dihydrogen phosphate (ADP) crystal, the potential V_0 required to give maximum transmission when used in the blue-green region ($\lambda = 0.5$ μm) is about 10,000 V. The effectiveness and ease of operation is enhanced by placing a quarter-wave plate in the optical path between the polarizer and the electrooptic medium. This is equivalent to introducing the bias in the transmittance curve. The transmitted intensity with a quarter-wave plate in the path is given by

$$\frac{I}{I_0} = \frac{1}{2} \left(1 + \sin \frac{\pi}{2} \frac{V}{V_0}\right)$$

$$= \frac{1}{2} + \frac{\pi}{4} \frac{V}{V_0} \quad \text{if } V \ll V_0 \tag{2.45}$$

If we apply a sinusoidal signal to the modulator, the transmitted intensity is modulated. Let the impressed voltage have a frequency w_n (modulation frequency) and depth of modulation m; then

$$\frac{V}{V_0} = m \sin w_n t \quad m \ll 1 \tag{2.46}$$

Thus the transmitted intensity is

$$\frac{I}{I_0} = \frac{1}{2} + \frac{\pi}{4} m \sin w_n t \tag{2.47}$$

The percentage of modulation is defined by $2J_1(\pi V_m/V_0)$, where $V_m = 0.380$ V_0 is the peak modulation voltage across the electrooptic material, which gives 100% modulation.

This modulator is called longitudinal effect device. A transverse-mode operation is also possible, where the electric field is applied normal to the light path in the material. It can be shown that the half-wave voltage V_0 is given by

$$V_0 = \frac{\lambda}{n_0{}^3\gamma} \frac{A}{L} \qquad (2.48)$$

where A is the electrode spacing. It is seen that V_0 can be reduced by choosing a proper geometry.

The Kerr Electrooptic Effect

The quadratic electrooptic effect is made use of in Kerr modulators. The phase retardation varies with the square of the modulation voltage rather than linearly with it. The quadratic effect may, however, be made linear if the dc electric field (the bias field) is large enough compared to the ac field (the modulation field). Let

$$E = E_0 + E_m \sin w_n t \qquad E_m \ll E_0 \qquad (2.49)$$

Therefore

$$E^2 = E_0{}^2 + 2E_0E_m \sin w_n t \qquad (2.50)$$

The phase change ϕ, being proportional to E^2, is given by

$$\phi = \phi_0 + \phi_m \sin w_n t \qquad (2.51)$$

The phase ϕ_0 is a constant, while ϕ_m varies with E_m, giving linear modulation.

Liquid Kerr cells using nitrobenzene and carbon disulphide have been used, but their applications are limited because of large power requirements. Potassium tantalate niobate (KTN) is a promising crystal in this sense, as the operating voltages are small (dc bias 300 V).

2.12.3 Acoustooptic Modulators

A moving grating placed in a beam modulates the beam. A traveling grating can be created in the medium by an ultrasonic field. The light beam passing through the medium is diffracted. The angle of diffraction θ is given by the grating equation

$$\sin \theta = N \frac{\lambda}{d} \tag{2.52}$$

where the integer N gives the diffraction order and λ is the wavelength of the light. The grating pitch d is equal to the wavelength of an acoustic field in the medium. Therefore $d = v/f$, where f is the frequency and v is the velocity of the acoustic field in the medium. Under low-angle diffraction, the angle of diffraction θ is given by

$$\theta = N \lambda \frac{f}{v} \tag{2.53}$$

We now define a parameter q as

$$q = \frac{2\pi\lambda Lf^2}{nv^2} \tag{2.54}$$

where L is the interaction length and n is the refractive index of the medium. When $q < 1$, we are in the Raman-Nath regime. The diffraction efficiency η is given by

$$\eta = J_1^2 \left(\frac{2\pi L \Delta n}{\lambda \cos \theta} \right) \tag{2.55}$$

where Δn is the refractive index change caused by the acoustic field and J_1 is the Bessel function of the first order and the first kind. The intensity of the diffracted wave varies with modulation. When $q \gg 1$, we are in the Bragg regime. The diffraction efficiency is given by

$$\eta = \sin^2 \left(\frac{\pi^2 LMP}{2H\lambda^2} \right)^{\frac{1}{2}} = \sin^2 \left(\frac{\pi L}{\sqrt{2}\lambda} \sqrt{MI_{aco}} \right) \tag{2.56}$$

where H is the acoustic field height, M the acoustooptic figure of merit, P the acoustic power, and I_{aco} the acoustic intensity. The figure of merit M is related to the other medium parameters through

$$M = \frac{n^6 p^2}{\rho v^3} \tag{2.57}$$

where p is the photoelastic constant and ρ the density of the medium.

2.12.4 The Magnetooptic Light Modulator

The magnetooptic light modulator presents an attractive alternative to the more familiar electrooptic devices for modulation in the wavelength band 1.1 to 5.5 μm. The modulator is based on the Faraday effect. When plane polarized light is passed through a medium in a direction parallel to the applied magnetic field, the plane of vibration is rotated. The amount of rotation θ in min. of arc is expressed as $\theta = VBL$, where B is the magnetic induction in teslas, L is the thickness in meters and V is the Verdet constant. The Faraday effect has been observed in a variety of gases, liquids and solids.

The rotation of the plane of polarization is a phenomenon quite different from the conversion of linear polarization to elliptical polarization in electrooptic devices. In this case, when linearly polarized light enters the medium and exhibits Faraday rotation, it divides into right- and left-handed circularly polarized waves of equal amplitudes. These waves travel with different velocities in the magnetized medium. The phase change thus introduced causes the plane of polarization to rotate on emergence. The amount of rotation is proportional to the component of magnetic field along the direction of propagation. The resulting rotation in the plane of polarization can be converted to amplitude modulation by passage through a linear polarizer.

The Faraday effect may be observed in all transparent materials, but its amplitude is usually small. Only materials of the ferromagnetic class exhibit large Faraday effects. Usually large Faraday rotation is accompanied by strong optical absorption. So an efficient magnetooptic modulator requires a material that gives the largest rotation per unit of optical loss from absorption. The yttrium iron garnet (YIG) crystal has high transparency in the near infrared region, and the ratio of Faraday rotation to the optical loss is 30 times higher than that of chromium tribromide. The region of transparency of YIG lies in the wavelength range from 450 to 1200 nm in the near infrared. A typical YIG modulator uses a rod 10 mm long and 0.5 mm in diameter. The light beam propagates along the rod axis (the x axis), and a transverse bias field (along the z axis) is applied by a permanent magnet. The value of the bias field is selected so that the rod is saturated. A small coil wound around the rod gives a small radio-frequency (RF) magnetic field along the x axis, thereby exhibiting the Faraday effect. The dc bias is such that the magnetic resonance frequency of the rod is outside the modulation bandwidth in order to avoid RF losses.

REFERENCES

1. Smith, W. J. (1966). *Modern Optical Engineering*, McGraw-Hill, New York.
2. Sirohi, R. S., and Kothiyal, M. P. (1987). Double wedge plate shearing interferometer for collimation test, *Appl. Opt., 26*, 4054–4056.
3. Murty, M. V. R. K. (1964). The use of a single plane parallel plate as a lateral shearing interferometer with a visible gas laser source, *Appl. Opt., 3*, 531–34.
4. Driscoll, W. G., and Vaughan, W. eds. (1978). *Handbook of Optics*, McGraw-Hill, New York.
5. Max, B., and Wolf, E. (1965). *Principals of Optics*, Pergamon Press, Oxford.
6. Noda, H., Namioka, T., and Seya, M. (1974). Design of holographic concave gratings for Seya-Namioka monochromators, *J. Opt. Soc. Amer., 64*, 1043–1048.
7. Silva, D. (1972). Talbot interferometer for radial and lateral derivatives, *Appl. Opt., 11*, 2613–2624.
8. Nakano, Y., and Murata K. (1985). Talbot interferometry for measuring the focal length of a lens, *Appl. Opt., 24*, 3162–3166.
9. Kothiyal, M. P., and Sirohi, R. S. (1987). Improved collimation testing using Talbot interferometry, *Appl. Opt., 26*, 4056–4057.
10. Clark, D. and Grainger, J.F. (1971) *Polarized Light and Optical Measurement*, Pergamon Press, Oxford.
11. Kapany, N. S. (1967). *Fiber Optics: Principles and Applications*, Academic Press, New York.
12. Rees, J.D. (1988) Office applications of gradient index optics, *SPIE, 935*, 27–51.
13. Stowe, D. W. ed. (1985). Fiber optic couplers, connectors and splice technologyII, *SPIE, 574*.
14. Tekippe, V. J. ed. (1986) Components for fiber optic applications, *SPIE, 722*.
15. Macleod, H. A. (1969). *Thin-Film Optical Filters*, Adam Hilger, London.
16. Schulze, R. (1967). Optische Messmethoden, *Handbuch der Physik, 39*, 755–834.
17. Yariv, A. (1976). *Introduction to Optical Electronics*, Holt, Rinehart and Winston, New York.
18. Goldstein, R. (1985). Electrooptic devices in review, *Lasers and Applications*, April, 67–73.
19. Lekavich, J. (1985). Basics of acoustooptic devices, *Lasers and Applications*, April, 59–64.

ADDITIONAL READINGS

Iizuka K. (1985) *Engineering Optics*, Springer-Verlag, Berlin.

Jenkins, F. A., and White H. E. (1976). *Fundamentals of Optics*, McGraw-Hill, New York, Kogakusha, New Delhi.

Longhurst, R. S. (1967) *Geometrical and Physical Optics*, Longmans, London.

Martin, L. C. (1960) *Technical Optics*, 2 vols., Pitman, London.

Mollet, Pol, ed. (1960). *Optics in Metrology*, Pergamon Press, New York.

Naumann, H. (1970). Optik für Konstrukteure, Wilhelm Knapp Verlag, Düsseldorf.
Van Heel, A.C.S., ed. (1967). *Advanced Optical Techniques*, North-Holland, Amsterdam.

CHAPTER 3

Basic Optical Systems

3.1 INTRODUCTION

This chapter deals with different optical systems in their basic forms. Variations for specific applications are dealt with whenever necessary. In this section various common features are discussed.

3.1.1 Aberrations

An optical system forms a perfect image within the paraxial region, i.e., when the object field and the lens aperture are small. In practice the paraxial conditions are not satisfied, and the location and size of the image is different from those given by the paraxial lens equations. We then say that the system suffers from aberrations. The amount by which the actual rays miss the paraxial image point is a measure of aberration. The primary aberrations were investigated and codified and analytical expressions for their determination were derived by Seidel.

There are two main categories of aberrations: monochromatic and chromatic. The monochromatic aberrations are spherical aberration, coma, astigmatism, field curvature, and distortion.

Spherical Aberration

A lens with spherical aberration brings rays from different parts of its aperture to different focal points as shown in Fig. 3.1. The longitudinal

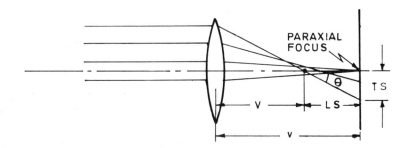

Figure 3.1   Spherical aberration.

spherical aberration of a ray is the distance from the paraxial focus to the axial intersection of the ray. Transverse spherical (TS) or lateral spherical aberration is measured at the paraxial image plane perpendicular to the optical axis. From Fig. 3.1 we have

$$LS = v - V \tag{3.1}$$

$$TS = LS \tan \theta \tag{3.2}$$

where θ is the angle that the ray makes with the axis. When the rays for the outer zones focus closer to the lens, we have undercorrected spherical aberration, which is usually associated with positive elements. Similarly when the marginal rays focus away from the lens, we have overcorrected spherical aberration, which is associated with negative elements.

The spherical aberration of a lens is a function of the object's position, the curvature of its surfaces for a given focal length, and the aperture. For an object at infinity, minimum spherical aberration is given by a nearly plano convex lens (the ratio of the curvatures being 1 to 6) provided that the more convex side is toward the object. When the lens is to be used for unit magnification, an equiconvex shape gives minimum spherical aberration. The image of a point formed by a lens with spherical aberration is a bright dot surrounded by a halo of light.

Coma

Spherical aberration is present even if the object point is on the axis (zero height). Other aberrations start appearing when the object point moves away from the axis (finite height). Coma is the first aberration to appear as we move off axis.

Coma arises from the variation of magnification with the aperture (Fig.

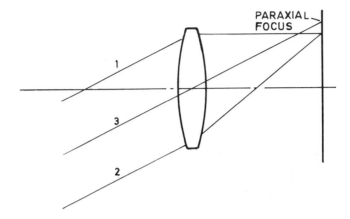

Figure 3.2   Coma.

3.2). Rays 1 and 2, passing through the lens at a larger aperture, come to focus at a height different from that of the central ray 3, which gives the paraxial image height. Coma is a nonsymmetrical aberration and produces an asymmetrical intensity distribution in the image of a point object. This makes the accurate determination of image position difficult, so that error in precision measurements occurs. Coma varies with the shape of the lens elements and is a function of the square of the aperture of the lens and the image height.

Astigmatism

Like coma, astigmatism is an off axis aberration. A point located away from the axis is imaged by the lens as two orthogonal lines separated along the optical axis as shown in Fig. 3.3. Between these two lines the image is an elliptical or circular patch. The distance between the lines is a measure of astigmatism. Astigmatism is a function of the image height and varies with the lens shape.

Field Curvature

When the focused image of a plane object does not lie on a plane but on a curved surface, the optical system is said to have field curvature (also called Petzval curvature). When astigmatism is present, two curved surfaces corresponding to the two astigmatic foci are produced that are away from the Petzval surface (Fig. 3.4). Petzval curvature depends on the index

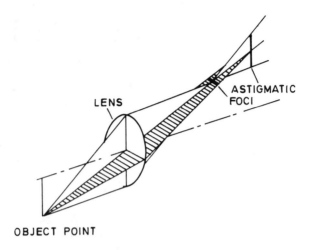

OBJECT POINT

Figure 3.3   Astigmatism.

of refraction, the curvatures of the lens, and the square of the image height. Positive lenses introduce inward curvature and negative lenses introduce backward curvature.

Distortion

Distortion is an aberration in which the image of an off-axial point is displaced from its position given by the paraxial equations, the displacement being a function of the image height. The effect is demonstrated when we consider the imaging of a square object (Fig. 3.5). Depending on the direction of displacement we have pincushion or barrel distortion. Distortion is a serious aberration if measurements are made on the image. If the object and image planes are interchanged, the type of distortion is also interchanged.

Chromatic Aberration

The refractive index varies with the wavelength of the light, and consequently the focal length of a lens is not the same for all colors. As a result, light of different wavelengths comes to focus at different points along the axis, so that longitudinal chromatic aberration results. Similarly, the height of an image varies with a change in wavelength, so that transverse chromatic aberration results.

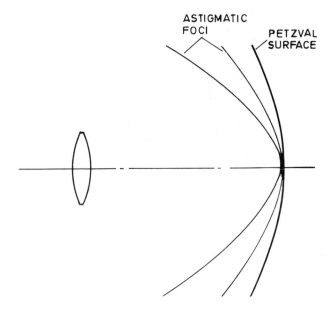

Figure 3.4   Field curvature.

Aberration Correction

Aberrations are corrected by combining more than one lens, by a proper choice of parameters and by playing the positive and negative aberration of the elements against each other. Lens systems are designed with specific applications in mind with given conjugates. For example, a telescopic objective lens is designed for objects at infinity and to accept a small field angle, whereas a microscope objective lens looks at a close object with a high numerical aperture.

3.1.2 Stop and Pupils in Optical Systems

The passage of light rays is limited by the stops in an optical system. These stops may be mountings of the lens system or intentionally placed diaphragms.

The stop that determines the maximum size of the cone of light from an axial point that the optical system will accept is called the *aperture stop*.

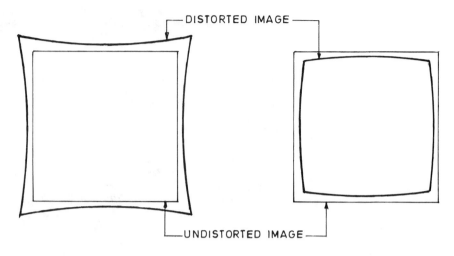

Figure 3.5   Distortion.

Fig. 3.6 shows the position of the aperture stop in a simple system. The aperture stop is not always located in front of the lens system. It may be anywhere within.

The field stop limits the size of the image that the system will image and is usually placed at one of the image planes in the optical system. The field angle ± α or the angular field of view is determined by the size of the field stop as shown in Fig. 3.6.

All the light that enters an optical system through the aperture stop leaves the system through the exit pupil. (see the microscope in Fig. 3.17 and the telescope in Fig. 3.23). Actually, the exit pupil is the image of the aperture stop formed by all the optics beyond the aperture stop (here the eyepiece). The entrance pupil is the image seen through the optics before the aperture stop. Often there is no optics before the aperture stop, in which case the aperture stop is also the entrance pupil. It is clear that the entrance and exit pupils are conjugates.

When an aperture stop is located at the focal point of an optical system, it is called the telecentric stop. This is an invaluable tool for minimizing the error due to poor focusing of an object or scales in metrological instruments such as the tool maker's microscope, the profile projector, etc.

The ray that passes through the centre of the aperture stop is known as the principal ray, and the cone of rays passing through the system

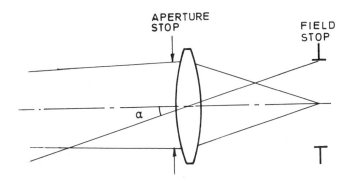

Figure 3.6   Stops in an optical system.

proceeds symmetrically about it. Fig. 3.7 (a) shows a projection of an object OA with a system with a stop at the lens. The image is BI at the image plane (a screen or a graticule plane). If the object is slightly defocused (A 'O'), the new image location would be B 'I' with a defocused projection BC at the image plane, which is different in height from the original image BI. Fig. 3.7 (b) shows the same situation with the aperture stop at the focus of the optical system. As a result, the principal ray is parallel to the optic axis in the object space and is the same for both object positions OA and O'A'. Although the exact image location B 'I' for the defocused location O 'A' of the object remains the same, its projection on the screen shows a symmetric defocusing around the original image BI. Hence a defocusing of the object will not be perceived as a change in the image size. Consequently, slight defocusing with a telecentric stop will not introduce measurement error.

Image Illumination

As mentioned earlier, the maximum size of the cone accepted by a lens system is governed by its clear aperture (aperture stop). The ratio of the focal length to the clear aperture of a lens system is called the relative aperture, f number, or speed of the lens. A lens with 200-mm focal length and 25-mm clear aperture has an f number of 8 which is also written as f/8.

For a lens working at finite conjugates (Fig. 3.8), the numerical aperture (NA) is defined as

$$NA = n' \sin \theta' \tag{3.3}$$

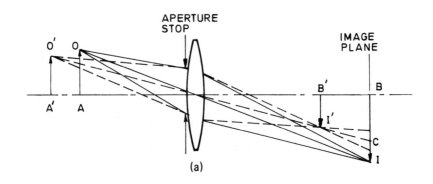

(a)

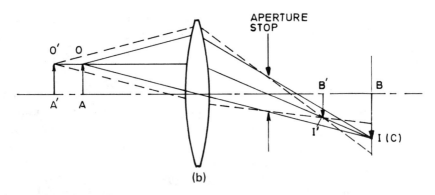

(b)

Figure 3.7   Telecentric stop. Image formation with (a) stop at the lens and (b) stop at the focal plane.

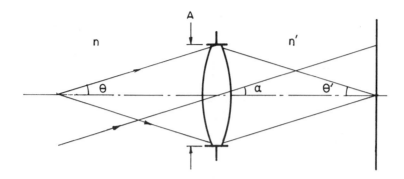

Figure 3.8   Numerical aperture of an optical system.

The number and the numerical aperture relate to the same characteristic of the f system, but the former is used with systems that image objects at large distances (the camera, the telescope), while the latter is more convenient with nearer objects (the microscope). For a corrected system imaging an object at infinity, we have the relation

$$f \text{ number} = \frac{1}{2\text{NA}} \tag{3.4}$$

The illumination (the energy per unit area) in the image of an extended object is inversely proportional to the square of the f number. The illumination is not, however, constant over the whole image. For off-axis image points, the illumination is usually lower than for the points on the axis. It decreases as $\cos^4\alpha$, where α is the angle subtended by the off-axis point at the aperture (the exit pupil).

3.1.3 The Resolution Limit and Diffraction at the Aperture

The image of an ideal point object imaged through an aberration-free optical system is not itself a point but an intensity distribution called the Airy distribution, as shown in Fig. 3.9. This is the result of diffraction at the aperture of the system. Most of the light is of course concentrated in the

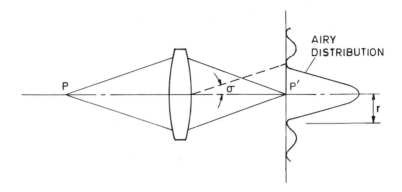

Figure 3.9   Airy distribution.

central maximum whose size (radius) depends on the relative aperture of the optical system and is given by

$$r = \frac{0.61\lambda}{n' \sin \theta'} = \frac{0.61\lambda}{NA} \qquad (3.5)$$

where λ is the wavelegth of the light. This can also be represented in terms of the angle σ subtended by the first minimum at the lens (the angular radius):

$$\sigma = \frac{1.22\lambda}{A} \qquad (3.6)$$

where A is the clear aperture.

Consider now the imaging of two equally bright incoherent point objects. Each point is imaged as an Airy distribution. The two intensity patterns will overlap if the points are close enough, and the total intensity will be the sum of the individual intensities. Fig. 3.10 shows the net intensity distribution (thick curve) for two separations of the object points. In Fig. 3.10 (a), the intensity distribution does not show any indication of the two object points being present, i.e., the object points are not resolved for this separation. In Fig. 3.10 (b) there is a dip in the center of the net intensity curve, and the object points will appear to be resolved. According to the Rayleigh criterion for resolution of images, the separation of the maxima of the two Airy patterns should be equal to the radius of the central maximum given by Eq. (3.5) or (3.6). Eq. (3.5) is useful in determin-

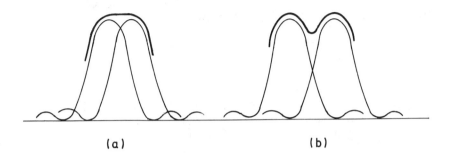

(a) (b)

Figure 3.10 (a) Nonresolution and (b) resolution of two point objects.

ing the resolution of systems such as microscopes working at close distances. To determine the minimum object separation for resolution, the NA on the object side (NA = n sin θ) is used. Eq. (3.6) is suitable for the evaluation of the resolution of systems working with objects at long distances, such as telescopes.

3.1.4 Image Evaluation

A perfect optical system imaging an ideal point object will send out a spherical wavefront coverging to the geometrical image point. We have seen that because of diffraction the image is not a point but an Airy distribution. In the presence of aberrations, the outgoing wavefront will depart from a spherical shape, resulting in a modfication of the Airy distribution. The aberrations of a system can be determined by measuring this departure from the reference shape (spherical or plane). There are several methods of doing this by geometrical or wave optics. These include simulation of ray tracing on an optical bench, the star test, the Foucault test, the Hartman test, the interferometer, etc.

The resolution of an optical system is limited by diffraction. There is a further deterioration of resolution in the presence of aberrations, because they modify the Airy distribution. The resolution is usually measured by examining the image of a pattern of alternating bright and dark lines or bars of equal width. A target with several sets of bars with varying spacing is used. The finest bar pattern that can be distinguished in the image of the pattern formed by the optical system is taken to be the resolution.

A more complete evaluation of an optical system is given by its modulation transfer function (MTF). In the image of a bar pattern the contrast or the modulation M_i is expressed as

$$M_i = \frac{I_{max} - I_{min}}{I_{max} + I_{min}} \tag{3.7}$$

where I_{max} and I_{min} represent the intensities of the bright and dark areas in the image respectively. The maximum value of modulation is unity when I_{min} is zero. In the same way, the modulation of the object M_o can be described. If the object pattern is sinusoidal, the MTF of the system is defined as

$$MTF = \frac{M_i}{M_o} \tag{3.8}$$

The MTF is a function of the frequency (the number of cycles per unit length). In an aberration-free optical system, the MTF is determined by the diffraction effects of the system aperture. Figure 3.11 (curve A) shows the MTF of such a system in terms of the limiting frequency (v_o) at which the

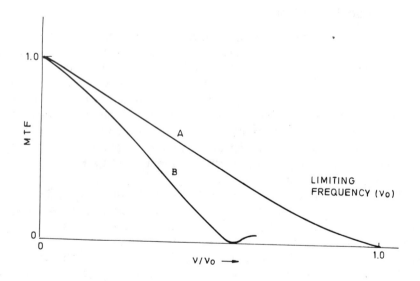

Figure 3.11   Modulation transfer function.

MTF is zero. The limiting frequency is given by

$$v_o = \frac{2NA}{\lambda} \tag{3.9}$$

In the presence of aberrations, the MTF curve is modified such as curve B. The MTF is experimentally determined using an object with variable frequency and measuring the modulation in the image.

The limiting resolution is reduced by the aberrations. In addition there is a minimum detectable modulation. Hence practical resolution may be reduced further. In any case, the limiting resolution is not a measure of the overall performance of an optical system. Two optical systems may have the same limiting resolution, but the one with higher modulation at lower frequencies will give a better contrast image.

The MTF is used to measure the optical performances of not only lenses but also films, image tubes, the eye, etc. The combined MTF of two systems can be obtained by simply multiplying the individual MTF's.

3.1.5 Depth of Focus and Depth of Field

In the image of an object, each image point is illuminated by a cone of rays with its base at the exit pupil of the optical system and its apex at the geometrical image point. On either side of the focused image plane, the point enlarges into a disc (a defocused patch). Even at the true image plane, the image is not a point but the Airy pattern as a result of diffraction. The depth of focus is the maximum displacement of the screen (film or reticle) from the ideal image plane without causing a serious deterioration of the image. The depth of field is the conjugate distance in the object space corresponding to the depth of focus.

The depth of focus is the displacement of the screen so that there is no marked change in the Airy pattern. Based on the Rayleigh criterion this can be expressed as

$$\delta v = \pm \frac{\lambda}{2n' \sin^2 \theta'} \tag{3.10}$$

This is the value for a well-corrected optical system. In the presence of aberrations this value can be much larger. The depth of field is determined by the law of axial magnification.

In practice the size of the defocused patch or blur spot that can be tolerated is larger than the Airy disc because of aberrations or the resolution of the detector (e.g., film), and a geometrical optics approach can be used. The size of the acceptable blur may be specified as the linear diameter B of the blur spot or as an angular blur β i.e., the angular subtense of the blur spot from the lens.

In Fig. 3.12 a point P in the object plane is imaged at P′ in the image plane. The receiving plane may be shifted from the image plane so that the blur spot diameter B′ is within the acceptable limit. On the other hand, there is a blur spot of diameter B in the object plane that will give an image of the size of blur spot B′ when the receiving plane is placed at the exact image plane. We may write

$$\beta = \frac{B}{u} = \frac{B'}{v} \tag{3.11}$$

From Fig. 3.12 it can be seen that with the receiving plane at the image plane the object point P may be displaced by δu_1 to P_1 outside or by δu_2 to P_2 inside to produce a blur spot of diameter B′ at the image plane. We may then write

$$\frac{B}{\delta u_1} = \frac{A}{u + \delta u_1} \quad \text{and} \quad \frac{B}{\delta u_2} = \frac{A}{u - \delta u_2} \tag{3.12}$$

It can be seen that the depth of field is not equal inside and outside the object plane. When δu_1 or δu_2 is small compared to u, however, the depth of field may be written as

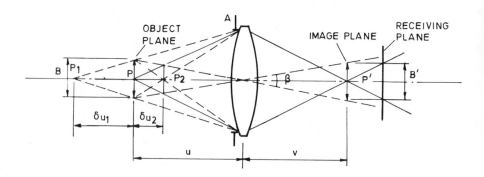

Figure 3.12 Depth of focus and depth of field.

$$\delta u = \frac{u^2\beta}{A} = \frac{uB}{A} \qquad (3.13)$$

On the image side, the depth of focus is given by

$$\delta v = \frac{v^2\beta}{A} = \frac{vB'}{A} \qquad (3.14)$$

If image is at the focal point of the system, then the depth of focus is

$$\delta v = \frac{v^2\beta}{A} = f\beta(f \text{ number}) \qquad (3.15)$$

The object distance for which the depth of field extends to infinity is called the hyperfocal distance of the optical system. δu_1 may be written as

$$\delta u_1 = \frac{u^2\beta}{A\text{-}u\beta} \qquad (3.16)$$

When $\delta u_1 = \infty$ we have

$$u \text{ (hyperfocal)} = \frac{A}{\beta} \qquad (3.17)$$

3.2 THE EYE

The eye is the final optical detecting system in visual optical instruments. It is therefore necessary to understand the function of a human eye as well as its capabilities and limitations.

Figure 3.13 shows a section of an eye. The eyeball is an approximately spherical body about 25 mm in diammeter. The front part is protruded and is called the cornea. The outer shell, called the sclera, is white and opqque except for the cornea, which is clear. Most of the power of the optical system of the eye is provided by the cornea. The lens of the eye is responsible for its variable power. The aperture of the lens is controlled by the iris, which is almost opaque and strongly colored. The iris is capable of expanding and contracting to control the amount of light entering the eye. The pupil size varies from about 2 mm in diameter in very bright light to about 8 mm in diameter in very dim light.

The inner surface of the rear part of the eyeball contains the receiving

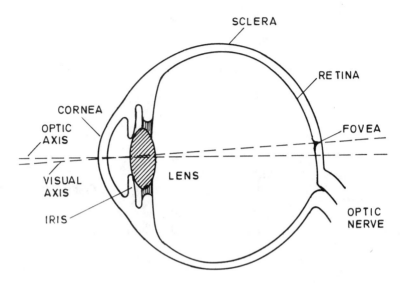

Figure 3.13   Section of a human eye.

screen known as the retina, which consists of blood vessels, nerve fibers, light-sensitive rod and cone cells and a pigment layer. The nerve fibers connect the retina to the brain. Slightly to the outer side of the optical axis of the eye is the macula, the center of which is called the fovea. The central 0.3 mm diameter of the fovea has the highest resolution. Only cones are present in this region. Some distance away from the fovea only rods are present. Cones are responsible for color vision and operate at normal light levels. Rods become active at low light levels but cannot distinguish colors.

The optical system of a relaxed normal eye brings parallel rays to a focus on the retina, and distant objects are clearly seen. The eye is focused to see near objects by changing the shape and hence the power of the lens. This process of varying the power is known as accommodation. The maximum change in power that the eye can produce is called the amplitude of accommodation. Its value for an average person varies with age, being 14 diopters at 10 years, 4 diopters at 40 years, and zero diopters at 75 years. When the eye receives parallel rays, its power is at a minimum. It is at the maximum when it is focused at 250 mm from the eye. This is the closest distance at which the eye can focus and is called the near point or the least distance of distinct vision.

To examine an object critically the eye is rotated to bring the image of the object onto the fovea. The line joining the fovea, the nodal point of the optical system of the eye, and the object point is called the visual axis; it makes an angle of 5° to 6° with the optical axis. The aperture stop of the eye is the iris. There is no field stop. The complete field of view of the eye is quite large—about 130° vertically and 200° horizontally.

3.2.1 Spectral Sensitivity of the Eye

The eye is sensitive to electromagnetic radiation of wavelengths between 0.4 and 0.7μm. The sensitivity varies with wavelength; equal amounts of energy of different wavelengths produce in general different sensations of brightness. The eye is most sensitive to yellow-green light at the wavelength of 0.55 μm. Figure 3.14 shows the relative wavelength sensitivity of the eye for a normal level of illumination (solid line). For lower levels of illumination, the peak shifts toward the blue end of the spectrum (broken line).

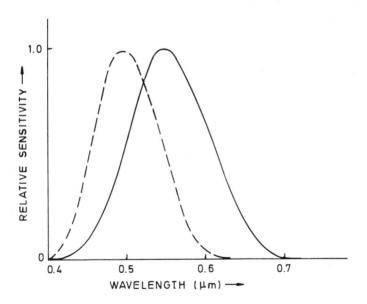

Figure 3.14   Wavelength sensitivity of the eye.

3.2.2. Resolution of the Eye

In order to see two independent neighboring object points, two conditions must be fulfilled: first, the optical system of the eye must produce resolved images (the Rayleigh criterion) and secondly, the light-sensitive receptors (cones) must have a structure fine enough so that the images are separated by one cone; otherwise the images will appear as one. Two object points that subtend an angle of about 1 min of arc at the eye satisfy these conditions. The resolution limit of the eye for two point objects is therefore conventionally assumed to be 1 min of arc (about 0.073 mm at 250 mm) in the design of optical instruments. The resolution here refers to the region of the fovea where it is highest. Resolution deteriorates rapidly as we move away from the fovea.

The resolution of the eye is much better in certain situations that we come across in measurement. Figure 2.20 (Chap. 2) shows several measurement situations: (a) two lines, typically reticle lines, can be superimposed with a resolution of 1 min of arc; (b) two straight lines can be aligned to 15 sec of arc; (c) a straight line can be aligned with the straight edge of an image with an error of 1 min of arc, and on the other hand a broken straight line can be aligned to 20 sec. of arc; (d) a straight line can be located symmetrically between a pair of inclined lines with a resolution of 15 sec of arc; (e) a straight line can be placed symmetrically between a pair of lines to 5 sec of arc. These properties of the eye are made use of in the design of optical measurement systems.

3.3 THE MAGNIFIER

When two object points placed at the normal viewing distance ($\approx$250 mm) subtend an angle of less than one min of arc at the eye, the eye cannot see the two points as separate. Optical systems are used to magnify this angle and present it to the eye, which then can see them as separate. A magnifier is the simplest optical system for achieving this.

The magnifier consists of a lens of focal length f with the object located at or within its first focal point. In Fig. 3.15 the object of height h placed at a distance u from the magnifier is imaged at a distance v with a height h'. The image is virtual, and both u and v are negative quantities. The lens equation is

$$\frac{1}{v} = \frac{1}{f} + \frac{1}{u} \tag{3.18}$$

From Fig. 3.15 we have

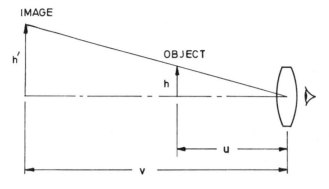

Figure 3.15   Magnifier.

$$h' = \frac{hv}{u} = \frac{h\,(f - v)}{f} \tag{3.19}$$

If the eye is located at the lens, the angle subtended by the image is given by

$$\alpha' = \frac{h'}{v} \tag{3.20}$$

For an unaided eye viewing the same object at a distance of 250 mm, the angle subtended will be given by

$$\alpha = -\frac{h}{250} \tag{3.21}$$

The magnifying power of the magnifer is the ratio between the two angles:

$$MP = \frac{\alpha'}{\alpha} = 250 \left(\frac{1}{f} - \frac{1}{v} \right) \tag{3.22}$$

If we adjust the object distance so that the image is at infinity, i.e., u = −f and v = ∞, we obtain

$$MP = \frac{250}{f} \tag{3.23}$$

On the other hand, if the focus is set so that the image appears 250 mm away, i.e., $v = -250$ mm, then

$$MP = 1 + \frac{250}{f} \tag{3.24}$$

The values of MP given by Eqs. (3.23) and (3.24) are those conventionally used to express powers of magnifiers and eyepieces. The magnifying power of a magnifier rarely exceeds 10×. When higher magnification is required, a microscope consisting of two lens systems is used.

3.3.1 The Magnifier with a Telecentric Stop

It is often necessary to measure an object against a scale with the help of a magnifier. If the object or its image and the scale happen to be in slightly different planes, the eye will see them under different magnifications because they are seen under different angles by the eye, as in Fig. 3.16 (a), so that measurement error occurs. If a telecentric stop is used with the magnifier, the principal ray is parallel to the axis in the object space, and both are seen under same angles; see Fig. 3.16 (b).

3.4 THE MICROSCOPE

The microscope consists of an objective lens and an eyepiece. The objective produces an enlarged real inverted image of the object. The eyepiece reimages the object and magnifies it still further. Figure 3.17 shows the optical system of a microscope. The distance between the eyepiece and the exit pupil is known as the eye relief or eye clearance. The eye is placed at the exit pupil for observation, and its diameter should be smaller than that of the eye pupil for all the light gathered by the microscope to enter the eye.

In Fig. 3.18 the lens systems have been replaced by their principal planes; f_0' and f_E' are the focal lengths of the objective and eyepiece, respectively. From the figure,

$$\frac{h'}{h} = \frac{g}{f_0'} \tag{3.25}$$

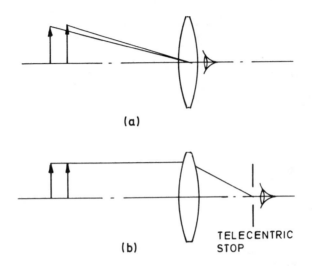

(a)

(b)

TELECENTRIC
STOP

Figure 3.16 Magnifier (a) without and (b) with telecentric stop.

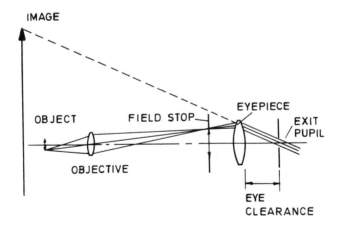

IMAGE

OBJECT

OBJECTIVE

FIELD STOP

EYEPIECE

EXIT
PUPIL

EYE
CLEARANCE

Figure 3.17 Optical arrangement of a microscope.

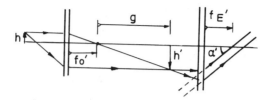

Figure 3.18   Microscope. Lenses are replaced by principal planes.

where g is the distance between the focal points of the objective and the eyepiece and is known as the optical tube length. The image is viewed under an angle α' given by

$$\alpha' = \frac{h'}{f_E'} \tag{3.26}$$

The angle α subtended at the unaided eye viewing the object h at a distance of 250 mm is given by Eq. (3.21). Hence the magnifying power is

$$MP = \frac{\alpha'}{\alpha} = -\frac{g}{f_0'} \frac{250}{f_E'} \tag{3.27}$$

which is the product of the magnifications of the objective and the eyepiece. The negative sign indicates that the final image is inverted. In a conventional microscope the value of g usually is 160 mm. Thus an objective with a 16-mm focal length has a magnifying power of 10.

The resolution of a microscope is limited both by diffraction and by the resolution of the eye. A point object is imaged by an optical system with a circular aperture as the Airy pattern (Sec. 3.1.3). For a microscope to resolve two object points, two conditions should be fulfilled: first, the Airy patterns (the images) corresponding to the two object points should be formed separately (resolved) in the image plane of the objective, and secondly the eyepiece should present these images to the eye with a separation that is more than the resolution limit for the eye. As explained in Sec. 3.1.3 the smallest separation r between two object points that will allow them to be resolved in the image plane of the objective in incoherent illumination is given by

$$r = \frac{0.61\lambda}{n \sin \theta} = \frac{0.61\lambda}{NA} \tag{3.28}$$

where NA = n sinθ, is the numerical aperture of the objective.

The eye has a resolution of 1 min of arc. At 250 mm, this corresponds to 0.073 mm. For the eye to resolve the separation r, the total microscope magnification (MP) should satisfy the relation

$$r \times MP = 0.073 \tag{3.29}$$

Substituting for r we obtain

$$MP = \frac{0.073\ NA}{0.61\lambda} \tag{3.30}$$

with λ in mm. This gives the magnification at which the diffraction and visual limits match. This is the minimum magnification required for the given numerical aperture. Setting λ = 0.55 μm. MP≈ 220 NA. Any mag-nifcation beyond this value is empty magnification, but a value several times this limiting value is used for comfortable viewing.

3.4.1 Microscope Objectives

The design of microscope objectives increases in complexity with increas-ing maginification and numerical aperture. The numerical aperture deter-mines the resolution limit. Figure 3.19 shows various types of microscopic objectives of different powers. These objectives are designed to work at specific conjugates and tube lengths. In the figure, the various objectives are (a) low-power achromatic doublet. (b) Lister objective, 10×, NA 0.25, (c) Amici objective, 20×, NA 0.5 to 40×, NA 0.8, (d) an immersion objective. The object is to the left.

3.4.2 The Eyepiece

The eyepiece used in microscopes (as well as in other instruments) consists of two lenses (or lens systems), the field lens and the eye lens. The action of the field lens is shown in Fig. 3.20. When placed exactly at the internal image, the field lens has no effect on the power of the eyepiece. It does however, bend the ray bundles back toward the axis so that they pass through the eye lens. In this way the field of view may be increased without increasing the diameter of the eye lens. The eye relief must be kept within reasonable limits (5 to 10 mm) for convenient viewing.

In practice, a field lens is rarely located exactly at the image plane, it is either ahead of or behind the image, so that defects such as scratches, dust, etc. on the field lens are not visible. Depending on the position of the

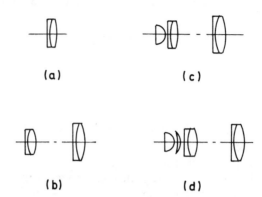

Figure 3.19   Microscope objective designs: (a) a low-power achromatic doublet, (b) a Lister objective, (c) an Amici objective, and (d) an immersion objective.

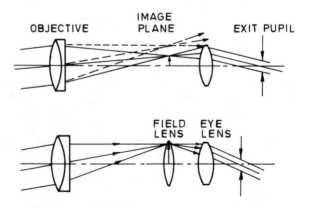

Figure 3.20   Eyepiece in an optical system.

field stop there are two types of eyepiece systems.

The Ramsden eyepiece consists of two plano convex lenses with convex sides facing each other; see Fig. 3.21 (a). If the two lenses have the same focal length and are separated by the distance equal to the individual focal length, then the focal length of the eyepiece itself is also the same, i.e., $f = f_1 = f_2$. If both lenses are made of the same glass, the chromatic aberration is reasonably well corrected. A field stop of appropriate size is placed in front of the field lens at the image plane.

The Huygens eyepiece consists of two plano convex lenses of the same glass oriented as shown in Fig. 3.21 (b). The usual selection of focal lengths f_1 and f_2 and their separation d for removing chromatic aberration is

$$f_1 = 2f_2 \text{ and } d = (f_1 + f_1)/2.$$

The total focal length is $f = (4/3) f_2$. The image plane is between the lenses. Any measurement graticule must be placed here. The graticule will therefore be seen by the eye lens alone for which the correction is not good. This eyepiece is therefore not suitable for use with a graticule. When a measurement graticule is to be used, the Ramsden eyepiece is employed.

An eyepiece is required to cover a fairly wide field. For a 10× eyepiece with a field stop of 20 mm, the field angle is over ±20°. Because of the extensive use of eyepieces for measurement and the necessity to use them to cover wide fields, the basic Ramsden eyepiece has been considerably improved.

The Kellner eyepiece, shown in Fig. 3.22 (a), is a Ramsden eyepiece with an achromatized eye lens to reduce the lateral color. This gives an improved color correction and can be used up to ±15° field angles. This is employed generally in measuring instruments.

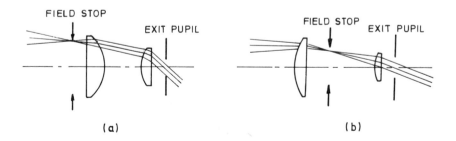

Figure 3.21   Eyepiece types: (a) Ramsden and (b) Huygens.

The orthoscopic eyepiece, shown in Fig. 3.22(b), is better than the Kellner eyepiece and can be used up to ±25°. This also has a very good distortion correction.

The symmetrical eyepiece shown in Fig. 3.22 (c) consists of two achromatic doublets. It covers a field of ±25° and is in general a superior eyepiece but for its distortion which is 30 to 50% greater than that of the orthoscopic eyepiece. The symmetrical eyepiece finds wide application in military instruments.

The Erfle eyepiece is a wide field eyepiece with a coverage of ±30°. Its distortion is similar to that of the orthoscopic for the same angular field and the curvature is less by about 40%; see Fig. 3.22 (d).

3.5 THE TELESCOPE

A telescope enlarges the apparent size of a distant object. This is accomplished by presenting to the eye an image that subtends a larger angle than does the object at the eye. A telescope usually works with both object and image located at infinity. It is referred to as an afocal instrument, since it has no focal length.

Telescope may be classified as astronomical (or inverting), Galilean,

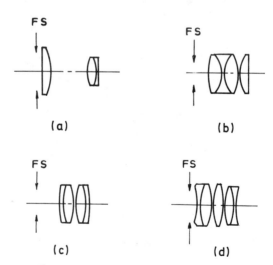

Figure 3.22 Eyepiece designs: (a) Kellner, (b) orthoscopic, (c) symmetrical, and (d) Erfle.

and terrestrial, or erecting. An astronomical telescope is composed of two positive lens systems with common focal points as shown in Fig. 3.23 (a). The objective lens forms an inverted primary image at its focal point. The eyepiece then reimages the object at infinity where it may be comfortably viewed by a relaxed eye. The final image is a virtual inverted image of the object.

If we replace the positive eyepiece of the astronomical telescope with a negative lens, we have a Galilean telescope with a negative lens, we have a Galilean telescope as shown in Fig. 3.23 (b). The lenses are placed so that the focal points again coincide. The object for the eyepiece is virtual, as an internal image is not formed. There is no inversion, and the final image presented to the eye is erect. Since there is no real image formed in a Galilean telescope, there is no location where a graticule may be inserted.

The terrestrial telescope is similar to the astronomical telescope except that an additional lens system, an erector, is employed between the primary image and the eyepiece to erect the image so that final image seen by the eyepiece is similar to the object; see Fig. 3.23 (c).

The magnification of a telescope can be determined with reference to Fig. 3.24.

$$\tan \alpha = \frac{h'}{f_0'} \quad \text{and} \quad \tan \alpha' = \frac{h'}{f_E} \tag{3.31}$$

The magnifying power is

$$MP = \frac{\tan \alpha'}{\tan \alpha} = \frac{f_0'}{f_E} = -\frac{f_0'}{f_E'} \tag{3.32}$$

because $F_E' = -f_E$. In the astronomical telescope the exit pupil is real; see Fig. 3.23 (a), and the eye pupil coincides with this when seeing through such a telescope. In the Galilean telescope, on the other hand, the exit pupil is virtual. The eye pupil of the observer acts as the exit pupil of the telescope, and its conjugate virtual image acts as the entrance pupil.

3.5.1 Exit Pupil Diameter

The diameter of the exit pupil of a telescope should match the eye pupil. The eye pupil varies in diameter from 2 to 8 mm depending on the brightness of the scene being viewed. In normal visual conditions the eye-pupil diameter is about 3 mm and hence an exit pupil diameter of 3 mm will give the optimum design. Variations from this value are common. For example, in surveying instruments an exit pupil of 1.0 to 1.5 mm is frequently used to keep the size and the weight of the instrument low. In ordinary binoculars and riflescopes a larger pupil diameter ($\approx$ 5 mm) is

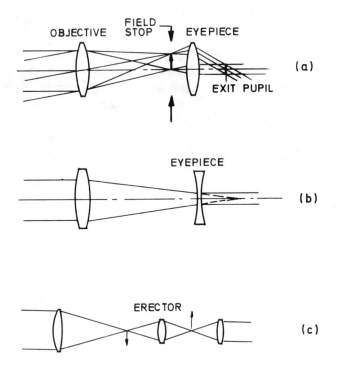

Figure 3.23 Various telescope arrangements: (a) astronomical, (b) Galilean, and (c) terrestrial.

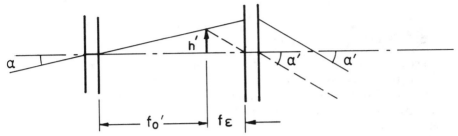

Figure 3.24   Astronomical telescope. Lenses are replaced by principal planes.

to keep the size and the weight of the instrument low. In ordinary binoculars and riflescopes a larger pupil diameter ($\approx$5 mm) is provided that makes it easier to align them with the eyes. In low-light-level instruments an exit pupil diameter of 7 to 8 mm is used to obtain maximum possible illumination at the retina.

3.5.2 Resolving Power

In the case of a telescope one speaks of angular resolving power because the object is generally at very large distance. The resolution limit of a telescope is the smallest angle made by the two object points at the objective that can be seen as separate points in the image. From the diffraction theory this angle is given by

$$\sigma = \frac{1.22\lambda}{A} \tag{3.33}$$

$$= \frac{6.7 \times 10^{-4}}{A} \text{ radians}$$

where λ is the wavelength of light ($\approx$0.55μm) and A is the diameter of the objective in mm. This angle is magnified by the MP of the telescope before being presented to the eye. If the eye has to see the image points as resolved, then MPσ should be equal to the resolution of the eye, i.e., 1 min of arc (2.9×10^{-4} rad), which gives

$$MP = 0.43A \tag{3.34}$$

This gives the minimum magnification required; but magnification of two or three times this value is used to minimize the visual effort.

3.5.3 Telescope Objectives

Most telescope objectives are ordinary doublets. Figure 3.25 (a) shows a telescope construction giving a long overall focal length but a short mechanical dimension. This type of construction is also used in engineering telescopes such as surveying instruments and theodolites, where focusing at various object distances is achieved by shifting the rear elements (see Chap.5). When it is desired to focus the telescope at a very close distance as in an alignment telescope, a positive rear lens close to the focal plane is used, see Fig. 3.25 (b).

3.6 RELAY SYSTEMS

An image can be carried through a long distance, without sacrificing the field angle and still keeping the diameter of the lenses low, with the help of relay lenses. Figure 3.26 shows an arrangement of relay lenses. A field lens A is placed at the primary image formed by the objective. A lens B,

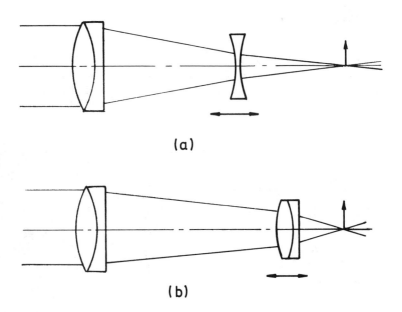

(a)

(b)

Figure 3.25   Telescope objective types: (a) long focal length and (b) short focal length.

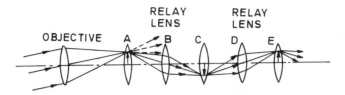

Figure 3.26 Relay lens system. (Courtesy of McGraw-Hill, Smith, W. J., 1966, Modern Optical Engineering.)

called the relay lens, forms a real image in the plane of another field lens C. A second relay lens D forms the image in the plane of the field lens E. This process can be repeated. It can be seen that the entrance pupil (here the objective) is successively imaged in the plane of the relay lenses without vignetting. The dashed rays coming from the first image indicate that large-diameter lenses would be required to cover the same field in the absence of relay lenses.

3.7 THE PROJECTOR

The projector is an optical system for reproduction of objects on a magnified scale on a screen. A projected image can be seen by more than one person at the same time. The basic components of a projector are the illumination system with light source and condenser, the object holder, the projection objective, and the screen.

Opaque (epiprojection) as well as transparent objects (diaprojection) can be projected depending on the type of illumination system. Interchangeable illumination systems are available in many instruments. Generally the image is projected on to an opaque screen. The objeserver and the projector are on the same side of the screen. In some cases, such as profile projection for engineering measurement, the image is formed on a translucent screen, so that the image can be seen from behind the screen. Figure 3.27 shows the optical arrangement of a projector. There may be minor variations of the optical arrangement depending on the light source, application area, etc., but the basic features are always retained. The illuminating ray bundle must be properly matched with the projection objective aperture.

The magnification of the projected image is obtained from the usual lens relations and is given by

$$M = \frac{f\text{-}v}{f} \tag{3.35}$$

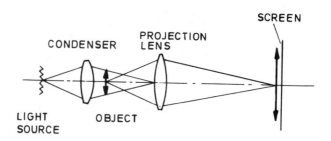

Figure 3.27   Optical arrangement of a projector.

where v is the distance of the screen from the projection lens and f its focal length.

The projection objectives for slide projection are similar to camera objectives in construction and image quality. The focal length is around 100 mm for general-purpose projection. In projection for applications in engineering measurement by tool maker's microscope or profile projector, a series of objectives of differennt focal lengths are available to vary the magnification. Further, these are equipped with telecentric stops to minimize focus error.

As mentioned above, there are two types of screens in use, opaque and translucent. An ideal opaque screen should reflect back all the light falling on it without any preferential direction. A screen is called a diffusing screen when it looks equally bright from all directions. A good diffusing surface can be prepared from oxides of metals, such as magnesium oxide.

Translucent screens, which are generally used in technical projectors for projecting scale lines of profile images, are made of ground glass. The diffusing ability of the glass increases with the increased size of the grains. The upper limit for the size of the grains is set by the resolution required in the image. Small angle scattering of the ground glass screen results in dropping of brightness at the edges of the image. Since the major portion of the light energy goes in and around the direction of the ray, the eye receives only a small percentage of light from the edges, as shown in Fig. 3.28. This can be overcome by placing a field lens next to the screen, which makes the scatttered rays bend toward the axis. The power of the field lens is so chosen that the aperture of the objective is imaged at the eye pupil. The rays from the rim of the screen are bent toward the eye, and the screen looks equally bright all over. A Fresnel lens can be effectively used as a field lens in such cases.

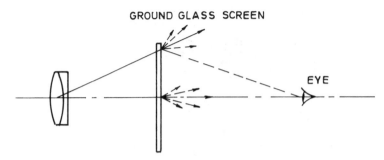

Figure 3.28   Screen of a projector. Intensity falls off at the edges.

3.8 CONDENSERS

Most of the objects imaged by optical instruments are not self-luminous and must be illuminated by artifical light sources. An optical system used for the illuimination of an object is known as a condenser. The actual condenser type depends on the requirements of the problem and the type of instrument with which it is to be used.

3.8.1 Condensers for Projection

Let us consider the projection of a thin object such as a film. Figure 3.29 (a) shows a light source S placed behind an object F that is to be projected by a projection lens L. Light rays can be drawn through various points (A, B,C) of the object starting from the light source and passing through the projection lens. The solid lines passing through the projection lens. The solid lines passing through the axial point A indicate the contribution at A by the whole source, but the projection lens aperture is partially utilized. The dotted lines through the point B indicate that the source contributes partially to this point and a still smaller aperture of the projection lens is made use of. No rays can be drawn through the point C that will enter the projection lens. It is obvious that illumination in the image will fall off rapidly. The situation can be marginally improved by bringing the source closer to the film. The essential requirements of an ideal condenser are to illuminate the object area uniformly and to fill the projection lens aperture through each object point. Two condenser arrangements are possible.

Figure 3.29 (b) shows the first arrangement, which is used when the light source has a structure, as a filament does. The filament is imaged

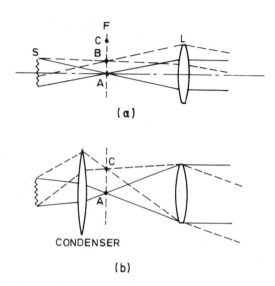

(a)

CONDENSER

(b)

Figure 3.29 Projection of a thin object (a) without a condenser and (b) with a condenser. (Courtesy of McGraw-Hill, Smith, W. J., 1966, Modern Optical Engineering.)

directly into the aperture of the projection lens in such a way that the image size is equal to the aperture size. It can be seen that in the present case even point C receives light from the entire source and fills the entire aperture of the illuminating lens. The minimum diameter of the condenser to illuminate optimally the point C is such that the extreme ray from it reaches the opposite side of the projection lens aperture.

The aberrations in condensers are not important from the point of view of imaging but may introduce nonuniformity of illumination. Spherical aberration is the main concern. Figure 3.30 shows a condenser with spherical aberration. It is obvious that there will be a fall in the illumination at the edge of the field as some of the rays from those areas will miss the projection objective. The effect of chromatic aberration is to color unevenly the image, as one end of the spectrum may miss the objective aperture. Chromatic aberration can be tolerable without any achromatization except in some cases such as microscope condensers. Spherical aberration can be considerably reduced by designing the condenser with two elements, as in Fig. 3.31 (a), or with three elements, as in Fig. 3.31 (b), or by using an aspheric surface, as in Fig. 3.31 (c).

The second condenser arrangement can be used when the light source

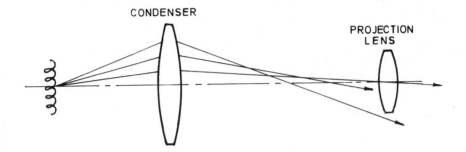

Figure 3.30   Effect of spherical aberration of a condenser.

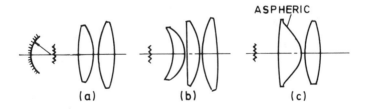

Figure 3.31   Variations of a condenser design: (a) two element, (b) three element, and (c) using an aspheric surface. (Courtesy of McGraw-Hill, Smith, W.J., 1966, Modern Optical Engineering.)

is uniformly bright, as with the carbon arc lamp, the discharge lamp, etc. In this arrangement the light source is directly imaged on the object so that the image size is equal to the object size as shown in Fig. 3.32. The condenser aperture must be large enough to send rays from the top of the condenser to the bottom of the projection lens through the top of the object. In motion picture projectors employing arc lamps, the condenser may be an ellipsoidal mirror.

A mirror placed behind the light source is generally used to improve the utilization of light in condensers. The source is at the center of curvature of the mirror effectively imaging the source on itself thereby sending more light in the desired direction; see Fig. 3.31 (a).

3.8.2 Condensers for Microscopy

Microscope condensers are usually required to give large variable numerical aperture of the illuminating beam to match the numerical aperture of the objective. Hence the correction for spherical and chromatic aberrations becomes important. One must also vary the area of the object to be illuminated.

Figure 3.33 shows the arrangement of a microscope condenser. The light source is imaged by a field lens into the first focal plane of the condenser, so that each point of this image produces a parallel beam passing through the object. Also an iris diaphragm I_1 at the field lens (the field iris) is imaged by the condenser at the object, thus giving a uniformly illuminated, sharply defined disc of light in the object plane. The actual area of the object to be illuminated is controlled by the field iris. The size of the light source is not important, because it is not imaged in the object

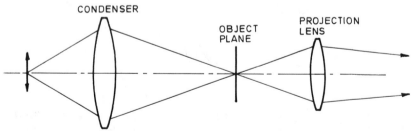

Figure 3.32   Condenser with the light source imaged on the object.

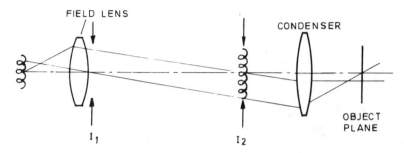

Figure 3.33   Microscope condenser.

plane. An iris diaphragm I_2 behind the condenser controls the angle of the illuminating beam.

When opaque objects are examined with a microscope, surface illumination is required. Since the area being examined lies directly beneath the objective, light must be directed normal to the surface; otherwise the bulk of the light is specularly reflected and does not enter the objective.

Figure 3.34 shows arrangements for the surface illumination of an object. In Fig. 3.34 (a) light from a tungsten filament lamp converges on the area of work beneath the objective. The lamp filament is not focused on the surface, for in this case the image of the filament would be seen in the microscope. Two such illuminators are required. A number of such lamps can also be arranged in an annular mount surrounding the microscope objective.

Figure 3.34 (b) shows an alternative method. BS is a beam-dividing plate that allows the light reflected from the object to pass through the objective. Such an arrangement cannot be used with objectives of high magnification, because in this case the clearance between the objective and the surface is very small. For high-magnification objectives, an arrangement of the type shown in Fig. 3.35 may be employed. Here the objective functions as a condenser as well. The iris I_1 is the field iris, and the iris I_2 controls the angle of the illuminating cone. This arrangement is identical to that of Fig. 3.33 for trans–illumination.

3.8.3 Collimated Beam Illuminator

Instruments such as spectrographs, tool maker's microscopes, profile projectors, and interferometers require collimated beams of light for their

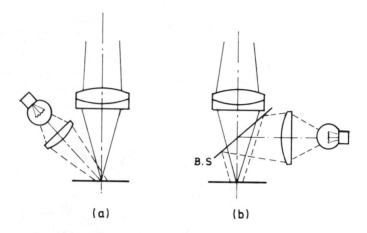

Figure 3.34 Condensers for viewing in reflected light: (a) light from a tungsten filament lamp converges on the area of work beneath the objective, and (b) a beam dividing plate allows the light to illuminate the object axially.

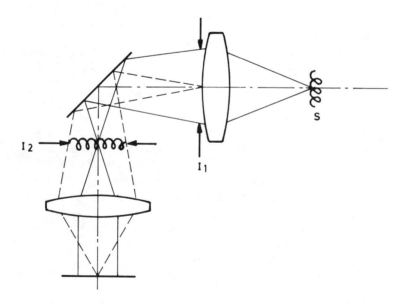

Figure 3.35 Microscope condenser for reflected light.

operation. A collimated beam can be obtained by placing a light source at the focus of an achromatic doublet. In practice an aperture such as a slit A is placed at the focus of a lens and is illuminated by a light source, see Fig. 3.36 (a). A solution of the problem is to place a broad source close to A in such a position that A appears filled by light as viewed from any point on the objective O. This solution can be applied to some problems where the available light source is large and the aperture small. A condenser system, however, is usually required between the source and the slit. There are two main ways of arranging a condenser, ways similar to those used in a projector.

Figure 3.36 (b) shows the first type of arrangement, in which the light source is imaged in the plane of the slit. The condenser size is chosen so that light reaches any point on the objective from all points of the slit. The dotted rays show the optimum size of the condenser. In another arrangement a field lens may be placed just before the slit, as in Fig. 3.36 (c); we choose its power so that the condenser aperture is imaged at the objective. This helps to keep the condenser small. Figure 3.36 (d) shows the second arrangement, which is to be used when a large aperture is to be illuminated by a small source.

3.9 INTERFERENCE AND INTERFEROMETER

Light is an electromagnetic wave, and the interference of light is a manifestation of its wave motion. The frequency of light is very high (10^{14} Hz); an observer therefore cannot sense the variation in amplitude. When two light waves meet at a point, the resultant amplitude depends on the phase difference of the two waves. Figure 3.37 (a) shows two waves meeting in the same phase; the resultant amplitude is the sum of two amplitudes. Figure 3.37 (b) shows waves meeting in phase opposition; the resultant is the difference of the two amplitudes. These two extreme cases are called constructive and destructive interference respectively. There can of course be intermediate phase relationships of the two waves; the resultant amplitude in any case is the algebraic sum of the individual amplitudes.

There are several ways of producing interference. As an example, Fig. 3.38 shows the Lloyd's mirror arrangement for producing interference of light. A slit source S is placed just above but parallel to the plane of mirror M. Light can reach the point A by two paths. It can go from S to A directly along the straight line SA; or it can go along the path SBA where B lies on mirror M. A ray such as BA seems to come from the point S', where the distance BS is equal to BS'. Light reflected from mirror M can be considered as coming from a source at S'. Light waves starting from S and S' at the same time are in the same phase disregarding a phase change of π on

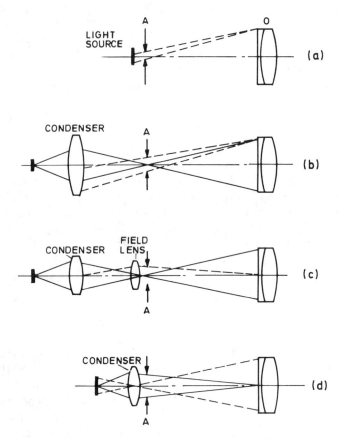

Figure 3.36 Condensers for collimated beam illumination: (a) aperture at focus, (b) light source imaged in the plane of the slit, (c) field lens before slit, and (d) when a large aperture is to be illuminated by a small source.

reflection. To reach A they have to travel different distances, hence take different times. The waves traveling the path SBA lag behind the waves traveling directly along SA. When this lag is equal to the time of a complete cycle or any whole number of complete cycles, the waves at A are in the same phase and hence interfere constructively at A. Points at which this condition is satisfied may, for example, be A_1, A_2, A_3, etc. At other points, such as C_1, C_2, C_3, etc., the lag between light from S and S′ may be an odd

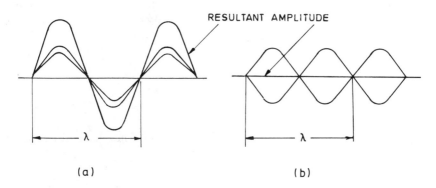

Figure 3.37 Interference of light: (a) waves meeting in the same phase and (b) in phase opposition.

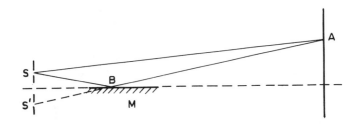

Figure 3.38 Lloyd's interferometer.

number of half cycles, bringing the waves to opposite phase so that they interfere destructively. A series of bright and dark bands, the so called interference fringes, are produced parallel to the slit on the screen.

Mathematically we can describe this phenomenon by considering the superposition of two harmonic waves of frequency w in the same plane. Let

$$Y_1 = a_1 \sin wt$$

$$Y_2 = a_2 \sin (wt + \phi) \tag{3.36}$$

be two waves with amplitudes a_1 and a_2 and phase difference ϕ. Resultant amplitude is the sum of Y_1 and Y_2. The resultant irradiance (intensity), which is the square of the amplitude, can be written as

$$I = a_1^2 + a_2^2 + 2a_1a_2 \cos \phi$$

$$= I_1^2 + I_2^2 + 2 \sqrt{I_1I_2} \cos \phi \tag{3.37}$$

where I_1 and I_2 are irradiances of the two waves. The phase difference is expressed as

$$\phi = \frac{2\pi}{\lambda} \Delta \tag{3.38}$$

where λ is the wavelength of the light and Δ the path difference. The irradiance I shows maxima (constructive interference) when

$\phi = 0, 2\pi, 4\pi$, etc. Thus

$$I_{max} = (\sqrt{I_1} + \sqrt{I_2})^2 \tag{3.39}$$

The minimum irradiance (destructive interferene) is obtained when

$\phi = \pi, 3\pi, 5\pi$, etc., giving

$$I_{min} = (\sqrt{I_1} - \sqrt{I_2})^2 \tag{3.40}$$

In terms of path difference this means that

$$\Delta = \frac{2m\lambda}{2} \text{ for maxima} \tag{3.41}$$

$$\Delta = \frac{(2m + 1)\lambda}{2} \text{ for minima} \tag{3.42}$$

where m = 0, 1, 2,... represents the order of interference.

3.9.1 Contrast of Interference Fringes

Contrast of interference fringes is defined as

$$C = \frac{I_{max} - I_{min}}{I_{max} + I_{min}} \tag{3.43}$$

where I_{max} and I_{min} represent the relative irradiances of the maxima and minima in the interference pattern. The contrast is a maximum, i.e., unity, when $I_{min} = 0$. I_{min} is zero when the interfering beams have equal magnitude ($a_1 = a_2$). When $a_1 \neq a_2$ the contrast will be less than unity. For quasimonochromatic sources, the contrast also reduces with increasing path difference, becoming zero when the path difference exceeds the coherence length of the light source. A laser source with a long conherence length gives good constrast fringes even with large path differences.

3.9.2 Conditions for the Interference of Light

A stationary interference pattern is obtained when (a) the two interfering waves are coherent; (b) the two interfering waves propagate in the same direction or intersect at a very small angle. Interference will take place in the region of superposition even for a large angle of intersection if the two beams are coherent; the fringes, however, will be too close to be seen with naked eye; (c) the waves are vibrating or have components vibrating in the same plane.

3.9.3 The Interferometer

The interferometer is an optical system that makes use of the interference phenomenon for measurement. Most of the interferometers used for technical applications make use of a system of two surfaces, or can be reduced to one, as shown in Fig. 3.39. AB and CD represent the two surfaces enclosing a small angle. A light wave traveling along LO is incident on AB. At O the incident wave is split into two waves traveling along OE and OM respectively. The OM wave is sent back from surface CD along MN and then NF. The waves along OE and NF are parts of the same wave along LO and are capable of interfering provided they superpose and the conditions of interference are satisfied. In the figure, OE and NF are seen widely separated; in practice, however, the conditions are such that they superpose.

The path difference between the two waves is the difference between

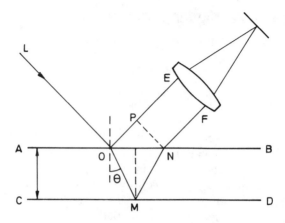

Figure 3.39 Path difference calculation for interference.

the path OP and the path OMN, because beyond P and N the two waves travel identical paths. This path difference can be calculated to be Δ = 2nh cosθ , where θ is as shown, h is the separation between the planes AB and CD at the point of interest, and n is the refractive index of the medium between them. As already discussed, at the points where the conditions

$$2nh \cos \theta = \frac{2m\lambda}{2} \qquad (3.44)$$

$$2nh \cos \theta = \frac{(2m + 1)\lambda}{2} \qquad (3.45)$$

are satisfied, bright and dark fringes respectively are formed. A fringe passes through all those points where the path difference is constant.

The path difference introduced between the two waves should be less than coherence length (the length of the wave train emitted by the light source) to produce interferences; see Fig. 3.40 (a). When the path difference is large, the waves do not superpose and no interference is produced, see Fig. 3.40 (b).

Generally, n is a constant quantity. Variation in Δ can be made by

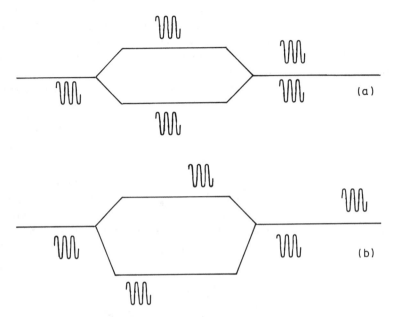

Figure 3.40   Effect of path difference in excess of coherence length: (a)interference and (b)no interference.

varying h or θ or both (in practice only one is allowed to vary).

If θ is kept constant, generally at 0° (normal incidence), h can vary if there is a small wedge angle between the surfaces AB and CD; h is constant along the lines parallel to the line of intersection of the two planes. The fringes, which are called fringes of equal thickness, are thus a set of straight, equally spaced bands parallel to the line of intersection of the surfaces. θ is kept constant by using a collimated beam of light. The separation between the two successive bright or dark bands is given by

$$\bar{x} = \frac{\lambda}{2n\alpha} \tag{3.46}$$

where α is the angle between the interfering beams. The well known Newton's rings are the interference fringes of equal thickness that occur when a spherical surface is placed on a plane surface. Here the loci of equal thickness are circles and hence the fringes are circular (rings).

In the expression 2nh cos θ, if h is kept constant, the fringes can be obtained by varying the angle φ. A fringe will now map all those points for which θ has the same value. These are, therefore, called fringes of equal inclination. They can be observed by illuminating a plane parallel plate by a point source and viewing it with a large lens.

For the majority of engineering applications, fringes of equal thickness are employed. A collimated beam at normal incidence is used. We discuss here some interferometer set-ups.

3.9.4 The Fizeau Interferometer

Figure 3.41 shows the basic arrangement of a Fizeau interferometer. A collimated beam coming from a lens L illuminates the system of two surfaces S and T. The light reflected from the lower surface of S and the upper surface of T is focused by lens L where the eye pupil is placed. The surface S is mounted on an adjustable mount so that the angle between the

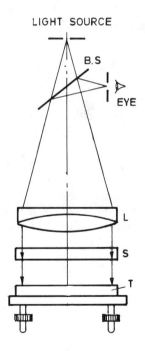

Figure 3.41   A Fizeau interferometer.

surfaces, and hence the width of the fringes, can be adjusted. In actual practice one of the surfaces, say S, is a standard surface and the other one is a test surface. The shape of the fringes tells us about the shape of the test surface.

If two surfaces of practically, the same curvature are placed in physical contact after being carefully cleaned, a thin air film a few micron thick is always trapped between them. The angle of the air film can be changed by pressing softly at one corner with a wooden stick. If such a system is illuminated by a broad source of light, the fringes seen are still of equal thickness, because the separation between the surfaces is very small and the variation in the angle of incidence has pratically no influence. Such an arrangement is a very convenient method for rapid testing of surfaces in a workshop.

3.9.5 The Twyman-Green Interferometer

The optical arrangement of the Twyman-Green interferometer is shown in Fig. 3.42. The incident wave is divided into two waves at the beam splitter BS. They are sent to the plane mirrors M_1 and M_2. The returning waves reunite at BS again and interfere as they proceed, toward the observer. A virtual image M_2' of M_2 in front of or behind M_1, depending on the distance of M_2 from BS, can be seen if one looks into the beam divider from the side of observer. M_1 and M_2' can now be assumed to be the two surfaces providing interfering beams. As in the case of the Fizeau interferometer, standard and test surfaces are used as $M_1$and M_2 respectively for testing work. M_1 and M_2 are mounted on adjustable mounts for adjusting the angle of the wedge formed between M_1 and M_2. There is also a provision for displacing M_1 parallel to itself to vary the effective distance between $M_1$and M_2' and hence the path difference between the interfering beams.

3.9.6 The Mach-Zehnder Interferometer

Figure 3.43 shows the arrangement of the Mach-Zehnder interferometer. The input beam is split at the beam splitter BS_1 and recombined at BS_2. M_1 and M_2 are mirrors. One of the two paths BS_1–M_2–BS_2 and BS_1–M_1–BS_2 serves as a reference path and the other as the test path. The object to be tested is placed in the test path and the beam passes through the object. The test beam passes through the test object only once, whereas there is a double passage due to reflection in the Fizeau and Twyman-Green interferometers. This interferometer is useful in heat transfer and aerodynamic studies.

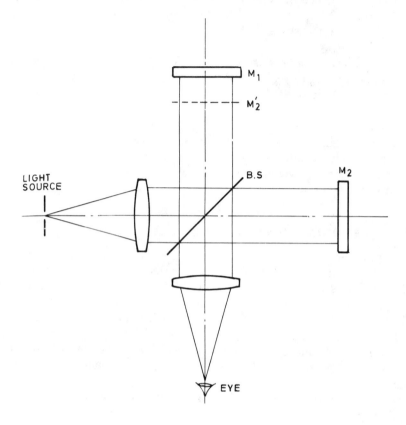

Figure 3.42   A Twyman-Green interferometer.

In the Fizeau interferometer the path difference will be zero when the two surfaces are in contact. This, however, introduces the risk of damage to the surfaces. On the other hand, in the Twyman-Green interferometer, the path difference can be made zero without any physical contact of the surfaces. By construction, the optical paths in the two arms of the Mach-Zehnder interferometer are equal.

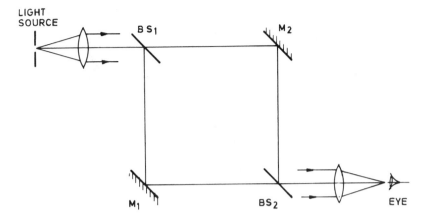

Figure 3.43   A Mach-Zehnder interferometer.

REFERENCES

1. Smith, W. J. (1966) *Modern Optical Engineering*, McGraw-Hill, New York.
2. Longhurst, R. S. (1967) *Geometrical and Physical Optics*, Longmans, London.
3. Jenkins, F. A. and White, H. E. (1976) *Fundamentals of Optics*, McGraw-Hill, New York.
4. Welford, W. T. (1962) *Geometrical Optics*, North-Holland, Amsterdam.

CHAPTER 4

Length Measurement Techniques

Optical length measurement techniques play an important role in metrology. Optical systems such as microscopes, telescopes, and interferometers have all been used for length measurement depending on the requirements of accuracy, range, etc. Interferometry, an accurate length measurement technique, was generally confined to the standards laboratory, but the use of the laser has brought it now to the shop floor.

4.1 THE TOOL MAKER'S MICROSCOPE

The tool maker's microscope (TMM) is an instrument for the measurement of the length of a variety of engineering objects. It is provided with cross-slides, appropriate fixtures for mounting the workpiece, and measuring devices such as the micrometer screw, glass or moiré scales, etc. A tool maker's microscope has two basic functions: to aim at the object with an optical or mechanical contact and to measure any displacement given to the slide to determine the length of the object. Figure 4.1 shows the schematic optical arrangement of a TMM. We shall discuss here its various features.

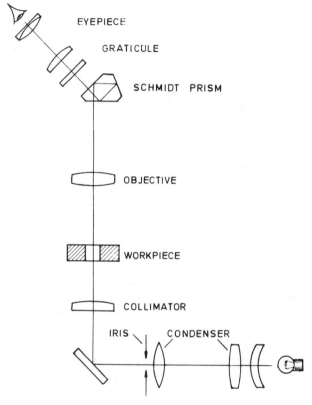

Figure 4.1   Optical arrangement of a tool maker's microscope.

4.1.1 Features of the Tool Maker's Microscope

Erect Image and Inclined Tube

A conventional microscope gives an inverted image of the object. It is always convenient to have an erect image from the measurement point of view. This is achieved in most case by placing a Schmidt prism between the objective and the front focal plane of the eyepiece (Fig. 4.1). This prism erects the inverted image of the object produced by the objective and at the same time bends the optical axis by 45° for convenient viewing by the observer.

Variable Magnification

The total magnification of the microscope is given by the product of the magnifications of the objective and the eyepiece. The eyepiece magnification is generally fixed at 10×. The desired magnification is obtained by selecting the objective of appropriate magnification from a set of objectives between 1× and 10×.

Work Clearance

The work clearance is the distance from the rim of the objective to the point at which the microscope focuses for a given objective. The work clearance decreases as the magnification of the objective increases. Typical values vary from 75 mm to 30 mm as the magnification varies 1× to 10×.

Work Diameter

The work diameter gives the size of the object that can be seen by TMM with a given objective. The diameter of the field stop placed at the focal plane of the eyepiece determines the size of the image that can be seen by the eyepiece. The usual value for the field stop diameter is 20 mm. On the object side, the work diameter is obtained by dividing 20 by the magnification of the objective. Hence the work diameter will be 20 mm and 2 mm for 1× and 10× objectives respectively.

Objectives

The objectives used in a TMM are required to give well-defined images for accurate measurement. In order that the lens should geometrically reproduce the object, it should be sufficiently corrected for distortion consistent with the accuracy requirements of the instrument. The objectives are provided with telecentric stops to minimize errors due to defocusing of the workpiece.

The Eyepiece

A 10× eyepiece is universally employed. The field covered by it, corresponding to 20 mm clear diameter of the graticule, is about 40°. A well-corrected eyepiece suitable for this is employed. The eyepiece can be axially adjusted to compensate for short and long sight over a range of ±5 diopters with the help of a scale on the eyepiece mount.

The Binocular Head

The usual eyepiece of a TMM can be replaced by a binocular head consisting of two eyepieces for binocular vision. Both eyepieces receive the same image of the object by suitable beam splitting. Its advantage is obvious: eyestrain might otherwise occur when the microscope were used for protracted periods.

4.1.2 Aiming Techniques

To measure length with a TMM, contact must be made with points between which the distance is to be measured. Various techniques are discussed in this section.

Graticules

Laying graticule lines (cross-wires for example) on the profile image of an object is a common aiming technique used in the TMM. The graticule is fixed at the front focal plane of the eye-piece and is sharply imaged by it along with the image of the object. A graticule line is set on the boundary of the image. Using a broken line instead of a continuous line enhances the sensitivity of setting (see Chap. 3).

The Double Image Prism

Double image prisms (DIP) are useful aiming devices in some situations. Two types of double image prisms have been used.

Figure 4.2 (a) shows the construction of an axially symmetric DIP. A beam of light from an object passing through such a prism is split into two

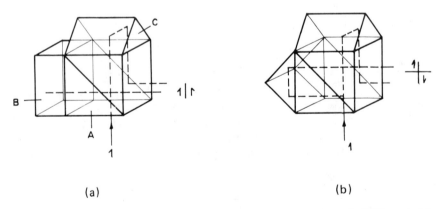

(a)   (b)

Figure 4.2   Double image prisms (DIP): (a) axis-symmetric DIP and (b) centrosymmetric DIP. (Courtesy of Carl Zeiss, West Germany.)

parts at the diagonal face of a cube beam splitter A. One of the beams is reflected back by a plane mirror B and the other one by a 90° prism C. These beams now travel in the same direction, which is perpendicular to the original direction, and they form two images of the object. It is found that one of the images is reversed compared to the other, and that there is symmetry about an axis that passes through the roof edge of the 90° prism. If the object is moved perpendicularly to the plane of symmetry, the two images move in opposite directions. If the object is rotated, the images rotate in opposite directions. This property is used to align the sides of the object perpendicular to the measurement direction.

As for the centro-symmetric DIP, if the plane reflecting surface B of the axis-symmetric DIP is replaced by a 90° prism, we have images one of which is completely inverted compared to the other, see Fig. 4.2 (b). There is symmetry about a point determined by the point of intersection of two roof edges of the prisms. Transverse motion of the object results in opposite movement of the images as before, but the rotary motion of the images corresponding to the rotary motion of the object is now in the same direction.

In the DIP the reference mark is the symmetry axis of the axis-symmetric DIP or the cross-section of such axes in the centrosymmetric DIP. A physical cross is not required: it is enough to overlap the images. The double image prism offers an advantage over a crosswire as the two images move in opposite directions, thereby doubling the setting sensitivity.

The centrosymmetric double image technique is particularly useful in aiming at (locating) the centers of holes, for example, to measure the center distances. In actual practice, red and green filters are used in the DIP such that the two images are colored. Superposing the two colored images gives a third color. Incomplete superposition is made clear by colored boundaries on a dark background. When the superposition is complete, the optical axis of the instrument coincides with the axis of the bore. Symmetry of a workpiece about an axis can be determined by examining it under an axially symmetric DIP.

The Mechanical Feeler

The mechanical feeler is an aiming device where contact with the workpiece is made with the help of a spring-loaded ball feeler. The contact force is only a few grams weight in order to avoid deformation of the measured surface.

Figure 4.3 shows the optical arrangement of a mechanical feeler. This optical arrangement is inserted into the tool maker's microscope in place

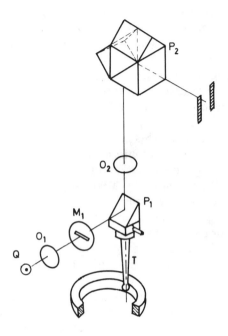

Figure 4.3   Optical arrangement of a mechanical feeler. (Courtesy of Carl Zeiss, West Germany.)

of the objective. The light source Q illuminates a slit M_1 through a lens O_1. The slit is imaged in the eyepiece plane by the objective O_2 over the mirror surface P_1 and the centrosymmetric DIP P_2. Two images, a red and green, are formed because of the presence of the DIP. To make contact, the workpiece is pressed against the spring-loaded feeler T attached to the reflector P_1. This tilts the optic axis of the image-forming beam, and consequently the separation of the two colored slit images changes. The workpiece is moved until the two images overlap, indicating contact with the surface.

A mechanical feeler is ideally suited for blind bores. Generally in profile imaging only the top edge of the workpiece nearest the objective is sharply focused. But with a mechanical feeler, contact can be made with the object anywhere along its side, and it is therefore also ideal for the measurement of gauges.

The Optical Feeler

The optical feeler is free from any contact pressure. A light cross makes the contact with the measured surface, which acts as a reflector in this arrangement. Figure 4.4 shows the mode of operation of an optical feeler. The light source Q illuminates a cross slit K_1 through a lens O_1. The luminous cross K_1 is imaged in the measuring plane K_2 through an optical arrangement consisting of the prism P_1 and the objective O_2. The microscope objective O_3 projects the image of K_2 at K_3 in the eyepiece image plane through the double image prism P_2. Initial adjustment of the cross K_1 is such that its center lies on the optical axis of the microscope. Hence only one image is seen in the eyepiece even in the presence of the DIP P_2. When a workpiece W is placed near the luminous cross K_2, a virtual image of K_2 is formed by the measured surface of the workpiece acting as a plane mirror. This virtual image appears to lie within the workpiece. As long as the measured surface is away from the optical axis (where the center of the actual cross lies), the virtual image will be away from the optical axis, and two colored images (one red and the other green) will be seen in the image plane

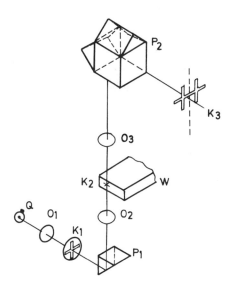

Figure 4.4   Optical arrangement of an optical feeler. (Courtesy of Carl Zeiss, West Germany.)

besides the direct image (in Fig. 4.4, the direct image is not shown for the sake of clarity). As the workpiece is brought nearer to the luminous cross K_2, the two images move closer to each other and finally coincide, giving a single colored cross when the measured surface exactly coincides with the optical axis of the optical system. This is the correct setting, which is obtained by touching the surface by a light cross without any measuring force.

It has been mentioned that the measured surface acts as a mirror. This is a sensitive aiming device and is used for measuring workpieces such as gauges. The surfaces of these workpieces are invariably ground or lapped and hence act like a mirror.

The Knife-Edge Technique

The knife-edge technique of aiming is used when the workpiece surface does not give a sharp profile image (cylinders, external threads). Knife edges are lapped to strict straightness. The knife edge is pressed on the work; correct setting is achieved when no light passes through the contact line between the work and knife edge as seen through the microscope. To make thread measurements, the knife-edge technique is the most accurate method.

4.1.3 Measuring Techniques

The object is placed on the cross-slide of a microscope and displaced for aiming at the points of interest. The measurement of displacement is done by using one of the several available techniques.

Micrometer Screws

The cross-slides of the usual TMM are displaced with the help of precision micrometer screws, which are also used for measurement. The micrometer screws give a least count of 5 μm or less. The measuring range usually available is 25 x 25 mm and can be extended with the help of gauge blocks.

Precision Scales

Precision scales, which are usually made of glass and read through a reading microscope or a projection system, give a least count of 1 μm or better. Very fine (approximately 0.005 mm wide) chromium lines are deposited on glass blanks. The line spacing is generally 1 mm and is accurate to a fraction of a micrometer. In the case of a reading microscope, the scale is magnified by the objective and an interpolation graticule or graticule system is arranged in the eyepiece. In the case of projection readout, the graticule is placed at the screen plane. Various interpolation devices have been employed.

The *optical vernier* is a graticule that divides the magnified distance between the scales lines into a number of parts (usually 100 for 1-mm scale spacing). The field of view of a reading microscope equipped with an optical vernier appears as shown in Fig. 4.5. For a reading microscope with total magnification of 100, 1 mm of scale will appear as 100 mm. The vernier can easily accommodate 100 divisions in this distance, giving a least count of 0.01 mm. Interpolation can be made up to 0.005 mm. It is difficult to use an optical vernier in projection read-out as this will require a large screen size and thus a large instrument.

When a least count of 0.001 mm or better is required, *optical micrometers* are used. Two such arrangements are the parallel plate micrometer and the diagonal scale micrometer.

The *parallel plate micrometer* consists of a fixed graticule G_1 with 10 double lines that divide the magnified main scale division into 10 parts. A plane parallel plate P is introduced into the beam of light coming from the objective; see Fig. 4.6 (a). When this plate is tilted in the light beam by rotating a cam K, there results a linear displacement of the main scale image in the fixed graticule plane. A circular scale G_2 carrying divisions

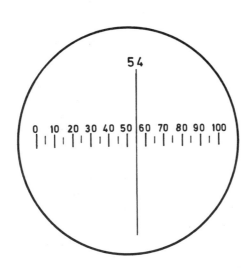

Figure 4.5   Field of view with an optical vernier.

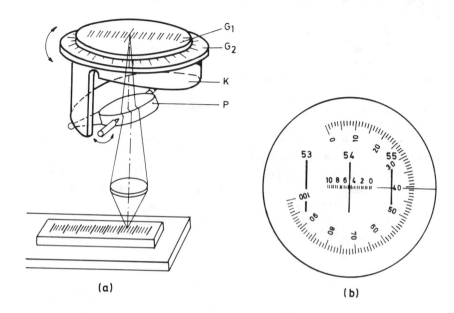

Figure 4.6   Parallel plate micrometer: (a) optical arrangement and (b) field of view. (Courtesy of Springer-Verlag, Berlin)

from 0 to 100 is attached to the cam. A rotation of the cam from 0 to 100 shifts the scale image by a distance equal to the distance between two double lines in G_1. Thus each division on G_2 is equivalent to $1/10 \times 1/100$ = $1/1000$ th of the main scale, i.e., 0.001 mm. To read any scale position, the cam is rotated by a knob until a main scale line is enclosed symmetrically between a double line on G_1. Figure 4.6 (b) shows the field of view in such a micrometer. The reading in the figure is 54.540 mm.

An alternative arrangement is possible in which the graticule G_1 is physically displaced with the help of a cam. The rotation of the cam is monitored by a graduated (100 parts) drum fixed to the cam. Both these micrometer arrangements can be used for projection readout.

The diagonal scale micrometer employs a single fixed graticule, shown in Fig. 4.7, to get a least count of 0.001 mm, which is a big advantage, as the reading of the scale is instantaneously available. There are no adjustments required for taking a reading. The magnified distance between the

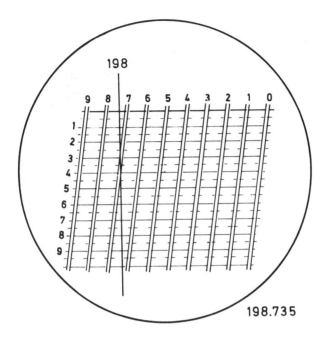

Figure 4.7   Diagonal scale micrometer. (Courtesy of Carl Zeiss, West Germany.)

main scale line is divided into 10 parts by 10 inclined double lines. The inclined distance of the double line is divided into 100 parts for a total of 1000 divisions. The reading procedure is obvious from the figure. The reading on the inclined double line is taken at the point where the main scale lies symmetrically between the double lines.

4.1.4 Illumination of the Object

The engineering objects viewed by a TMM are not self-luminous and therefore must be artifically illuminated using condenser systems (Chap. 3). Thin objects such as transparencies are best projected by illuminating them with a convergent beam. Solid objects with appreciable thickness along the optical axis pose some difficulties in a convergent beam.

In Fig. 4.8 (a), an objective P is shown imaging an object of circular cross-section illuminated by convergent light. The section that can be

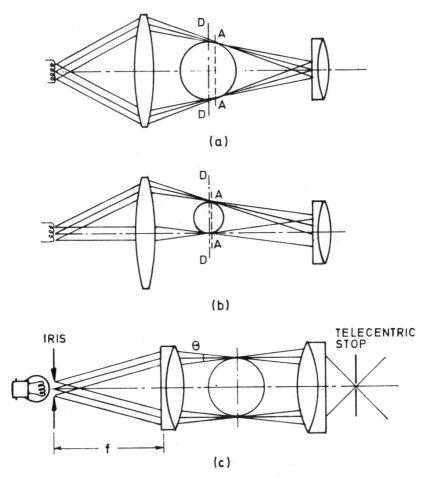

Figure 4.8   Illumination of solid objects: (a) and (b) with convergent light and (c) with collimated light.

sharply focused on the screen is the one where the illuminating cone of light is tangential to the object. Thus AA and not the diametric section DD will be focused. With the object placed symmetrically on the axis, the image is truly circular but does not correspond to the diametric section. The discrepancy is not constant but varies with the size of the object projected.

When the object is displaced from the axis, the illuminating cones are not tangential to the object at the same section. In Fig. 4.8 (b) the object is placed with one end on the axis. Since the cones are tangential at D on one side and at A on the other, which do not lie on a section perpendicular to the optical axis, both points cannot be focused by the projection lens simultaneously. Intermediate points focus at positions between those of A and D. A sharp image is therefore not obtained, and the best compromise form does not give a circular image.

A collimated beam condenser alone provides a symmetrical illumination of the object wherever it is placed in the beam; see Figure 4.8 (c). For a point source at the focus of a collimating lens, all light rays will leave the lens parllel to the optical axis adn must be tangential to the spherical object about a diametric palne. In fact, the source is of finite size (diameter s), and a set of parallel bundles coming from different points of the source leave the collimator at different angles. An illuminating cone of angle $\theta = s/f$ is available at all points of the object to the right of the collimating lens.

4.1.5 Illumination of Screw Threads

We have seen that for a proper projection of a profile the illuminating cone should be tangential to the profile section in the plane normal to the projection lens axis. Further, the light reaching the profile section should proceed to the projection lens without any obstruction. Special arrangements are required for a satisfactory projection of threads in view of these two conditions. Screw threads can be projected either by inclining the threads to the optic axis at an angle equal to that of their helix, or by setting their axis square to the optical axis and inclining the illuminating beam through the helix angle. In the first case the image projected is that of a section normal to helix, while in the second case a true axial section is projected.

Figure 4.9 (a) shows the thread inclined at the helix angle to the optical axis so that the illuminating cone may be tangential to the thread being projected. FF denotes the focal plane of the projection lens. With such an arrangement, one thread form only can be imaged sharply in the image plane, the thread whose center lies on the line of intersection of the plane FF and a plane normal to that of the diagram and containing the thread axis. Since the thread is inclined, the plane FF is normal to the helix, and an image of the corresponding section of the thread form is formed.

In Fig. 4.9 (b),the thread is shown normal to the projection lens axis, while the illuminating beam is inclined to this axis by helix angle α, so that the illuminating rays are tangential to the thread forms in the focal plane

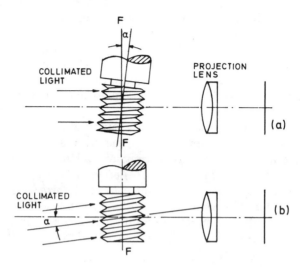

Figure 4.9   Illumination of screw threads: (a) screw inclined and (b) illuminating beam inclined.

FF of the projection lens. In this case all the thread forms lying within the field of the projection lens are focused together sharply. It is clear that the correct axial section is projected in this arrangement.

Most microscopes, however, use an arrangement in which the thread is kept normal to the optical axis, and the illuminator and the microscope tube are tilted through the helix angle simultaneously. They are actually mounted on a common axis and can be tilted by about 15° on either side. This case is similar to the first case of thread projection, and a helical section of the thread is imaged. To obtain information regarding the correct axial section, a cosine correction factor must be applied to the measurement on the helical section. Figure 4.10 shows a thread of helix angle α. P denotes the true pitch. When an image is projected by inclining the illuminating beam through the helix angle, it is the true axial pitch P that is projected and the true flank angle β that is measured. If p' denotes the pitch measured normal to the helix, then $p = p' \cos \alpha$. The apparent flank angle β' is also different from β. The depth d is common to both sections. Hence

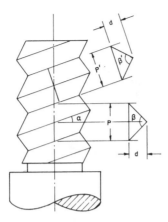

Figure 4.10   Corrections applied to thread measurement with inclined screw.

$$2d = \frac{P}{\cos \alpha} = \frac{P'}{\cos \alpha'}$$

This gives

$$\tan\left(\frac{\beta'}{2}\right) = \tan\left(\frac{\beta'}{2}\right)\cos \alpha$$

4.1.6 The Reflection of Light

The objects examined by a TMM generally have a good surface finish and specularly reflect the light falling on them for their illumination. If such reflected light, after its passage through the projection lens, arrives in the image plane outside the shadow of the object, no deterioration in definition results. In certain circumstances, however, the light falls on the screen inside the shadow and reduces the contrast between dark and bright areas, and the edge of the image may not be focused with certainty.

An object with appreciable thickness, such as a rectangular block with one face parallel to the optical axis, gives a sharply defined image of the edge nearest to the projection lens. Figure 4.11 (a) shows an object A of rectangular section with its front face in the focal plane FF of the projection lens. The ray 1 parallel to the axis grazes the surface of the block. The

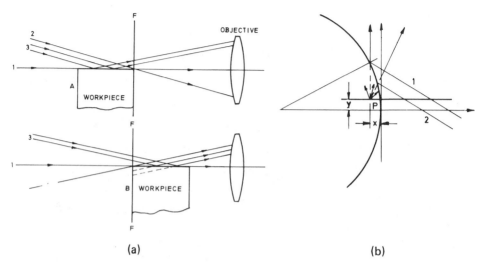

(a)   (b)

Figure 4.11   Reflection of light from solid objects: (a) plane objects and (b) cylindrical objects (courtesy of Carl Zeiss, West Germany.)

limiting ray 2 grazes the front edge. Rays parallel to 2 incident upon the surface are reflected from it and cross the focal plane outside the object. These rays (3) illuminate the image plane outside the shadow. Similarly, all the rays intermediate between 1 and 2 falling upon the block are reflected to cross the focal plane outside the object. These rays will also merely add to the illumination of the bright areas in the image. The front edge of the block is thus imaged sharply without loss of contrast.

However, if a block B is placed with its rear edge in the focal plane, the reflected rays 3 , when produced back intersect this plane inside the object. These rays will fall inside the shadow in the image plane and reduce the contrast. The boundary lines of the rear edge are therefore not focused sharply. It can be seen that no other plane except the front edge is sharply focused.

Cylinders (particularly of large diameters including screw threads) provide much the same difficulty in projection as rectangular blocks. Light reflected from the surface of a cylinder facing the lens appears to originate from inside the object, as shown in Fig. 4.11 (b), so that there is reduced contrast in the screen image. Light reflected from the surface on the condenser side merely adds to the illumination on the bright area. This accounts for a halo around the shadow. A variable aperture iris diaphragm is provided at the focus of the collimating lens of the illuminating system in a TMM so that the angle of the illuminating cone can be controlled depending on the diameter of the cylinder.

4.2 OPTIMETERS

Optimeters are optical instruments for the measurement of length by the optical lever principle. An optical lever consists of a light beam incident on a mirror and received on a screen after reflection. A tilt β to the mirror introduces a 2β tilt in the reflected beam, which results in its linear movement on the screen. Large linear movement can be obtained corresponding to small angular tilt if the screen is placed at a large distance from the mirror. These instruments have a very small measuring range and a least count of 0.001 mm or less and are intended for linear measurement by the comparative method.

There are two main classes of instruments employing the optical lever. In the first, a spot of light moves over a scale on which its displacement is measured: in the second, light starting from an illuminated scale is brought to a screen after reflection at a mirror so that as the latter rotates, an image of the scale moves past a fiducial mark on the screen. The former class suffers from the disadvantage that for a reasonable magnification and range of measurement the scale tends to be inconveniently large. This is avoided in the second method, in which a magnified image of a small scale is formed in the plane of the screen, only part of the image being seen for a given position of the mirror. In this case, however, a more complicated optical system is necessary in order to obtain a well-defined scale image, whereas in the other type a simple optical system is adequate.

Most of the optimeters used in length measurement are based on the second method. The optical arrangement takes the form of an autocollimator, i.e., both the scale and its reflected image are situated in the focal plane of a collimating objective. Figure 4.12 shows the optical arrangement of such an optimeter. Light from the external source L illuminates a graticule 3 that carries a scale. The graticule is placed at the focal plane of the objective 5. The collimated light from 5 illuminates a mirror 6 that is tilted by a small amount under the action of the measuring plunger 7. The bottom end of the measuring plunger contacts object 8 being measured. The mirror is pivoted at 9.

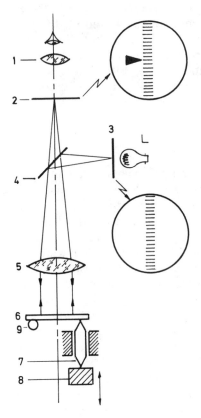

Figure 4.12   Optical arrangement of an optimeter.

The parallel beam of light reflected by the mirror passes back through the objective and forms an inverted image of the scale on the graticule 2. The scale reading is taken with reference to an index mark on 2. The reflected image of the scale is seen by an eyepiece or can be projected on a screen.

Displacement of the plunger provides a tilt of the mirror 6 and causes a displacement of the scale image with reference to the index mark. If X is the distance between the point of contact of the plunger and the axis of rotation of the mirror, the mirror rotates through an angle x/X, for a small displacement x of the plunger, and the reflected beam is deviated by $2x/X$. The corresponding displacement of the scale image in the graticule plane is 2xf/X, where f is the focal length of the lens. The magnification of the displacement is given by 2f/X. If the eyepiece (or the projection lens) magnification is m, the total magnification is given by 2fm/X. For f=250, m=10, and X=5, the magnification is 1000 times, which means that a displacement of 1 μm will be observed as 1 mm in the instrument. The range of these instruments is very small (≈0.1 mm), and they are generally used for accurate comparative measurements.

With one reflection at the tilting mirror, a magnification factor of 2 is obtained for the deviation of the reflected beam. By means of more than one reflection, the magnification may be increased. For n reflections at the mirror, the magnification is 2n. In the optimeter shown in Fig. 4.13, two reflections take place at the mirror M_1 for light passing from the collimator into the telescope. Figure 4.14 shows another mirror arrangement whereby still higher magnification for the deviation of the reflected beam can be achieved. This makes it possible to use objectives of smaller focal length and reduce the overall instrument dimension.

4.3 INTERFEROMETERS FOR LENGTH MEASUREMENT

In interferometric length measurement, the length is measured in terms of the wavelength of light. For a long time the primary standard of length was the International Prototype Metre, a bar of platinum-iridium alloy with two fine lines inscribed on it at a spacing of 1 m. In 1960 the metre was redefined as 1,650,763.73 wavelengths of an orange radiation arising by the transition between two specified levels of [86]Kr. Subsequent developments

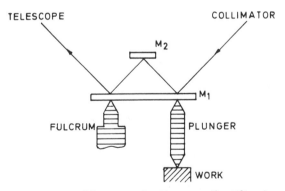

Figure 4.13   Optimeter with two reflections on the tilt mirror.

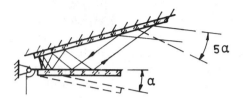

Figure 4.14   Multiple reflection tilt mirror arrangement for an optimeter.

in the frequency stabilization of lasers (Chap. 1) made it possible to determine their wavelengths with an accuracy higher than that of ^{86}Kr. Hence a stabilized laser wavelength appeared a natural choice to repalce the existing standard. But in view of the advances in time and frequency measurement, the definition of the meter was changed [1-3] in 1983 to length traveled by light in free space during 1/299792458 s. This definition relates to wavelength via laser frequency measurement and the speed of light. Several stabilized laser sources (Chap. 1) can be used to realize this definition.

Two interferometric techniques have been used in length measurement: Multiple wavelength interferometry and single wavelength interferometry.

4.3.1 Multiple Wavelength Interferometry

Multiple wavelength interferometry is used to determine the absolute length of slip gauges in terms of wavelength of monochromatic light. Slip gauges are used as length standards in workshops, and their lengths are measured up to an accuracy of 0.1 μm or better. All gauge block interferometers are variations of the Fizeau interferometer or of the Twyman-Green interferometer.

Measurement Principle

To measure the length of a gauge it is fastened to an optically flat base plate, and interference fringes are produced between a reference surface R and the surface of the gauge G and the base plate B as shown in Fig. 4.15. R' is the image of reference surface R in the Twyman-Green system. Figure 4.15 (c) shows a typical intereference pattern with two displaced fringe systems, one each on the gauge and the base plate. The only information that is readily determined by a glance at the fringe pattern is that the gauge

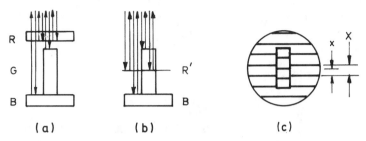

Figure 4.15   Interference arrangement for the excess fraction technique of length measurement: (a) Fizeau system, (b) Twyman-Green system, (c) typical interference pattern.

length is not an integral multiple of a half wavelength. The ratio $x/X = e$ is the fraction of $\lambda/2$ by which the gauge is bigger than an integral multiple of $\lambda/2$. The fringes in the two patterns will fall on the same line if the gauge length is an integral multiple of a half wavelength. A 1 cm slip gauge contains approximately 30,000 half wavelengths, and this number is to be determined. The excess fraction method is used for this. This requires two things: (1) that approximate length (within few half wavelengths) should be known in advance, which is always possible by using accurate length comparators, and (2) that the fraction should be measured at more than one wavelength.

Suppose that for the radiations λ_1, λ_2, λ_3, and λ_4 the measured fractions are e_1, e_2, e_3, and e_4. The length of the gauge is

$$L = \frac{(N_1 + e_1)\lambda_1}{2} = \frac{(N_2 + e_2)\lambda_2}{2} = \frac{(N_3 + e_3)\lambda_3}{2} = \frac{(N_4 + e_4)\lambda_4}{2} \qquad (4.1)$$

where N_1, N_2, N_3, and N_4 are numbers of half wavelengths corresponding to λ_1, λ_2, λ_3, and λ_4 respectively. From these expressions, the length L can be calculated by the method of coincidences. A first approximation to N_1 is obtained by dividing the nominal value of L by the known value of $\lambda_1/2$ and ascribing to N_1 the whole number of the quotient. Addition to N_1 of the measured fraction gives the first approximateion to L, which is now divided by the known values $\lambda_2/2$, $\lambda_3/2$, and $\lambda_4/2$, so that we obtain whole numbers and fractions $N_2 + e_2'$. N_3+e_3' and N_4+e_4' for each wavelength. If these fractions do not agree within experimental error with those observed, another value of N_1, differing from the first by 1 is tried. This process is continued until an agreement between the measured fraction and the calculated fraction is obtained, giving the correct value of N_1. A few trials will lead to the correct result. The process of calculation is avoided by use of a special slide rule developed for the purpose.

Gauge Block Interferometers

Figure 4.16 shows a Fizeau gauge block interferometer. Light from a cadmium source 1 is collected by a lens 2 and focused on a pinhole 3. Light from 3 is collimated by a lens 4, from where it goes through a constant deviation prism 5 whose rotation determines the wavelength that will passes through the reference flat 6 to the upper surface of the gauge block 7 and base plate 8 to which it is fastened. Light retraces nearly the same path to the beam splitter 9 close to 3. An eyepiece (not shown) placed close to 10 is used to see the interference pattern. Two fringe systems are formed due to interference between the wavefront reflected from the reference flat and the wavefronts coming from the top surface of the gauge block and the base plate. The spacing and orientation of the interference fringes can be adjusted by operating the adjusting screws provided with the reference flat mount. If the end faces of the gauge are plane and parallel, the fringes on the gauge will be parallel to those on the base. The fraction may be estimated by eye to one-tenth or even one-twentieth of the fringe spacing.

Figure 4.17 shows a gauge block interferometer based on the Twyman-Green design. The input beam is a collimated beam from a constant-deviation prism. The technique of measurement is the same.

In the Fizeau instrument the minimum separation between the reference surface and the base plate is governed by the size of the gauge. This

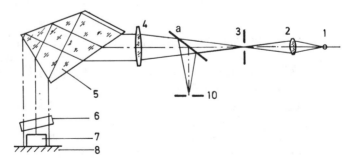

Figure 4.16   Gauge block interferometer based on the Fizeau system.

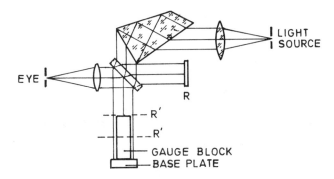

Figure 4.17   Gauge block interferometer based on th Twyman-Green system.

leads to two things. First, interference fringes between the base plate and the reference surface can be produced only if the gauge length is less than half the coherence length of the light; second, the two intereference fringe systems will have different contrasts because the path differences in the two are not equal.

On the other hand, in the Twyman-Green interferometer, the virtual reference surface R' can be adjusted in the middle of the slip gauge length. Therefore in this interferometer the contrast of both fringe systems can be made equal, and a gauge of length equal to the coherence length can be measured.

Figure 4.18 (a) shows another type of gauge block interferometer similar to the Twyman-Green interferometer. This one can be quickly transformed into an arrangement for comparative measurement as shown in Fig. 4.18 (b). Here the test guage is compared against any standard gauge block (not necesarily of the same nominal length). This avoids the necessity of fastening the surface to a base plate. There is interference between G_1 and G_4 and between G_2 and G_3; see Fig. 4.18 (c). Two interference patterns are formed for which the fractions can be determined for different wavelengths and evaluated for the difference between the test gauge and the standard gauge. The standard gauge can be zero-thickness gauge as shown in Fig. 4.18 (d). This is constructed with two elements A and B with holes for the light to pass and joined together with a common face C.

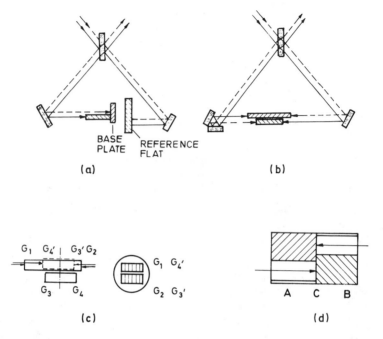

Figure 4.18 Alternate arrangement of a gauge block interferometer based on the Twyman-Green system: (a) normal configuration, (b) and (c) comparative measurement, and (d) zero thickness gauge.

4.3.2 Gauge Block Interferometry with Laser Light

Gauge block interferometry has been conventionally carried out using quasimonochromatic sources such as cadmium or krypton. These sources cannot be applied to the measurement of length greater than a few tens of centimeters because of their poor coherence length. Laser light can be used to measure longer length; a frequency-stabilized laser at several wavelengths or a stabilized tunable laser is required.

Visible-wavelength interferometry, however, is too sensitive, and the measurement at long lengths becomes difficult because the interference fringes of short wavelength are affected by air turbulence, mechanical vibrations, and wavelength instability. Long-wavelength (infrared) interferometry, on the other hand, is insensitive to all these. There has been considerable interest recently in the use of infrared lasers for long length measurement [4–6]. Two lasers, He-Ne at 3.39 μm and He-Xe, have been found useful. Because of the long wavelength of light, the excess fraction method can be employed even at a single wavelength, provided the length of the gauge block is known within certain limits in advance. With long wavelengths these limits can be quite reasonable. From Eq. (4.1), the length L of the gauge is given by

$$2L = (N + e)\lambda \tag{4.2}$$

If the length is known within ± ΔL, this makes the right-hand side of the above equation uncertain up to

$$u = \pm \frac{2\Delta L}{\lambda} \tag{4.3}$$

If $|u| < 1/2$ then there will be no uncertainty in the integer N due to the uncertainty ΔL in the initial estimate of L. We need only determine accurately e consistent with the desired accuracy in the length measurement. It is obvious that we must have

$$|\Delta L| < \frac{\lambda}{4} \tag{4.4}$$

for this technique to succeed. For λ = 3.39 μm (He-Ne laser), λ/4 is approximately 0.9 μm. This accuracy is achievable with present-day optical comparators.

An infrared interferometer with a He-Ne laser at 3.39 μm with electronic scanning of fringes has been developed for the measurement of gauge blocks at single wavelengths [4]. Figure 4.19 shows the schematic arrangement of the intereferometer. It is aligned using a visible laser L_1 (a He-Ne laser at 0.633 μm). The beam fron an infrared laser L_2 (a frequency-stabilized He-Ne laser at 3.39 μm) is converged by a spherical mirror C_1 and collimated by an off-axis paraboloidal mirror C_2. M_1 is the base plate and GB the gauge block fastened to it in one arm of the interferometer. The mirror M_2 on a piezoelectric transducer (PZT) forms the other arm. A mask F with 3 holes allows light from the base plate fringe as shown in Fig. 4.20 (a). Light from the fringe on the gauge is deflected by a prism S. The lens O converges light on two separate PbS photodetectors. To determine the excess fraction, the fringes are scanned by applying a low-frequency sawtooth signal to PZT. The typical output is shown in Fig. 4.20 (b). The fringes from the base plate and the gauge block are simultaneously recorded, and

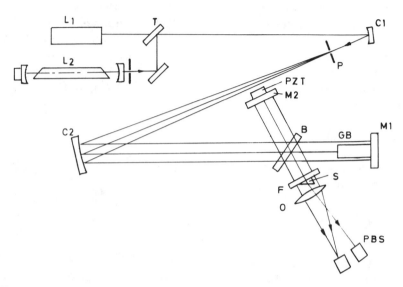

Figure 4.19   He-Ne IR gauge block laser interferometer. (Modified from Ref. 4, used with permission.)

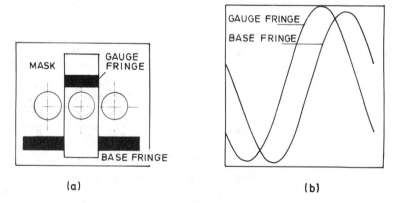

Figure 4.20   (a) Fringes in the IR interferometer. The holes on the mask separate the gauge and the base fringes. (b) Signals due to the base and gauge fringes. (Modified from Ref. 4, used with permission.)

their phases are determined by

$$\phi = \sin^{-1} \frac{2(I(\phi) - I_0)}{I_M - I_m}$$

(4.5)

where $I(\phi)$ is the intensity at the point of measurement. I_M and I_m are maximum and minimum intensities respectively and I_0 is the mean intensity. Thus ϕ_1 and ϕ_2 can be determined for the base and gauge fringe, respectively, giving a phase difference of

$$\Delta\phi = \phi_1 - \phi_2$$

(4.6)

Eq. (4.2) for the length of the gauge can also be written as

$$L = \frac{(N + \Delta\phi/2\pi)\,\lambda}{2}$$

$\Delta\phi$ can be determined to an accuracy of about $1°$, which will result in an uncertainty of $0.01\mu m$ at this wavelength. The other important parameters that affect the accuracy are the refractive index of air and the influence of temperature on the length of gauge block. These quanities are determined and correction applied to the length measurement.

For multiple wavelength interferometry, the He-Xe laser has been used [5,6]. This laser gives two closely spaced lines, namely 3.37 μm and 3.51 μm, that can oscillate simultaneously and can be stabilized. The interference order equation for length can be written as

$$2L = 2(L' \pm \Delta L) = (N_1 + m_1 + e_1)\lambda_1 = (N_2 + m_2 + e_2)\lambda_2$$

(4.7)

where L' is the approximate length determined by other methods, ΔL the uncertainty in the length measurement, and

$$N_1 = \frac{2L}{\lambda_1}, \quad N_2 = \frac{2L}{\lambda_2}, \quad m_1 = \frac{2\Delta L}{\lambda_1}, \quad m_2 = \frac{2\Delta L}{\lambda_2}$$

(4.8)

each of which is rounded to the nearest whole number. If ΔL is small, so that $m_1 = m_2 = m$ except for a fraction, then we may write

$$m = \frac{(N_2 + e_2) - (N_1 + e_1)\,f}{f - 1}$$

(4.9)

where $f = \lambda_1/\lambda_2$. The uncertainty in m as a function of small errors in e_1, e_2, and f is given by

$$\delta m = \frac{2f}{f-1} \, \delta e + \frac{2L}{\lambda_1} \, \frac{\delta f}{f-1} \tag{4.10}$$

where $\delta e = |\delta e_1| = |\delta e_2/f|$, and δe and δf are errors on fractions e and f respectively ($f \approx 1$). The number m will be uniquely determined, provided $|\delta m| < 1/2$. As f approaches 1, the required accuracies of reading the interference order and the ratio between the laser wavelengths become severe.

Using a He-Ne laser, for L = 3 m the two terms on the right-hand side of Eq. (4.10) give $\delta e < 0.01$ if $\delta f = 0$, and $\delta f = 1.2 \times 10^{-6}$ if $\delta e = 0$. With the scanning fringe technique discussed earlier, $\delta e < 0.01$ can be achieved. The condition on δf is rather severe and cannot be achieved if the two wavelengths vary independently with time. However, if the two wavelengths oscillate simultaneously in the same cavity, the condition can be satisfied.

As mentioned, ΔL is such that the whole numbers $m_1 = m_2 = m$. The variation in actual value of m in going from m_1 to m_2 should be less than $1/2$ if it has to be uniquely defined. Hence

$$\frac{2\Delta L}{\lambda_1} = m \quad \text{and} \quad \frac{2\Delta L}{\lambda_2} = m \pm \frac{1}{2}$$

This gives

$$|\Delta L| < \frac{\lambda_1 \lambda_2}{4 \, (\lambda_1 - \lambda_2)} \tag{4.11}$$

For the He-Xe laser, $\Delta L \approx 21 \ \mu m$. Hence the preliminary measurement becomes simple, as an uncertainty of 21 μm can be achieved over long lengths.

The optical arrangement of the interferometer using the excess fraction method with the He-Xe laser is shown in Fig. 4.21. The arrangement is similar to that in Fig. 4.19. Since both the He-Xe lines are simultaneously available, a grating is used to separate the signals due to the two wavelengths.

Another interesting possibility with the He-Xe laser is two-wavelength simultaneous interferometry (TWSI) for measurement of long lengths [6].

The combined intensity in the inference pattern of an interferometer using wavelengths λ_1 and λ_2 is given by [7]

$$I = I_1 \cos \frac{2\pi L}{\lambda_1} + I_2 \cos \frac{2\pi L}{\lambda_2} + \text{const} \tag{4.12}$$

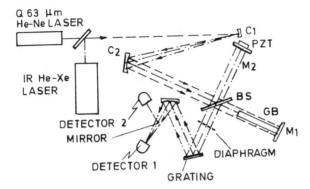

Figure 4.21   He-Xe IR gauge block laser interferometer. (Modified from Ref. 5, used with permission.)

where L is the difference in the lengths of two arms of interferometer and I_1 and I_2 the intensities of the respective beams. This signal can be high-pass filtered and squared giving a signal

$$I^2 = I_1^2 \cos^2 \frac{4\pi L}{\lambda_1} + I_2^2 \cos^2 \frac{4\pi L}{\lambda_2}$$

$$+ I_1 I_2 \cos \frac{2\pi L (\lambda_1 + \lambda_2)}{\lambda_1 \lambda_2}$$

$$+ I_1 I_2 \cos \frac{2\pi L (\lambda_1 - \lambda_2)}{\lambda_1 \lambda_2} + \frac{I_1^2 + I_1^2}{2} \qquad (4.13)$$

Passing the signal through a low-pass filter, we can eliminate the first three terms so that

$$I^2 = I_1 I_2 \left\{ \cos \frac{2\pi L}{\lambda_s} + C_0 \right\} \qquad (4.14)$$

where $\lambda_s = \lambda_1 \lambda_2 / (\lambda_1 - \lambda_2)$ and $C_0 = (I_1^2 + I_2^2)/2I_1 I_2$. The simultaneous interference with λ_1 and λ_2 therefore becomes equivalent to interferometry at a synthetic wavelength of λ_s. This value can be sufficiently large when λ_1 and λ_2 are close. For example, for He-Xe lines, $\lambda_s = 84.15$ µm. TWSI is therefore useful in measuring long lengths.

4.3.3 The Laser Interferometer For Displacement Measurement

In a gauge block interferometer absolute length is measured indirectly by the excess fraction method using one or more wavelengths. Length (displacement) can also be measured at a single wavelength by actually counting the number of half wavelengths or fraction thereof in a given length.

In a Twyman-Green interferometer, if a crosswire is set on the center of a fringe and a $\lambda/2$ displacement given to one of the mirrors, a movement of one fringe will occur, and the next immediate fringe will take the place of the first one at the crosswire. If a mirror is mounted on the bed of a machine, the movement of the bed can be determined by counting the number of fringes moving past the crosswire and multiplying the number by $\lambda/2$. The direction of movement of the bed governs the direction of movement of the fringes. Since the number of fringes to be counted is very large (about 30,000 per cm), the counting in an actual interferometer is done electronically and a digital display is obtained.

Retroreflectors

When a plane-mirror Twyman-Green interferometer is used for the measurement of displacement, the mirror should move parallel to itself within a very close tolerance, because any tilt will result in a change in fringe spacing and give a false signal that will be recorded as a displacement. This puts very stringent conditions on the manufacture of the guides on which the moving member is guided. The use of retroreflectors overcomes this difficulty. The retroreflector is an optical component that sends back an incident ray parallel to itself irrespective of the orientation of the reflector, as against the case of a plane mirror, where a change in orientation by an angle α results in a 2α change in the direction of the returning beam. Cube corner retroreflectors are generally used.

Figure 4.22 shows a Twyman-Green interferometer with cube corners. It can be seen that the incident and reflected beams are widely separated and that no light goes back to the source. In ordinary spectral lamps this is of no significance. But if a laser source is used, no light should go to the source, as it can disturb the output characteristics of the laser. This is an additional advantage of cube corners when a laser is to be used. Further, two inteference patterns 1_s and I_D are separately accessible. In conventional interferometers the interference pattern I_s is lost to the source unless a beam splitter is interposed between the source and the beam splitter.

Reversible Counting

For counting the fringes, the interference pattern is projected on a narrow slit. A photodetector behind the slit receives the light flux passing through

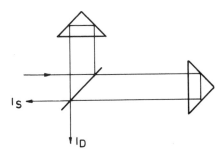

Figure 4.22 Interferometer with cube corners as retroreflector for displacement measurement.

it. The photodetector signal will show a sinusoidal variation as a function of the displacement of the moving reflector, one cycle for every $\lambda/2$ displacement. A photodetector signal will show sinusoidal variation irrespective of the direction of displacement, and the count of the cycles will always be added. A reversible counting technique is employed to overcome this difficulty.

In order to achieve reversible counting, the interferometer must be capable of providing two outputs, one varying sinusoidally and the other cosinusoidally (i.e., with a 90° phase difference between them) with the displacement. Two such signals are called quadrature signals.

The simplest technique is to use two slits with separate detectors; see Fig. 4.23 (a); instead of one, whose separation is adjusted so that signals produced in the two detectors have the desired phase difference when the interference pattern moves with the displacement of the reflector. One can now study the phase relationship of the two signals in the two cases, namely when the fringes move to the left and when they move to the right. Taking the signal A as reference, the other signal leads or lags behind by 90° depending on the direction of motion; see Fig. 4.23 (b), (c) and (d). An electronic circuit can distinguish this difference and produce a command signal for the bidirectional counter whether to add or subtract the incoming pulse.

The above method is applicable in a plane-mirror interferometer in which the fringes can be adjusted by suitably tilting the mirror. When retroreflectors are used in Twyman-Green interferometer, however, the incident wavefront is sent back precisely in the same direction, and fringes are actually not seen. Since the two interfering wavefronts are parallel to

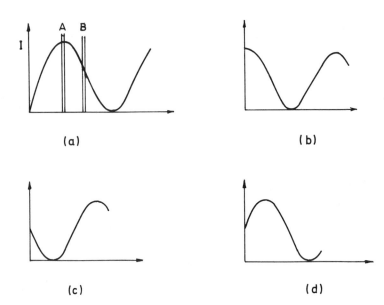

Figure 4.23 Signals for bidirectional counting. (a) Generation of two Signals in quadrature using a pair of slits, (b) reference signal, (c) signal leads by $\pi/2$ and (d) signal lags by $\pi/2$.

each other, the fringe width becomes infinite, and the field of view appears uniformly bright or dark depending on the path difference. The brightness of the whole field now varies sinusoidally as a function of the reflector displacement. This light taken on a photodetector (even without a slit) will produce a sinusoidal signal. To obtain the second quadrature signal, the polarization technique may be used.

Figure 4.24 shows an interferometer using the polarization technique to obtain signals in quadrature. The input light is polarized at 45° to the plane of the figure. After reflection from the moving cube C_1 and reflection from the beam splitter, the light is still linearly polarized at 45°. Its vertical (X) and horizontal (Y) components are in same phase. In the reference arm a quarter-wave plate is placed through which the light beam passes after reflection from the corner cube C_2. The light after the quarter-wave plate becomes circularly polarized, so that its vertical and horizontal componet have a phase difference of 90°. The X and Y components of the two beams are separated by a polarizing beam splitter or a Wollaston prism. The signals produced by the interference of the XX and YY components have a 90° phase difference. It can be seen that once the desired phase relation-

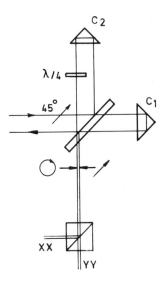

Figure 4.24 Interferometer arrangement for producing quadrature signals using the polarization technique.

ship is achieved it will not change. When slits are used, only a small amount of light is transmitted at a time, and the signals produced are weak. In the present case, there are no slits and hence strong signals are available.

This description of the polarization transformation through the interferometer does not consider the polarization properties of the cube corners and the beam splitter. Exact calculation can be carried out using the Jones calculus, and the phase relationship of the outputs can be obtained. The phase relationship of the outputs can be varied by manipulating polarization components.

Processing of the Signals

Figure 4.25 shows the processing of the quadrature signals. The sinusoidal signals are put into rectangular form by Schmidt triggers. The shaped signals are represented by the two digital states 1 and 0. The sequence of the states of the two signals is representative of the direction of motion. One, two, or four pulses per cycle are produced, depending on the least count requirement. When four pulses are produced, the least count is $\lambda/8$.

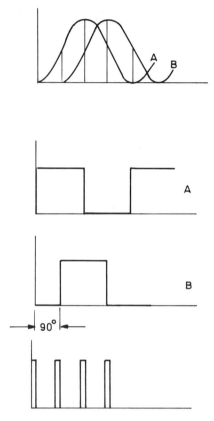

Figure 4.25   Signal processing of quadrature signals.

The total count is converted into displacement by multiplying it by the least count.

This laser interferometer presents a difficulty. Figure 4.26 shows the output of one of the photodetectors as one of the reflectors is moving. The intensity variations are centered around the triggering levels of the counter. If the intensity of either the beam or the source changes, the variations in the intensity may not cross the triggering levels, thus there will be malfunctioning of the instrument until the trigger levels are readjusted. The changes in intensity may occur as the laser ages or as turbulence deflects the laser beam slightly. The triggering levels can be adjusted for

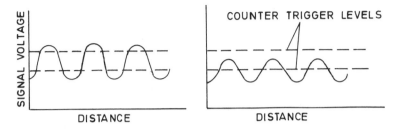

Figure 4.26   Disabling of trigger levels due to loss of signal strength.

long term changes. For fast changes in intensity, however, one usually finds that automatic adjustment is difficult in shop conditions. Therefore, for optimum performance two signals with a constant average dc level and sinusoidal components related to the optical path difference are required.

The problem of variable dc level and sinusoidal components that disables the counting circuits can be overcome by removing the average dc level. This can be done by an electronic subtraction process that removes the average signal level from the photodetector signals. In the interferometer designed to achieve this objective, four signals are generated instead of the normal two [8]. Figure 4.27 shows the optical arrangement. The usual two outputs of a Michelson interferometer are each divided into two with the help of polarization beam splitters giving a total of four signals spaced in phase by $\pi/4$. Using three of the signals, two signals in phase quadrature can be generated by subtraction, as shown in Fig. 4.28. These two signals have zero average levels. Although their amplitudes may change depending on light source intensity and fringe contrast, the position at which the signals cross the zero level have a fixed relationship to the optical path difference and are unaffected by the presence of dc components in the original interferogram signals.

4.3.4 Two-Frequency Laser Interferometer for Displacement Measurement

The problem associated with a variable dc level of the fringe-counting interferometer signal is completely avoided in a two-frequency laser interferometer. This operates on a heterodyne principle (Fig. 4.29). The laser gives out two optical frequencies f_1 and f_2 a few MHz apart. They are orthogonally polarized. A polarizing beam splitter B_2 sends f_1 toward the measuring cube corner C_1 f_2 and to the reference cube corner C_2. The

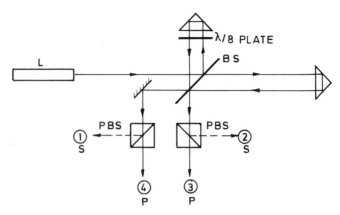

Figure 4.27 Interferometer with four outputs. L, frequency stablized laser; BS, beam splitter; PBS, polarizing beam splitter; P and S, orthogonal polarized components of light. (Modified from Ref. 8, used with permission.)

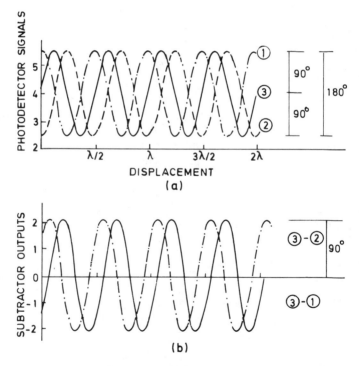

Figure 4.28 (a) Output signals of the four-output interferometer and (b) subtractor outputs. (Modified from Ref. 8, used with permission.)

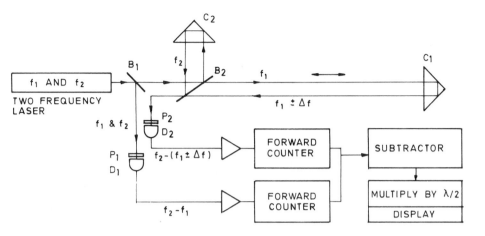

Figure 4.29 Two-frequency laser interferometer for displacement measurement.

reflected beams are combined on a detectors D_2 and produce interference. Since the frequency of the two beams is different, the resultant intensity fluctuates at the rate of $f_2 - f_1$ just as beats are produced when two tuning forks of slightly different pitches are struck simultaneously. Similarly a signal of frequency $f_2 - f_1$ is produced on detector D_1 when light of two frequencies falls on it via the beam splitter B_1. This serves as the reference signal. Since the interfering beams are orthogonally polarized, polarizers P_1 and P_2 are placed in front of the detectors to cause interference.

When the cube corner C_1 is moved, the reflected beam frequency is Doppler shifted $f_1 \pm \Delta f$ depending on its speed and the direction of movement. The frequency of the Doppler signal produced on D_2 is then $f_2 - (f_1 \pm \Delta f)$.

The two frequencies from D_1 and D_2 are counted by two separate counters, and their total counts are then subtracted to produce a net count. If there has been no motion, the frequencies of the two signals are equal, the net count will be zero, and the subtractor will not accumulate any count. Motion of the cube corner, on the other hand, raises or lowers the Doppler frequency depending on the direction of motion, producing a net positive or negative count corresponding to the distance traveled.

When the measuring cube corner moves, the reflected light has a new Doppler-shifted frequency given by

$$f_1' = f_1 \left(1 - \frac{2v}{c}\right)$$

(4.15)

where v is the velocity of the cube corner and c the velocity of light. Thus

$$\Delta f = f_1' - f_1 = -\frac{2v}{\lambda}$$

where λ is the wavelength of the laser light.

The distance L traveled in a given time t is $L = vt$. The net accumulated count will be

$$N = \Delta ft = \frac{-2v}{\lambda} \ t = \frac{2L}{\lambda} \quad \text{or}$$

$$L = \frac{N\lambda}{2}$$

(4.16)

Thus the total displacement is obtained by multiplying the accumulated count by $\lambda/2$, which is performed electronically and the result displayed. The displacement is measured with a least count of $\lambda/2 \approx 0.3$ μm. The resolution can be improved by electronic splitting of the sinusoidal signal.

The resolution can be doubled optically by double passing the interfering beams through the interferometer as shown in Fig. 4.30. The additional optics for double passing is common to both reference and measurement beams [9]. Hence movement of these optics does not introduce measurement error or instability. It is possible to make more than two passes through an interferometer, but the size and configuration of the optics imposes limitations. Further light losses and wavefront distortion with multiple passes can be excessive.

In this inteferometer, the detectors continuously generate an ac signal even when the reflector is stationary and can be ac coupled to the following amplifiers. This interferometer is therefore also referred to as an ac interferometer. Since the counting pulses are not generated by triggering counter levels, loss of light in the measurement beam for whatever reason does not stop the functioning of the instrument. A loss of light up to 90% can be tolerated.

4.3.5 The Plane Mirror Interferometer

A useful variant of a laser interferometer is a plane mirror interferometer,

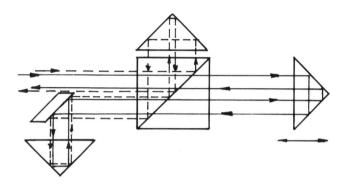

Figure 4.30   Double-pass interferometer. (Modified from Ref. 9, used with permission.)

in which plane mirror are used as tilt-insensitive reflectors in place of cube corners. Figure 4.31 shows the arrangement. It is essentially a Michelson interferometer in which the beams pass through the interferometer twice. This double passing is achieved by using a polarizing beam splitter (PBS) and a quarter-wave plate in each arm. The incoming laser beam is split into orthogonally polarized beams. The plane of polarization of the return beams from the mirror is rotated through 90°. Both the beams now proceed to the cube corner reflector C. The retroreflected beams again proceed to the same mirrors as before. A double pass through the $\lambda/4$ plates restores the original polarization condition of they beams, and the leave the interferometer parallel to the incident beam. An analyser in the output beams is required to produce interference, because the beams are orthogonally polarized. The interfering beams are parallel, and the fringe width is infinite. The tilt of the mirrors does not influence this parallelism except in an offset of the return beam, since tilting of the first reflected beam is exactly compensated by the second reflection. The resolution of this inteferometer for displacement measurement is doubled, because the measurement beam is reflected twice from the mirror reflector. Another configuration of the plane mirror interferometer is shown in Fig. 4.32. Here only the retroflector of the measurement beam is replaced by a plane mirror and has been used with a two-frequency laser source [9].

The plane mirror interferometer is particularly useful in a two-axis system. The plane mirror as retroreflector for x-axis measurement can be allowed to move in the y direction without affecting the signal strength or

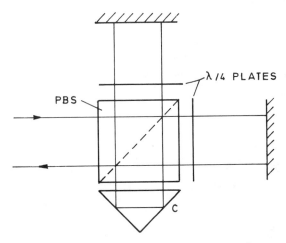

Figure 4.31   Double-pass plane-mirror interferometer.

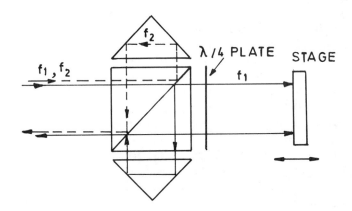

Figure 4.32   Plane-mirror interferomter for use with a two-frequency laser. (From Ref. 9, used with permission.)

the y measurement. Therefore both mirror reflectors of a two-axis system can be mounted on the same moving part. This is an important considera- tion in reducing the Abbe error in measurement. Usually the measuring point is placed where the two axis beams cross; see Fig. 4.33. If we use a standard cube corner interferometer, the x and y reflectors are to be mount- ed on two separate stages that move in the corresponding directions. This will result in an offset of the x and y measurement axes and thus a geometric error. High-performance stages are valuable tools in integrated circuit lithography and inspection equipment as well as in metrology.

Another configuration of the plane mirror interferometer useful for precision measurement of thermal expansion is shown in Fig. 4.34. The specimen and the base plate on which it is placed act as two mirrors of the interferometer. The input beams are obtained at the beam splitter BS. PBS is a polarization beam splitter. After a double passage through the inter- ferometer, two interference outputs are generated at the beam splitter B'S' for bidirectional counting [10]. Other arrangements have also been re- ported for thermal expansion measurement [11, 12].

Two double-pass variations of the plane mirror interferometer shown in Fig. 4.32 are illustrated in Fig. 4.35. Arrangement (a) doubles the optical resolution, as four measurement beams are produced. Arrangement (b), which is obtained by incorporating a λ/4 plate in the double pass optics, does not have the advantage of increased resolution but is suitable for thermal expansion measurement and high performance metrology stages [9].

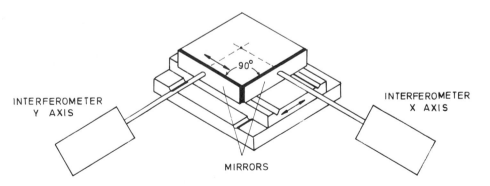

Figure 4.33 Two-axis plane-mirror interferometer system. (Courtesy Hewlett Packard, United States.)

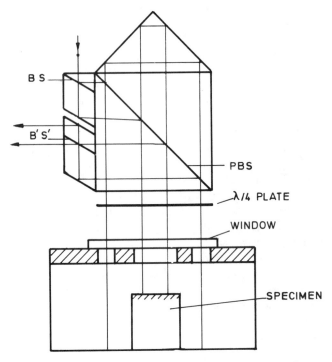

Figure 4.34   Measurement of thermal expansion with a plane-mirror interferometer configuration. (Modified from Ref. 10, used with permission.)

The laser interferometer is a fast, accurate and high-precision length measuring instrument. It has been used for the calibration of machine tools as well as for variety of length, angle, straightness, squareness, and related measurements.

4.3.6 Wavelength and Correction for Air Refractive Index

In interferometers the wavelength is the multiplying unit. The accuracy of the measurement will depend on the accuracy with which the unit is known. If an accuracy of 0.1 μm per meter (1 part in 10^7) is required, then the wavelength accuracy should be of the same order or better than 1 part in 10^7. Stabilized lasers provide such accuracies.

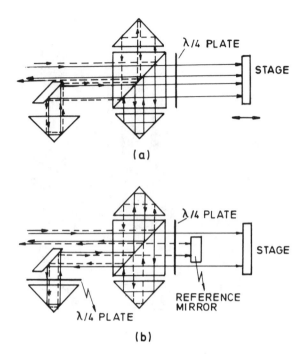

Figure 4.35 Two variations of the double-pass plane-mirror interferometer. [9]

The wavelength of light in air is given by λ_0/n where λ_0 is the vacuum wavelength and n is the refractive index of air. The index n generally fluctuates within 1×10^{-5}. Hence a correction is required in the wavelength to meet the requirement of accuracy as mentioned above. The refractive index can be calculated from Elden's formula [13]. In an actual measurement situation, the temperature, pressure, and vapor pressure of the air on which the refractive index mainly depends are continuously monitored. To obtain a correction accuracy of about 1×10^{-7}, these three quantities are measured with accuracies of 0.05°C, 13 pa, and 133 pa, respectively.

4.4 THE MOIRÉ TECHNIQUE FOR DISPLACEMENT MEASUREMENT

When two gratings are superimposed, we obtain moiré fringes. Figure 4.36

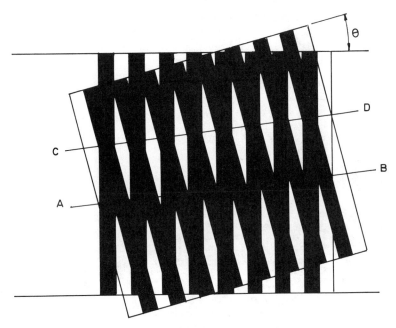

Figure 4.36   Generation of moiré fringes with two linear gratings.

shows the superposition of two gratings of equal dark and clear spaces
with their lines inclined at a small angle. It is seen that the area around the
line AB is predominantly dark and actually looks like a dark band. Similarly, the area around the line CD appears as a bright band. These alternate
bright and dark bands appear over the whole area of the overlap and have
the appearance of fringes. These fringes are known as moiré fringes (or a
moiré pattern). In this particular case of linear gratings, these fringes are
straight and run almost perpendicular to the grating lines.
   The spacing of the fringes is given by $P=d/\theta$, where d is the pitch of
the grating and θ the angle between the grating lines. Thus the spacing of
the fringes can be controlled by the angle θ. When θ is zero, the spacing is
infinite, that is the entire overlap area becomes uniformly dark or bright
depending on whether the bright areas of one grating fall on the dark or
bright areas of the other grating, respectively. It can be easily seen that if
one of the gratings is displaced perpendicular to the grating line, the moiré
fringes also move normal to the direction of displacement. For a movement

of one pitch of the grating, the moiré pattern displaces by one fringe. If we count the number of fringes passing across a reference mark we can determine the displacement of the moving grating in multiples of grating pitch. Thus, if one of the two gratings is attached to a moving member, displacement can be measured digitally as in an interferometer. At the same time, this technique is less sensitive to environmental conditions than are interferometers.

An actual measurement system consists of a long main grating attached to the moving member and a small fixed grating of the same pitch, the index grating, as shown in Fig. 4.37. The two gratings are mounted with a small gap of a fraction of a millimeter (0.2 mm) determined by both geometrical and diffraction considerations [14]. A collimated light beam illuminates the overlap area, and transmitted light is received on a photodetector. As the main grating moves, the photodetector receives varying amounts of light, and a periodic signal is produced; one complete cycle for one pitch movement. Counting the periods and multiplying the count by the pitch gives the displacement. A similar arrangement can be made with a reflecting main grating.

As in the case of the interferometer for displacement measurement, a single photodetector cannot determine the direction in which the grating is moving. The 90° displaced signals are essential for bidirectional counting. In the present case, these two signals are obtained by modifying the index grating. The index grating is divided into two parts as shown in Fig. 4.38. The two parts have the same grating pitch as the main grating, but

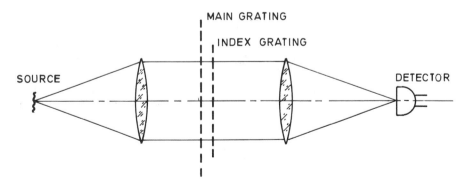

Figure 4.37 Optical arrangement of a moiré linear measurement system.

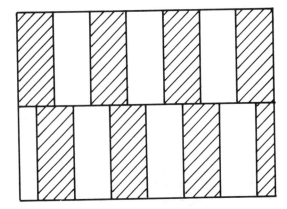

Figure 4.38   Two-field index grating for bidirectional counting.

they are displaced with respect to each other by a quarter of the pitch. The lines of the main and the index gratings are aligned parallel, and the light from the two areas of the index grating is collected and delivered to two separate detectors. The signals of these two detectors have the required 90° phase difference for bidirectional counting. These signals can be processed in the same way as in the case of the fringe counting interferometer.

There is, however, the same problem in this scheme as in the fringe counting interferometer, namely the change in the average dc level of the signal due to fluctuations in the light intensity. To overcome this difficulty, four signals are derived, at 0°, 90°, 180°, and 270°. This is achieved by having four sets of grating lines instead of two on the index grating, each set shifted by a quarter of the pitch. Four detectors receive light from the four areas, giving four signals as required. The alternate signal pairs are combined and amplified in a push-pull amplifier stage, so that the modulations add and the dc levels cancel, resulting in a pair of quadrature signals symmetrical about a mean level.

If the variation in light intensity is gradual because of overall reduction in the light transmission of the optical surfaces, this four-signal system operates satisfactorily. Since the four signals are not derived from the same area of the grating surface, local nonuniform contamination can still affect the mean dc level. Proper sealing of the gratings can overcome this difficulty over small to medium lengths. Alternatively, the problem can be overcome by generating the four signals from a common area of the grating [14, 15].

Fringe Subdivision

The two quadrature signals may be treated electronically to obtain four counting pulses per period, as in the case of the fringe counting interferometer shown in Fig. 4.25. Thus with a grating having 25 lines per mm, a resolution of 0.01 mm can be obtained. Further subdivision is required to enable coarser gratings to be used or to improve the resolution of the system. This may be done by subdividing the signal period into 10 to 100 parts. This will give a 1-μm resolution with 10 lines per mm. A 0.1-μm resolution can be obtained by dividing the signal from a grating having 100 lines per mm into 100 parts [14,15].

Metrological Gratings

High-accuracy gratings on glass having a few lines per mm to 100 lines per mm have been used for displacement measurement. For use in reflection readout, staineless steel gratings have been used. For long-length applications up to 10 m multiple glass segments can be used, but flexible stainless steel tapes on which grating lines are obtained by an etching process have been found more convenient.

The moiré technique of displacement measurement is the most widely used method in a variety of machine tools, numerical-control machines, and coordinate measuring machines as well as in routine inspection instruments such as micrometers, comparators, etc. The moiré fringe technique can be used equally well in angle measurement, for which radial gratings are used.

4.5 THE PHOTOELECTRIC LINE AND EDGE DETECTOR

Scales and graticules used in measuring instruments are well-defined objects. In visual instruments these are seen by the human eye through a microscope or an eyepiece. A setting is done visually using the unique properties of the human eye such as symmetry setting (one line enclosed between two lines). Essentially we try to locate the center of the line.

Again, detecting the edge of an image is the usual problem in visual measurements. The visual process is time consuming and has limited sensitivity. In both line detection and edge detection, the human eye may be replaced by a photodetector. This provides an electrical signal that can be used for an indicating instrument or a digital display or to control certain other processes. Photoelectric line and edge detectors can provide precision in the range of 0.01 μm.

4.5.1 The Photoelectric Line Detector

Photoelectric line detection is achieved by placing a slit in the plane of the image of the line as shown in Fig. 4.39. The width of the slit is equal to the width of the line L. A photodetector D is placed to receive the light passed by the slit. The image of the line is usually dark and the area around it bright. When L and S perfectly align, no light will be received on D, and a minimum signal will be obtained. This provides the simplest line detector. The signal obtained here is dc, however, and is easily influenced by stray light. An ac detection signal is a desirable ideal. This is obtained by allowing either L or S to oscillate. The output of the detector is fed to an amplifier tuned to the frequency of the slit oscillation. When the line to be detected approaches and aligns itself symmetrically about the oscillating slit, an ac signal at double the frequency f of oscillation of the slit is obtained. A tuned amplifier at the slit frequency will show a null, which is the criterion for line detection. For any other position of the line, both f and 2f frequencies are available. Figure 4.40 illustrates this.

Photoelectric line detection is widely applied in precision measurement of line standards such as scales [16]. The line detector here takes the form of a photoelectric microscope, schematically shown in Fig. 4.41. The image of the scale line is projected by a microscope objective O in the plane of a slit S placed in front of a photomultiplier tube P. A scanning mirror M_1 placed in the image-forming path scans the line image across the slit. The scale is illuminated with a light source (not shown), and an eyepiece EP is provided for visual observation. A laser interferometer is used to

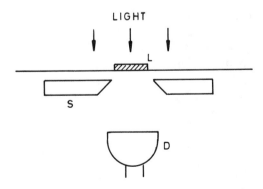

Figure 4.39 Simple arrangement of a photoelectric line detector.

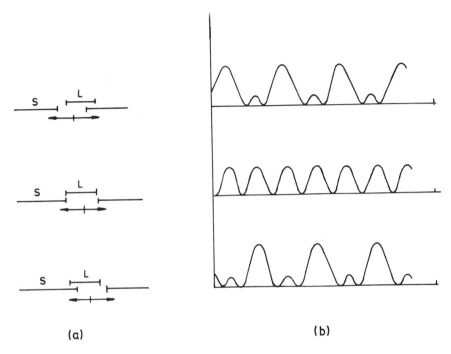

(a)   (b)

Figure 4.40 Generation of signal in an oscillating slit line detector. (a) Line L and slit S with double-ended arrow showing the amplitude of oscillation. (b) Detector signal for three positions of the line.

measure the displacement from one line to another. Measurement accuracy of better than 0.1 μm can be achieved.

4.5.2 The Photoelectric Edge Detector

The frequency of the mechanical scanning as discussed in the previous section cannot be very high, and this limits the speed at which the measurement can be made. If large numbers of measurements must be made, fast measurements are essential, such as in integrated circuit masks, where hundreds of patterns, whose exact dimensions are important, are packed into a small area. Moreover, fast measurements suppress the influence of environmental changes, mechanical deformation due to temperature drift, etc.

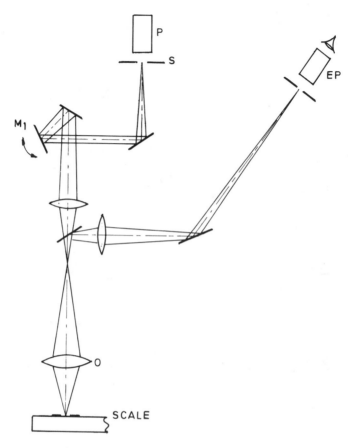

Figure 4.41   Photoelectric microscope for line standard calibration.

Figure 4.42 shows the arrangement of a photoelectric edge detector employing high frequency oscillation [17, 18]. It employs an inertia less electrooptic oscillating mechanism at a frequency in the MHz range. The light from a He-Ne laser 1 is focused into a diffraction-limited spot onto the test surface 7 with the help of an eyepiece 3 and an objective 6. The test surface may be a mask consisting of a chromium coating on glass; reflected light may be used for detection. In other situations, transmitted light may be used. The mask can be moved within the horizontal plane. The light reflected by the mask is detected by a photodetector 8. Visual inspection

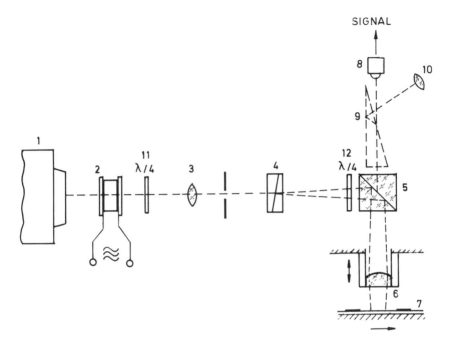

Figure 4.42   Electrooptic edge detector. (Modified from Ref. 18, used with permission.)

of the mask can be done through an eyepiece 10 after inserting a deviating prism 9. A wollaston prism 4 produces two beams that are orthogonally polarized and gives two closely adjacent spots on the mask. The separation depends on the split angle of the Wollaston prism as well as its location on the optical axis. A typical size for the light spots about 1.5 μm and the sweep about 0.7 μm. Actually, the spot is made to alternate continuously between two positions by periodically switching the plane of polarization of the linearly polarized light incident on the Wollaston prism between two orthogonal states. This results in only one of the two beams being emitted by the prism at a time, depending on the plane of polarization of the incident beam. The modulation is achieved by a KD*P polarization modulator 2 in combination with a quarter-wave plate 11. While the latter introduces a constant bias retardation, the KD*P modulator is actuated by a high frequency ac voltage of quarter-wave voltage amplitude. Given

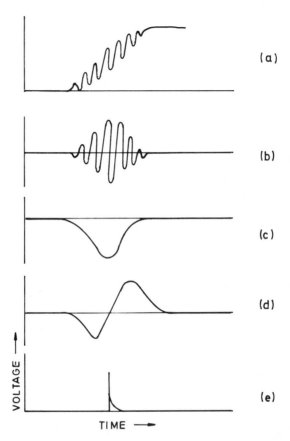

Figure 4.43 Edge detector signal processing. The detector produces a dc signal of low average value (a). As an edge approaches the flying spot, an ac signal is produced, and the dc signal increases. The signal is passed through a resonance amplifier (b), a demodulator (c), a differentiator (d), and a zero detector (e). (From Ref. 18, used with permission.)

appropriate orientation of the components, the combination of the KD*P crystal and the quarter-wave plate produces a periodical 90° rotation of the plane of polarization. The beam splitter 5 may have different reflection coefficients for the two orthogonal polarization states, so that a difference in the intensity of the two light spots occurs. To avoid this problem, a quarter-wave plate 12 is introduced after the Wollaston prism. The plate transforms the linearly polarized beam components into counterrotating circular components.

When the two light spots fall on bare glass, the detector will produce a dc signal of low average value; see Fig. 4.43 (a). As an edge approaches the flying spot, an ac signal is also produced along with an increase in average dc level, as one of the spots may fall on the reflecting surface and the other on bare glass. The maximum ac signal is produced when the edge is centered to the symmetry line of the double light spot configuration. The signal is passed through a resonance amplifier, Fig. 4.43 (b), a demodulator, Fig. 4.43 (c), a differentiator, Figure 4.43(d), and finally through a zero detector, Fig. 4.43 (e). The zero detect pulse represents the maximum ac signal and hence the edge to be detected. The pulse is used to command the displacement transducer to record the position. To achieve high measurement accuracy, the test object may be mounted on an xy stage equipped with plane mirror laser interferometers as displacement transducers.

Measurements may be carried at 10 cm/sec with this system. The minimum object width whose edges can be detected is governed by the size and sweep of the spot. A microscope objective at NA=0.60 gives a spot of about 1.3 μm. The sweep may be about half this value, giving a 2 μm area of overlap. The minimum object width will be more than this value. A high NA microscope gives a low depth of focus (1 μm for NA=0.60). If the test surface has a waviness in excess of this value, an automatic refocusing of the objective is required during the scanning of the object.

REFERENCES

1. Matsumoto, H. (1984). Recent interferometric measurements using stabilized lasers, *Precision Engineering, 6:* 87.
2. (1984) Documents concerning the new definition of metre, *Metrologia, 19:* 163.
3. Giacomo, P. (1983). Laser frequency measurements and redefinition of the metre, *IEEE Trans. Instrum. Meas.,* IM-32: 244.
4. Matsumoto, H. (1980) Length measurement of gauge blocks using a 3.39 μm He-Ne laser interferometer, *Jpn. J. Appl. Phys., 19:* 713.
5. Matsumoto, (1981) Infrared He-Xe laser interferometry for measuring length, *Appl. Opt., 20:* 231.
6. Matsumoto, H. and Seino, S. (1982) Infrared two wavelength interferometry for measuring long length, *Annals CIRP, 31:* 401.
7. Polhemus, C. (1973). Two wavelength interferometry, *Appl. Opt.,* 12:2071.
8. Downs, M. J. and Raine, K. W. (1979) An unmodulated bidirectional fringe counting interferometer system for measuring displacement, *Precision Engineering, 1:* 85.

9. Siddal, G.I. and Baldwin, R.R. (1984). Development in laser interferometry for poition sensing, *Percision Engineering*, 6: 175.
10. Bennett, S. J. (1977). An absolute interferometric dilatometer ,*J. Phys. E.Sci., Instrum*, 10:525.
11. Roberts, R. B., (1981). Absolute dilatometry using a polarization interferometer, *J. Phys. E, Sci, Instrum*. 14: 1386.
12. Okaji, M. and Imai, H., (1985) Precise and versatile systems for dilatometric measurement of solid materials, *Precision Engineering*, 7: 206.
13. Elden, B. (1966) The refractive index of air, *Metrologia*, 2: 71.
14. Shepherd, A. T., (1979) 25 Years of moire fringe measurement, *Precision Engineering*, 1: 61.
15. Zeiss Phocosin, A new photo-electric digitization system for length measurement with digital increments of 0.1 μm, CARL ZEISS, West Germany.
16. Beers, J. D. and Lee, K. B., (1982). Interferometric measurement of length scales at the National Bureau of Standards,*Precision Engineering*, 4: 205.
17. Schedewie, F., (1970). Electrooptic laser microscope for high speed, high precision edge detection,*Israel J. Tech, 9:* 263.
18. Kallemeyer, M., Kosanke, K., Schedewie, F., Solf, B. and Wagner, D., (1973). Rapid, precise, computer controlled measurement of X-Y coordinates, *IBM J. Research and Development*, 17: 470.

ADDITIONAL READINGS

Encyclopedia of Physics,Vol. 29, Springer-Verlag, Berlin.
Habbel, K. J. and Cox A., (1961).*Engineering Optics*, Pittman, London.
Luxmoore, Ed., (1983). *Optical transducers and techniques in engineering measurement*, Applied Science Publishers.
Rose, M., ed. (1971). *Laser Applications*, Academic Press, New York.

CHAPTER 5

Alignment and Angle Measurement Techniques

In construction of any type, alignment is almost invariably required. An experienced technician can achieve reasonable alignment with his eye using simple devices such as straight edges, squares, and scales. Stretched string or wire can provide a reference axis for alignment. Plumb lines and spirit levels produce alignment relative to gravity, which means the local horizontal plane. Optical alignment techniques enhance measuring capabilities and improve accuracy.

Telescope instruments play an important role in alignment practice. The extension of the optical axis of a telescope provides an absolutely straight line (a line of sight) in space that cannot sag. In optical alignment, lateral displacements or tilt angles of objects are measured with respect to the line of sight.

5.1 THE TELESCOPIC SIGHT

A telescopic sight is essentially a terrestrial telescope provided with a cross-hair graticule. The line joining the center of the cross-hair and the optical center of the objective provides the line of sight of the instrument. In other words, the object points lying on the line of sight are imaged at the center of the graticule. A telescope sight will encounter objects at different distances from it, and for critical measurement the sight should be focused on the object.

5.1.1 Focusing on the Object

When a telescope is focused on an infinitely distant object, an image of this object is formed at the back focal plane of the objective. For a nearer object at a distance D from the front focal plane of the objective, the image is formed at a distance d from the back focal plane so that $d = -f^2/D$, where f is the focal length of the objective. If d is less than the range covered by the depth of focus of the telescope, the image appers to be as sharply defined as was that of the infinitely distant object. This property is utilized in fixed-focus telescopes such as gun sites.

In telescopic sights for measurement, a graticule is fitted in the image plane for setting or measuring. Here the sharpness of definition is not the only consideration; the image of the object (the target) should also lie in the graticule plane or should be close enough so that there is no parallax between the image and the graticule. When the parallax is present, the coincidence of the image depends on position of the eye; measurement error is the result. The problem is shown in Fig. 5.1. To remove parallax, the eyepiece is adjusted first until the graticule is seen clearly. One then focuses on the object by the appropriate focusing mechanism. Parallax is tested by the movement of the eye behind the eyepiece, and focusing is continued until relative movement between the object and the graticule lines is removed..

Telescopic sights have to be focused both at infinity and at very near objects. One of the obvious methods is to move the eyepiece in or out along

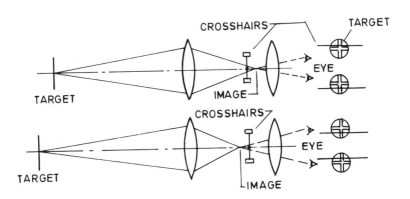

Figure 5.1   Test for parallax in a telescope sight. (Courtesy of McGraw-Hill, Kissam, P., 1962, Optical Tooling.)

the graticule until the image of the object is seen sharply in focus on the graticule. Taking an object range from infinity down to 1m, the extent of refocusing required for the objective of a 250 mm focal length is about 62 mm. The line of sight should remain fixed throughout the focusing range. This requires that the center of the graticule should not be laterally shifted during focusing. It should also be unaffected by vibrations. It is mechanically difficult to meet these two conditions, particularly because the eyepiece is supported only at one end of the main body of the telescope. The most satisfactory solution is the use of internal focusing, in which a lens system mounted on a separate tube is moved inside the main body of the telescope tube. If close tolerances are provided, the movement of the focusing lens can be made tilt free within acceptable limits.

A positive or negative lens placed between the objective and the graticule can be used for internal focusing, as shown in Fig. 5.2. The negative lens has the advantage that the equivalent focal length (EFL) of the objective increases, giving a larger image, and so the eyepiece is not required to magnify the image as much as it otherwise would. The positive focusing lens, on the other hand, makes the EFL shorter, thereby making it possible to focus the telescope on very near objects. It introduces slightly more error in the direction of the line of sight than a negative lens when displaced from the optical axis during focusing. Telescopic sights used in optical levels, theodolites, etc. use negative lenses for focusing, whereas an alignment telescope uses a positive lens for focusing.

5.2 THE ALIGNMENT TELESCOPE

An alignment telescope is a telescope sight for observing and measuring vertical and horizontal linear displacement from its line of sight. It is capable of being focused on targets very close to the objective lens and to the maximum practical working range. Since close-range focusing is re-

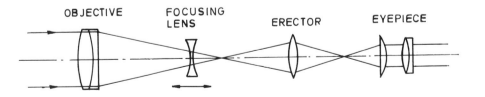

Figure 5.2   Optical arrangement of a telescopic sight.

quired, internal focusing with a positive lens is used. The precision and trueness of the axial guidance mechanism is crucial for accuracy. Most instruments have telescopic magnification between 30X and 50X at infinity. Magnification diminishes, and angular field correspondingly increases, as the telescope is focused on nearer objects. In some instruments the alignment telescope can be converted in an autocollimator when desired.

An alignment telescope is usually packaged in a barrel $2\frac{1}{4}$ in (57 mm) in diameter to fit in standard optical tooling mounts. The optical components are so mounted that the optical axis of the telescope and the mechanical axis of the barrel coincide. For measurements, the line of sight is to be displaced laterally (horizontally and vertically) without changing the direction. This is usually accomplished by an optical micrometer, as discussed in the next section. Figure 5.3 shows the optical arrangement of an alignment telescope. The effective focal length of the objective is small because of the presence of a positive focusing lens. This makes the primary image small. The erector system must provide enough magnification to compensate for this.

5.2.1 The Optical Micrometer

Figure 5.4 shows a plane parallel plate optical micrometer. It is mounted in front of the objective in the parallel beam. Rotation of the plate results in the shift of the line of sight. The displacement of the line of sight can be calibrated as a function of the angle of rotation of the plate. If the rotation angle is large, the calibration will be nonlinear. The micrometer is usually employed in the linear region and can measure only a limited offset from the line of sight.

5.2.2 Targets and Scales

The line-of-sight instruments require targets to align with or make offset

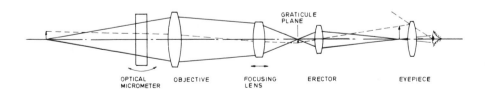

Figure 5.3   Optical arrangement of an alignment telescope.

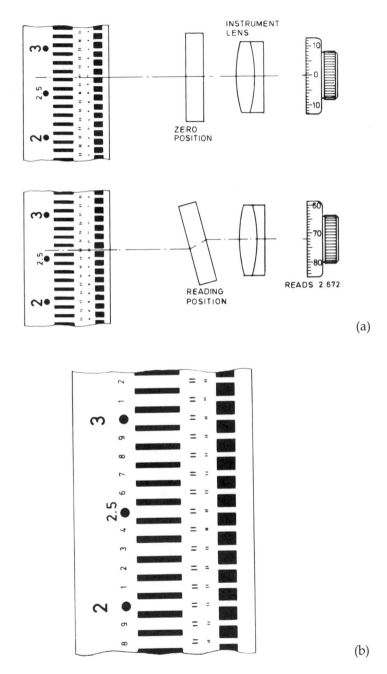

Figure 5.4 (a) Optical micrometer to measure offset from the line of sight. (Courtesy of Cubic Precision, K&E Electro Optic Products, USA.) (b) An optical tooling scale, (Courtesy of Cubic Precision, K&E Electro Optic Products, USA.)

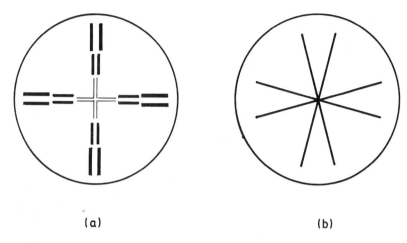

(a)   (b)

Figure 5.5   Alignment targets: (a) line pair and (b) double V.

with the line of sight. Figure 5.5 shows two typical target patterns. Figure 5.5 (a) is a target consisting of a series of line pairs of variable spacing. As mentioned in Chapter 3, the eye can place a line symmetrically between a pair of lines with very high accuracy. Here the graticule cross-hair of the line of sight instrument is placed between the line pairs of the target. Line pairs with variable spacing are provided, so that the target can be used at different distances. Only one pair will be best suited for alignment at a given distance; the others will be seen either too wide or too close. Figure 5.5 (b) shows a double V target. The setting here is achieved as shown in Fig. 5.6. Two white wedges are formed when the crosshair line super- imposes on the apex of V. These wedges should be aligned opposite each other for correct setting. The paired line target is found to give better accuracy than the V target. Both targets can be used for alignment or measurement in two axes.

For alignment of the target (which is eventually fixed to the object being aligned) with the line of sight of the instrument, the optical microm- eter is set for the zero position and the target is adjusted for the alignment. In another situation it may be necessary to measure the offset of the target with respect to the line of sight. The micrometer can now be rotated for alignment with the target and the offset read from the calibrated scale.

The optical micrometer has a limited range, and to measure offset beyond that limit scales are used. An optical tooling scale consists of a

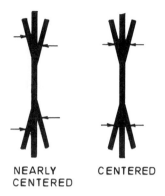

NEARLY   CENTERED
CENTERED

Figure 5.6   Alignment of a V target. (Courtesy of McGraw-Hill).

series of paired lines repeated at precise intervals that are less than the off-center measuring range of the micrometer, as shown in Fig. 5.4 (b). The gap between different series of paired lines is progressively increased as in case of the paired line target. Measurement with a tooling scale is shown in Fig. 5.4 (a).

5.2.3 The Collimator

A collimator is a device for projecting parallel rays. It consists of an objective lens, a cross-hair graticule placed at the focal plane, and a low-power light source that illuminates the cross-hair. The light rays from the cross-hair will be parallel as they leave the collimator, effectively projecting them at infinity.

For optical tooling, collimators are built in steel barrels in standard sizes; as in the alignment telescope, the optical axis is kept coincident with the axis of the cylinder. These are provided with two targets, a tilt target and a displacement target. The tilt target is used in place of the cross-hair graticule. It is graduated to read tilt angle in two axes. The target is essentially a cross-hair pattern whose center coincides with the optical (mechanical) axis of the collimator. The cross-lines are divided at regular intervals in both axes to represent angles in min of arc based on the relation to be discussed in Sec. 5.4. The displacement target is a standard alignment pattern mounted in front of or behind the objective.

5.2.4 The Optical Square

The optical square is essentially a pentaprism that bends the optical axis by 90° for alignment perpendicular to the optical axis. A plane normal to the optical axis can be generated by rotating the prism about the line of sight. By using a beam splitter on one of the reflecting surfaces of the pentaprism, two orthogonal light beams can be generated, as shown in Fig. 5.7. This modification is more useful with laser alignment, where a strong light beam is available.

5.2.5 Applications of the Alignment Telescope

The alignment telescope with its accessories is used for a variety of alignment problems in manufacturing and construction. Some typical alignment problems are discussed below.

Lateral Alignment

Lateral alignment requires lateral adjustment of the object (in two axes) on the line of sight or at a certain distance from it. The objects to be aligned generally do not have any reference marks, so scales or targets, as discussed earlier, are fixed on suitable receptacles on the objects. These scales or targets are then used for alignment with reference to the line of sight or measurement of displacement from it.

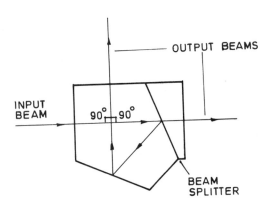

Figure 5.7   Generating two alignment beams at 90° with a pentaprism (optical square.)

Angular and Lateral Alignment

In several alignment problems, such as alignment of a series of bearings for a long shaft, both lateral and angular alignment is required. This is achieved using an alignment telescope and a tooling collimator. The tooling collimator is successively held in the bearings with the help of suitable adapters. Figure 5.8 shows the steps involved in the alignment procedure. When the alignment telescope is focused at infinity, it will focus the tilt target of the collimator in its cross-hair plane; see Fig. 5.8 (a). The displacement target D at the collimator objective is not seen, as it is out of focus. If the bearing in which the collimator is mounted is tilted with respect to the axis of the alignment telescope, the collimated beam coming from the collimator will enter the alignment telescope at an angle equal to the angle of tilt. Therefore the telescope will bring the center C of the tilt scale of the collimator to C' displaced from the center P of the telescope cross-hair. This tilt angle, if required, can be measured from the tilt scale of the collimator. To remove the tilt, the bearing is adjusted so that P and

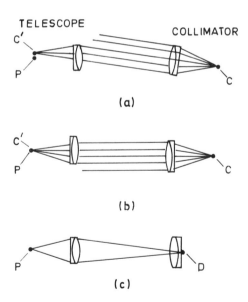

Figure 5.8   Alignment of a telescope and a collimator with (a) collimator axis tilted, (b) collimator axis parallel but not coincident with telescope axis, and (c) collimator axis coincident. (Courtesy of McGraw-Hill.)

C′ coincide; see Fig. 5.8 (b). The two are now parallel but may be laterally shifted. This lateral shift is removed by focusing the telescope on the displacement target of the collimator and making the necessary lateral adjustments. The tilt target is out of focus now and is not seen; see Fig. 5.8 (c). Repeating the above two steps a few times ensures both lateral and angular alignment.

Autoreflection

Autoreflection is a technique of angular alignment. An annular target is placed over the front end of the telescope as shown in Fig. 5.9. The target again consists of line pairs on two axes that intersect at the optical axis of the telescope when mounted. The image of the autoreflection target is observed in a mirror, and the mirror is adjusted until the target and the telescope cross-hair coincide. The autoreflection target may also be placed on the front glass surface of the telescope objective. The reflecting mirror used above may be fixed on the part to be aligned so that its reflecting surface is parallel to the proper reference plane in the part. This type of

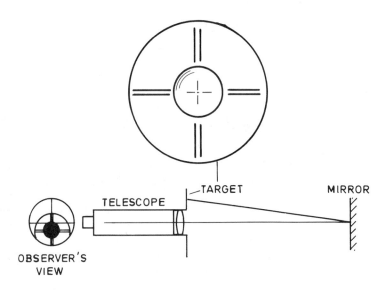

Figure 5.9 Alignment with an autoreflection target. (Courtesy of McGraw-Hill.)

alignment can also be done with autocollimation as will be discussed later. The autoreflection technique is not as accurate as autocollimation. There are two main sources of error. First, the autoreflection target may not be exactly centered on the axis. Secondly, refocusing of the telescope at different distances may introduce line-of-sight error. In autocollimation the sighting telescope is always focused at infinity.

5.2.4 Sensitivity and Accuracy

The sensitivity of the alignment telescope to target displacement from the line of sight is a function of target distance. With increasing target distance, the image size reduces, thereby increasing the error in linear measurement in the target plane. Generally an error of $5\mu m$ per m is to be expected. The overall accuracy for linear measurement is affected by the deviation in line of sight as a result of focusing. This error may be of the order of 1 sec of arc.

5.3 THE PHOTOELECTRIC ALIGNMENT TELESCOPE

A photoelectric alignment telescope produces an electrical error signal as a function of the lateral displacement of a remote body rather than the displacement of a target image as in visual instruments. Figure 5.10 (a) shows the optical arrangement of a photoelectric alignment telescope. A light source L illuminates the nose of the sensing prism P through a condenser C. The nose of the prism is in turn projected by an objective O into a cube corner prism CC, which is fixed to the object to be aligned. The prism P is actually a truncated pyramid with a clear nose and reflecting sides. Four detectors are placed around the prism at 90° intervals, two in teh plane of the paper and two perpendicular to it. These receive light reflected from the sides of the prism. Light is retroreflected by the cube corner, and an autoreflected image of the nose is formed on itself, provided the cube corner center coincides with the optical axis. In this situation, the autoreflected image is symmetrical at the nose, and all four detectors receive equal amounts of light; see Fig. 5.10 (b).

The autoreflected nose image displaces laterally when the cube corner moves off the telescope axis—Fig. 5.10 (c)—producing an imbalance in signal output. The detectors 1/1' and 2/2' detect positive and negative displacements along the x and y axes respectively. In order that ac error signals be produced, a modulated light source L is used.

The relation between the image displacement Δr and the cube corner displacement Δl at working distance R with objective focal length f is

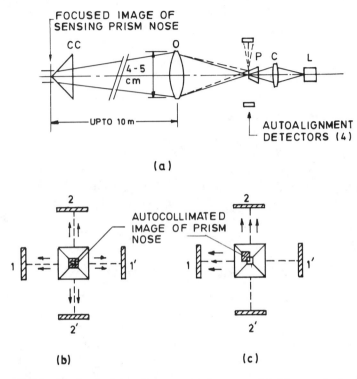

Figure 5.10 Photoelectric alignment telescope. (a) The optical arrangement. (b) The cube corner coincides with the optical axis. (c) The cube corner is off the optical axis. (Courtesy of McGraw-Hill.)

$$\Delta r = 2f \frac{\Delta l}{R}$$

A sensitivity of 10^{-6} to 10^{-7} m per meter can be obtained.

5.4 THE AUTOCOLLIMATOR

An autocollimator is a collimator with an eyepiece for viewing the auto-reflected image of its graticule for measurement of small angles. In Fig. 5.11, G is the illuminated graticule projected by the objective O. Light reflected back from a mirror M set normal to the collimated bundle is

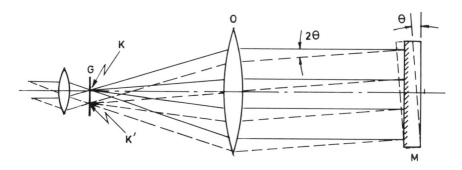

Figure 5.11 Principle of an autocollimator.

focused by the objective lens and forms an image of the graticule directly superimposed on the graticule itself. If the mirror is rotated slightly about any axis perpendicular to the autocollimator's optical axis, the return image of a point K will be formed at K' on the graticule displaced by a linear distance s proportional to the angle rotation θ and given by

s = 2fθ

where f is the focal length of the objective. If we measure the linear distance s, we can determine θ.

Figure 5.12 shows an optical arrangement of an autocollimator. The projected graticule is usually a cross-hair (1). The reflected image of the cross-hair is received on another graticule (2), which is here a tilt scale. We call it a tilt scale because the linear displacement of the image of the cross-hair, as a result of the tilt of the reflecting mirror, is indicated here directly as the tilt of the mirror. The maximum angle of the tilt of the mirror that can be measured is usually less than half a degree with graduation every one or one-half min of arc. Resolution of 1 sec of arc can be achieved with suitable micrometer devices such as a parallel plate micrometer with eyepiece (3). Autocollimators can be designed to measure tilt angles in one or two axes. In Fig. 5.12 the optical path is folded with the help of mirrors (4 & 5) to make the instrument compact. The eyepiece is placed at 60° to the optic axis for convenient viewing with the help of a deflection prism (6). The cross-hair graticule is illuminated by a light source (7).

An alignment telescope can be converted into an autocollimator if a graticule illumination device is included. In this case the instrument is adjusted for infinity focus.

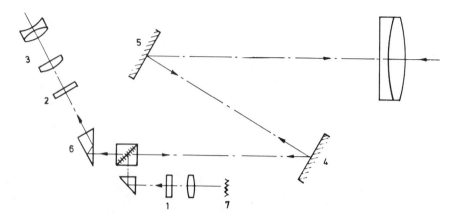

Figure 5.12   Optical arrangement of an autocollimator.

5.4.1 Applications of the Autocollimator

An autocollimator measures the tilt of a plane reflector with respect to its own axis. It is routinely used for alignment with the help of a plane reflector. With an appropriately mounted mirror, an autocollimator is used for checking the flatness or parallelism of machine tool ways, for measuring the flatness of surface plates, etc. The mirror is mounted on a flat base with three-point support. As the mirror is moved over the surface or guideway, any departure from flatness or parallelism tilts the mirror. The tilt is recorded over the entire object and analysed for surface errors. Rotary angular indexing and measuring devices can be calibrated with an autocollimator against optical cubes and polygons as standard angles. Alignment at 90° to the optical axis can be done with the help of a pentaprism. In the optical workshop, an autocollimator is used for precision measurement of wedge plates and a variety of prisms.

Not all the rays of the light beam projected by the autocollimator are parallel to the optical axis, because of the finite size of the graticule. The output is composed of an inifinity of collimated bundles emanating from different points on the graticule. As the distance between the autocollimator and the reflecting mirror is increased, some of the edge bundles that are not normal to the mirror begin to miss the objective lens, as shown in Fig. 5.13. This results in a reduction of illumination, followed by total extinction of the edge points as the distance increases. The angular range of the instrument is accordingly reduced. In the extreme case, when the

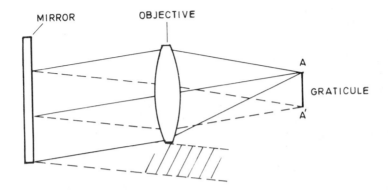

Figure 5.13   Vignetting of the reflecting beam in an autocollimator as the mirror moves away from it. (Courtesy of McGraw-Hill.)

reflected axial bundle of collimated light completely misses the objective aperture, the axial point is extinguished. The limiting angle θ of mirror tilt that can be measured at a distance l is given by

$$\theta = \frac{d}{2l}$$

where d is the diameter of the objective lens.

5.4.2 Sensitivity and Accuracy

An autocollimator measures angles by measuring linear displacement via the relation $s = 2f\theta$. A small change $\Delta\theta$ in θ results in a change Δs in s given by

$$\Delta s = 2f \, \Delta\theta.$$

If we use the symmetry setting criterion using double lines (a bifilar graticule pattern) in the micrometer eyepiece, Δs has an empirically determined value of 2.5μm. For an objective focal length of 250 mm this gives 1 sec of arc for $\Delta\theta$.

In order to achieve the limiting accuracy of the autocollimator, its full entrance aperture should be employed. The reflector should be large enough to return to the entrance aperture all the light that the geometry of the setup permits. With a smaller reflector the mirror itself acts as the entrance aperture. This results in error in measurement depending on how much and which part of the aperture is used. If only a part of the aperture is used there is a shift of image because of the obstruction, as will be explained below. This may result in error of up to a few sec of arc across the entrance pupil.

Lens systems usually have uncorrected aberrations. As a result, the image of an object point is a circular area in which the maximum light from the object is concentrated at the center as shown in Fig. 5.14. The center is recognized as the image of the object point; but if the aperture is partly obstructed, unless the obstruction is symmetrical with respect to the axis of the lens, the photometric center of light is not at the same place as before. Accordingly, the whole image will shift position slightly. The shift may be quantatively assessed by moving a card with an appropriate hole across the objective of instrument and by observing the effect on the reflected

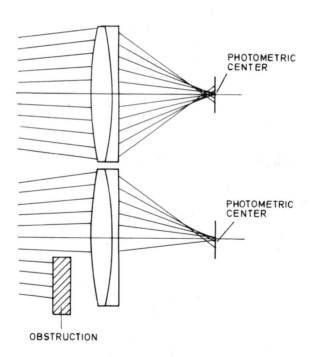

Figure 5.14   Effect of beam obstruction in an autocollimator. (Courtesy of McGraw-Hill.)

5.4.3 The Electronic Autocollimator

There are several versions of the electronic autocollimator, in which the human eye is replaced by a photoelectric detector for measurement. In one version, the reflected image of the graticule line is sensed by a slit vibrating symmetrically about the image line, as in the photoelectric microscope (Chap. 4). The vibrator is essentially a linear motor. A dc bias current maintains the symmetrical setting; the dc bias current is proportional to the angular tilt of the reflector.

Figure 5.15 shows another arrangement of the electronic autocollimator. Here the projected graticule is replaced by a sensing prism, which is essentially a truncated pyramid on a square base with a clear nose and reflecting sides. The nose is placed at the focal plane of a lens and is illuminated by a light source. Around the prism four detectors are arranged at 90° intervals, each receiving light reflected from one of the sides of the prism. When the reflecting mirror is perpendicular to the optical axis, the autoreflected image of the prism is formed symmetrically on the nose. All four detectors receive an equal amount of light. When the reflecting mirror is tilted, however, the autoreflected image is displaced from the nose, and the sides of the nose reflect more light on the appropriate detector, depending on the amount and the axis of the tilt θ as shown in Fig. 5.10 (b) and (c). The excess light gives an increase in detector signal, which is a measure of the tilt angle. The light source can be modulated to get an ac signal. An LED or a nonmoving electrooptic modulator is preferred.

As in the usual autocollimator, the displacement s of the nose image is given by

$$s = 2f\theta$$

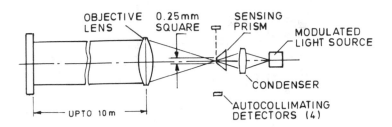

Figure 5.15 Photoelectric autocollimator. (Courtesy of McGraw-Hill, Driscoll, W. G. and Vaughan, W., Eds., Handbook of Optics, 1978.)

For a θ of 1 sec of arc, s is 2.5 μm for an objective of 250 mm focal length. This displacement is easily detected.

In yet another version of the electronic autocollimator, the eyepiece graticule, relative to which the displacement of the proejcted graticule is measured, is replaced by a detector array (Fig. 5.16). The displacement of the reflected graticule line, and hence the angle, is measured by scanning the detector array. Such a system is ideal for automatic evaluation, computation, display and printing the result at a rapid rate. Commercial instruments have a resolution of 0.1 sec of arc and an accuracy of 0.3 sec of arc over 60 sec of arc. The range is about 1°.

5.5 ALIGNMENT VIA A LASER

The He-Ne laser gives a strong beam of light that diverges very little as it travels in space. Because of its special features it can be used to determine all the six motions of a body (three of rotation and three of translation). The motion in the direction of the beam on the z axis is measured by a laser interferometer as discussed in Chap. 4. The two motions Δx and Δy perpendicular to the laser beam, the tilts θ and φ about the x and y axes, respectively, and the rotation γ about the laser beam form the alignment

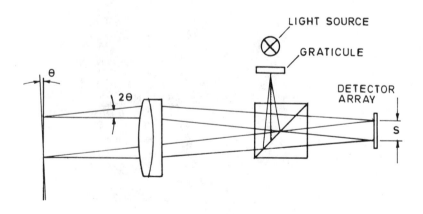

Figure 5.16   Electronic autocollimator.

problem and are discussed in this section.

5.5.1 The Tooling Laser

The tooling laser is a He-Ne laser of special construction as shown in Fig. 5.17. The plasma tube is enclosed in a metal tube housing finished to a standard diameter. The whole construction is extremely rugged to withstand the conditions on the shop floor. It gives out a collimated beam that is precisely centered and accurately parallel relative to the mechanical axis of the housing. The output from the laser is passed through a beam expander of low power to produce a beam that is about 10 mm in diameter and is collimated to the diffraction limit. The laser tube and the beam expander are enclosed in the same housing.

Beam Expander

An afocal system is used as a beam expander as shown in Fig. 5.18. The ratio of the focal lengths of the lenses L_1 and L_2 gives its magnification. The diameter of the output beam is given by $2d_2 = (f_2/f_1)\,2d_1$, where $2d_1$ is the diameter of the input beam and f_1 and f_2 are the focal lengths of the two lenses.

Reduction in Divergence

The beam expansion is accompanied by a reduction in the divergence of the laser: the divergence of the beam after the beam expander is $\alpha_2 = (f_1/f_2)\alpha_1$, where α_1 is the input divergence. It is not possible, however, to

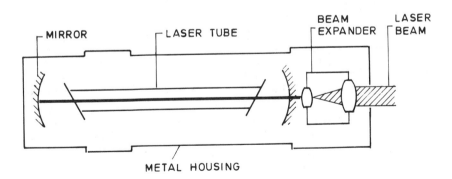

Figure 5.17   A tooling laser.

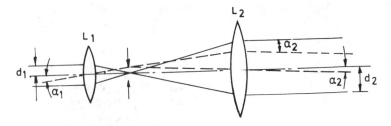

Figure 5.18   Laser beam expander.

reduce the divergence arbitrarily; the minimum limit is set by diffraction. In the case of conventional collimators, the limit of divergence is set by the smallest source size that can be used from energy considerations.

Increase in Stability

In a tooling laser, stability means the stability of the direction of the beam. The direction may change because of the instability of the resonator. It is easy to see that if the change in the direction of the laser beam is $\Delta\alpha_1$, then the beam leaving the beam expander will change in direction only by an amount $\Delta\alpha_2 = (f_1/f_2)\,\Delta\alpha_1$.

5.5.2 Visual alignment

The laser beam provides a straight reference in space. The objects to be aligned can be adjusted with respect to this beam. The simplest technique of course is to align the objects visually by receiving the laser beam on a ground-glass screen with scales marked on it.

5.5.3 Photodetectors

The laser output has a Gaussian distribution of intensity. This permits precise detection of the photometric center of the beam and accurate measurement of the displacement from the center by means of photodectectors. Two types of detectors have been used.

In the first type, the photosensitive surface is divided into two or four parts. Each of the halves or quadrants produces a signal. These signals are added in such a way that indicators show zero when light falls symmetrically about the line dividing the halves or the point of interesction of its quadrants.

If the beam diameter is small, a small displacement between the sensor and the beam produces a large output signal. But then the beam soon leaves the region where the displacement is linearly related to signal. Thus with a small beam diameter we obtain high sensitivity but a smaller measuring range. For larger beam diameters it is the other way around.

The small measuring range associated with high sensitivity is not a handicap, because these detectors are used as targets. The targets are mounted on the objects to be aligned. The object is aligned until the signals from the detectors show zero on the indicator.

When a relative displacement between the detector and the laser beam is to be determined, a plane parallel plate can be employed before the detector, and the plate is rotated until the detector shows a null, as in Fig. 5.19. The angle of the parallel plate can be calibrated in terms of the displacement from the center of the detector. The detectors are mounted on standard adapters that are used in conventional alignment techniques using alignment telescopes.

The second type of detector has a surface sensitive to the position of the laser beam. The detector shows the position of the photometric center of the incident light in one or two coordinates from the middle point of the detector. Here the range of measurement is independent of the size of the beam, only the whole beam must fall on the detector: but there may be nonlinearity when used over the maximum range.

5.5.4 Autoreflection

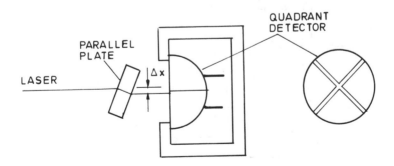

Figure 5.19   Laser alignment with a quadrant detector.

If a mirror, as shown in Fig. 5.20, is attached to a body, its tilt perpendicular to the laser beam can be measured by measuring the shift of the laser beam on a quadrant detector placed close to the laser head and far from the mirror. The displacement of the reflected spot, when the mirror is not perpendicular to the laser beam, is a function of the distance of the mirror. Usually this method is used to indicate null as the objects are aligned, i.e., the mirror becomes perpendicular to the beam when the alignment is complete. In such a case the distance of the object has no influence. However, if the mirror is replaced by a cube corner retroreflector, the detector will sense a small displacement along two axes.

5.5.5 Autocollimation

An arrangement of the type shown in Fig. 5.21 measures only the angle of tilt and is indpendent of the distance of the object, because the shift of the spot on the detector is a function the angle of the returning beam because of the presence of the lens L. A position-sensitive detector and not a quadrant detector is used because the spot on the detector is very small. The arrangement is insensitive to the motion of the mirror perpendicular to the beam.

5.5.6 The Collimator

A laser can be used as a collimator and a photosensor as a telescope, as shown in Fig. 5.22. The photosensor and the lens are mounted in a tube suitable for adoption in standard alignment fixtures. The displacement of the laser spot on the position-sensitive detector is a function of the tilts of

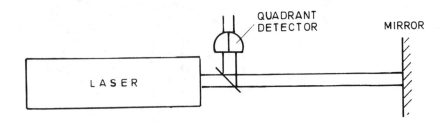

Figure 5.20   Autoreflection with a tooling laser and a quadrant detector.

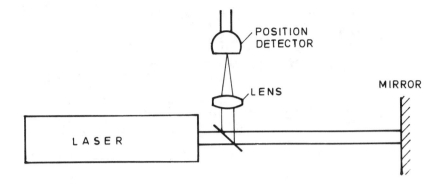

Figure 5.21   Autocollimation with a tooling laser and a position sensitive detector.

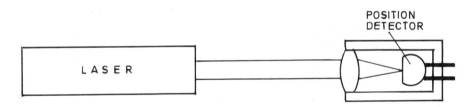

Figure 5.22   Collimator with a position detector for use with a laser for alignment.

the detector assembly about the two axes perpendicular to the laser beam. This technique has given a reproducibility of 0.2 arc sec, which is much better than that of the usual collimator and telescope techniques (1 sec).

5.5.7 Combined Measurement of Displacement and Tilt

Figure 5.23 shows a detector head combining a quadrant detector and a position-sensitive detector. The detector head is mounted on the job to be aligned. Displacements Δx and Δy are indicated by the quadrant detector, and tilts θ and ϕ are given by the position detector.

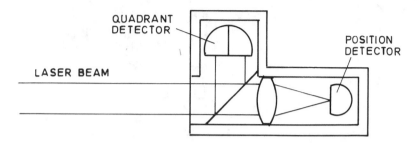

Figure 5.23 Combined measurement of tilt and displacement with a laser.

5.5.8 Rotation About the z Axis

A tooling laser giving a linearly polarized output can be used to measure the rotation of an object about the laser beam with the help of a polarizer. The intensity of the light transmitted by a polarizer is given by

$$I = I_0 \frac{(1 + \cos 2\gamma)}{2}$$

where I_0 is the input intensity and γ the angle between the transmission axis of the polarizer and the plane of vibration of the laser output. Figure 5.24 (a) shows the intensity variation with γ. The linear part of this curve can be used for the measurement of rotation.

When the laser beam passes through a Wollaston prism, two ortho-gonally polarized beams are produced. If the direction of vibration of the laser light is at 45° to the transmission directions of the two output beams, their intensities are equal; see Fig. 5.24 (b). The two outputs are received on separate photodetectors, and their signals are subtracted to give a null position. This position is used as the reference point to measure rotations about the laser beam axis. Since the two signals vary in opposite directions, the sensitivity is doubled. Rotations in the range of a few sec of arc can be measured in this way.

5.6 THE ASYMMETRY METHOD FOR HIGH-PRECISION ALIGNMENT VIA LASER LIGHT

In the methods discussed above, the center of the beam with its maximum intensity is used as a reference for the alignment of a system. By introdu-

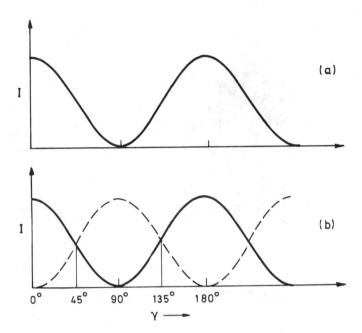

Figure 5.24   (a) Transmitted intensity from a rotating polarizer for linearly polarized input, and (b) transmitted intensities of the two beams from a rotating Wollaston prism for linearly polarized input.

cing a phase plate it is possible to obtain a minimum of intensity at the center of the beam. The asymmetry method [1] makes use of this minimum principle.

5.6.1 The Phase Plate

If a phase plate consisting of four sectors of parallel and homogeneous layers such that the adjacent parts of the plate differ in optical thickness by $\lambda/2$, where λ is the wavelength of light, is placed in the beam of laser, the output has orthogonal dark bands over the beam. The beam can be said to be carrying its own cross-wire. Figure 5.25 (a) shows the phase plate and Fig. 5.25 (b) the intensity pattern across the beam. The theoretical calculations for intensity distribution after the phase plate can be treated on the basis of diffraction theory. The intensity distribution across the beam is shown in Fig. 5.25 (c).

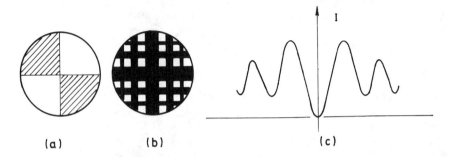

(a)   (b)   (c)

Figure 5.25   Effect of using a phase plate in a laser beam. (a) Phase plate, (b) intensity pattern, and (c) intensity distribution.

5.6.2 Detection of the Optical Axis

The dark cross in the center of the laser beam is clearly visible at any distance and can be used for making preliminary adjustments. To use the asymmetry method a rectangular aperture is inserted in the PA plane as shown in Fig. 5.26. The diffraction pattern is observed in the PB plane. When the aperture PA is exactly centered on the axis, the deffraciton pattern is exactly symmetrical and appears like the first diffraction pattern,; see Fig. 5.25 (b). A displacement between the aperture and the axis results in characteristic asymmetries in the diffraction pattern in the plane PB. The asymmetry is found in the four intensity maxima surrounding the central dark cross-band and increases with the displacement of the aperture from the axis. Both the amount and the direction of such a displacement can be measured by monitoring and comparing the intensities of the four maxima. In actual practice, rectangular apertures are mounted on alignment fixtures in the same way as conventional alignment aids, and the alignment is carried out until symmetry of the diffraction pattern is achieved. It has been estimated that a shift of about 0.01 mm of the aperture PA from the optical axis can be visually detected. Using four detectors to sense the change in intensity, misalignments much smaller than 0.01 mm can be detected and measured.

Telescopic methods of alignment can be used up to 10 m. The tooling laser extends this limit up to 100 m. The asymmetry method is best suited for larger distances.

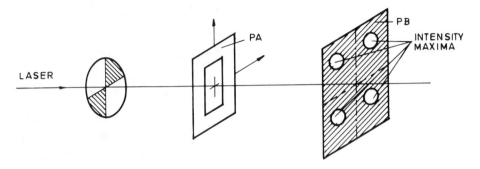

Figure 5.26   Asymmetry method of alignment with laser.

5.7 THE LASER INTERFEROMETER FOR ALIGNMENT

The use of the laser interferometer for displacement measurement has been discussed in Chap. 4. This technique can be adopted for alignment or straightness measurement using the arrangement shown in Fig. 5.27. A Wollaston prism splits the laser beam with a small included angle ϕ. A double mirror assembly reflects the beams back to the Wollaston prism where they are combined again. If the double mirror assembly or the Wollaston prism is laterally shifted by an amount Δx, an optical path difference equal to $2\Delta x \sin \phi/2$ is introduced beetween the beams. This path difference is converted into fringe count by the interferometer and Δx is determined.

5.8 THE OPTICAL DIVIDING HEAD

The optical dividing head and similar instruments such as the theodolite level and the goniometer, which perform measurement over 360°, incorporate a precision divided circle (usually of glass) that is read with the help of microscopic systems and optical interpolation scales. Alternatively the glass scale is replaced by a moiré measurement system consisting of a radial moiré grating and a small radial index grating. These give a least count of 1 sec of arc or less.

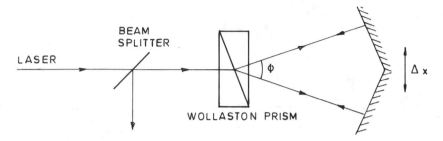

Figure 5.27   Laser alignment interferometer.

5.8.1 Divided Circle Readings: The Coincidence Principle

For high accuracy, the circles are read at two positions 180° apart, and average value is found, to cancel eccentricity errors [2]. This process of reading at two places and averaging is considerably simplified by the coincidence principle.

In coincidence techniques of reading circles, two parts of the scale circle 180° opposite each other are imaged together so that they move in opposite directions when the circle is rotated; see Fig. 5.28 (a) and (c). In (b), the scale reading (24) is displaed by an amount x_1 from the index line in the field of view, whereas the corresponding 180° opposite reading (204) is displaced by x_2. When there is no eccentricity error, $x_1 = x_2$. Conventionally the reading can be obtained by measuring x_1 and x_2, and the actual reading will be 24+ $(x_1 + x_2)/2$. In the coincidence principle, the lines of the two scales are made to coincide with the help of optical micrometer devices (plane parallel plates), and the average reading is found directly from the interpolation scale. In actual practice, the two parts of the scale are imaged together with a sharp dividing line, and images are moved in opposite directions by equal amounts for coincidence using the optical micrometer. The actual movement of the images for coincidence is half of $x_1 + x_2$, and hence an average is automatically obtained. A sharp dividing line between the two images is obtained by suitable prism systems. Optical dividing heads are designed for angular measurement and indexing in machining and layout jobs.

5.9 THE INTERFEROMETER FOR ANGLE MEASUREMENT

Figure 5.29 shows the Dowell's angle gauge interferometer. The mirrors M_1, M_2 and M_3 can be adjusted so that the two wavefronts produced at the

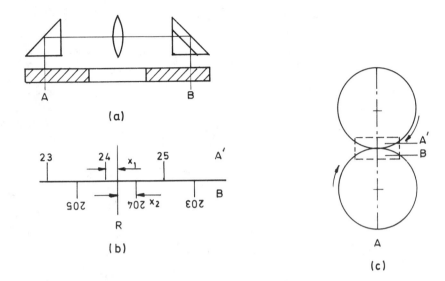

Figure 5.28 Coincidence principle of reading a circular divided scale: (a) A, B: 180° opposite parts of a divided circle, (b) A': image of A brought into coincidence with B, and (c) typical field of view of a reading system. R is the index mark.

beam divider B come out parallel to each other after traveling their respective paths, $BM_1M_3M_2B$ and $BM_2M_3M_1B$. In this case, no fringes will be seen in the field of view. If a plane parallel plate is now inserted at G, the wavefronts reflected from its two faces will remain parallel when they come out of the interferometer, irrespective of the orientation of the plane parallel plate, as is clear from the passage of the dotted lines. If, however, a workpiece with a small angle between its faces is placed at G, fringes will be seen in the field of view. The spacing of the fringes is a function of the angle between the faces. This interferometer is used for checking the parallelism of the slip gauge surfaces and for measuring small angles.

Figure 5.30 shows the arrangement of the Dowell's interferometer for comparing angles of a workpiece with master angles. The interferometer is first set with the standard angle for no fringe condition. The standard angle is then removed and the workpiece is inserted. Fringes will show if the workpiece angle deviates from the standard angle. The spacing of fringes gives the amount of deviation.

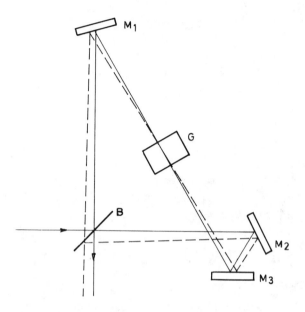

Figure 5.29 Dowell's angle gauge interferometer for small angle measurement.

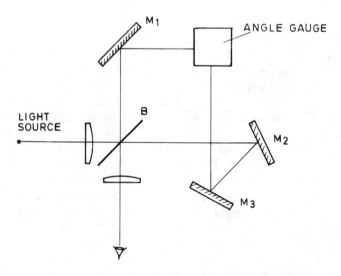

Figure 5.30 Dowell's interferometer for comparison of angles.

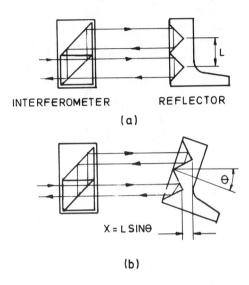

INTERFEROMETER   REFLECTOR

(a)

X = L SINθ

(b)

Figure 5.31   Laser angle interferometer.

Figure 5.31 shows a version of the laser interferometer suitable for angle measurement. Two cube corner reflectors are used in a single mount with their optical center displaced by an amount L. Interference is produced by the two beams reflected off the two reflectors. When the mount holding the reflectors is titled by an angle θ, the path difference between the two interfering beams is x = Lsin θ. Thus the angle θ is found by determining the path difference interferometrically by counting the fringes as in displacement measurement. This device is used with a laser interferometer for displacement measurement as a mirror as in an autocollimator for measurement of small tilt angles and associated applications.

REFERENCES

1. Betz, H. D. (1969). An asymmetry method of high precision alignment with laser light, *Appl. Opt., 8:* 1007.
2. Deumtich, F.. (1972). *Instrumentenkunde der Vermessungstechnik*, VEB Verlag für Bauwesen, Berlin, Chap. 2.

ADDITIONAL READINGS

Kissam, P. (1962). *Optical Tooling,* McGraw-Hill, New York.

Luxmoore, A. R. ed. (1983). *Optical transducers and techniques in Engineering Measurement,* Applied Science Publishers, London.

Driscoll, W. G. , and Vaughan, W. ed. (1978). *Handbook of Optics,* McGraw-Hill, New York, Chap. 16.

Dagnall, R. H. and Pearn, B . S. (1967). *Optical Alignment,* Hutchinson, London.

CHAPTER 6

Heterodyne and Phase Shifting Interferometry

Interferometry has been used in problems such as length and displacement measurement (Chap. 4), testing of optical components and systems, surface profiling, roughness measurement, etc. The advantage of interferometry is that the measurement unit is the wavelength of light, so that there is an inherently high resolution. Resolution of a fraction of a wavelength is easily obtained.

Interferometry measures the optical path difference (OPD) between two wavefronts. The path difference may be introduced by displacement, as in displacement measurement or by deviation of wavefront from an ideal wavefront, as in optical testing and profiling. The result in an interferometer is an interference pattern. In displacement measurement, the fringes crossing a point in the intereference pattern are counted, and no analysis of the interference pattern is required. In other cases, an interpretation of the fringe pattern is necessary and is more complicated.

The fringes in an interference pattern represent contours of equal path (or phase) difference. The path difference between two consecutive fringes in a fringe pattern is usually $\lambda/2$. From the locations of the centers of the fringes an OPD map is generated. This procedure is known as fringe interferometry. The intensity distribution in a fringe pattern is sinusoidal. In such a pattern the detection of the fringe center usually introduces an error in the range of $\lambda/10$. Thus, in fringe interferometry the accuracy of OPD maps is limited to about $\lambda/10$.

Presently we must test optical components and systems more accurately than before to be able to manufacture high-performance systems. Extremely smooth surfaces are required for high-power-lasers, laser gyrooptics, x-ray optics, magnetic storage discs, etc. where resolution in the subnanometer range of roughness measurement is expected. At the same time, the results are required rapidly. To achieve this, new interferometric techniques coupled with microcomputers have been developed recently. Heterodyne and phase shifting interferometry are two such techniques. These techniques eliminate the need for photographing a fringe pattern and then determining the position of fringe centers. Either phase at a point in the interferogram is measured electronically or the intensity at the point is measured and the phase calculated. These techniques have several advantages: the measurement accuracy is high; the measurements are done rapidly, avoiding slowly varying influences; low-contrast fringes can be used without sacrificing accuracy; intensity variations across the interference pattern do not influence the result; and measurements are carried out at a fixed grid of points. The use of computers gives an added advantage. The errors of the interferometer system can be measured, stored, and subsequently subtracted from the result. In this way, error-free results can be obtained, even in the presence of instrumental errors.

6.1 HETERODYNE INTERFEROMETRY

We have heterodyne interferometry when the two interfering waves have different frequencies. Let the two waves be represented as $A_1(t)$ and $A_2(t)$ given by

$$A_1(t) = a_1 \exp(-iw_1 t) \tag{6.1}$$

$$A_2(t) = a_2 \exp[-i(w_2 t + \phi)] \tag{6.2}$$

where ϕ is the phase difference, $w_1 = 2\pi f_1$ and $w_2 = 2\pi f_2$ the angular frequencies of the waves, a_1 and a_2 their amplitudes, and f_1 and f_2 their frequencies. The superposition of the two waves

$$A(t) = A_1(t) + A_2(t) \tag{6.3}$$

gives an electrical field at a detector. The output of the detector is a current proportional to the intensity $I = A(t) \cdot A(t)^*$ given by

$$I(t) = \tfrac{1}{2}a_1^2 + \tfrac{1}{2}a_2^2 + \tfrac{1}{2}[a_1^2 \cos(2w_1 t) + a_2^2 \cos(2w_2 t + \phi)]$$

$$+ a_1 a_2 \cos[(w_1 + w_2)t + \phi] + a_1 a_2 \cos[(w_2 - w_1)t + \phi] \tag{6.4}$$

The third and fourth terms on the right side of this equation correspond to components at angular frequencies of $2w_1$, $2w_2$, and $w_1 + w_2$, which are too high to be followed by practical detectors. Hence the output of the detector will be proportional to

$$I(t) = I_1 + I_2 + 2\sqrt{I_1 I_2}\, \cos[(w_2 - w_1)t + \phi] \qquad (6.5)$$

The first two terms here represent the intensities of the single beams and are constant. The third term represents an electronic signal with an angular frequency $w_2 - w_1$ giving $f_2 - f_1$ typically in MHz range. This term also contains the phase term ϕ, which was once carried by an optical carrier of 10^{14} Hz but is now carried by an electronic signal in the range of 10^6 Hz. Direct phase measurement of the electronic signal and hence the phase difference between the interfering beams is possible by comparing it with a suitable reference signal. As shown in Fig. 6.1, a time measurement of the zero crossing of the test and reference signals gives the optical path difference as $(t_1 / t_2)\, \lambda$. Obtaining a $\lambda / 100$ accuracy would require a time measurement accuracy of 1 μsec for a 10 KHz beat signal.

In heterodyne interferometry, no fringe pattern is actually seen, as the intensity at any point sinusoidally varies at the rate of $f_2 - f_1$, which cannot be followed by the human eye. An appropriate detector, however, can follow these variations. At any given point the phase of the detector signal is the same as the optical phase ϕ. To generate an OPD or phase map, two detectors are used. The signal from a fixed detector in the interference

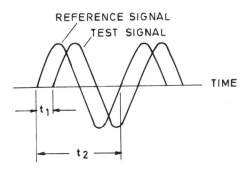

Figure 6.1   Signals carrying optical phase information in heterodyne interferometry.

pattern serves as the reference signal. A scanning detector scans the pattern, and the zero crossing time of its signal is compared with that of the reference signal, giving the phase at the point being scanned. After the scanning is complete, the measurement data is analysed by a computer, which provides an OPD map and other useful information. Heterodyne interferometry can give values of phase difference with a precision of a fraction of a degree.

6.1.1 Frequency Shifting

Heterodyne interferometry requires interfering beams of two slightly different frequencies. This is achieved by introducing a frequency shift in one or both beams. Several frequency shifting techniques are available.

One technique makes use of rotating polarization components, namely quarter and half-wave plates and polarizer [1-6]. For example, if circularly polarized light is transmitted through a rotating half-wave plate, the orthogonal components of the transmitted light show a frequency difference equal to twice the rotation rate of the plate. The principal difficulty in this method is that for reasonable rotation rates it is not possible to obtain a frequency shift much larger than a few KHz.

A moving diffraction grating can also be used as a frequency shifter, since it shifts the frequency of the Nth diffracted order by an amount Nvf, where v is the velocity component of the grating perpendicular to the grating lines and f is the spatial frequency of the grating. In other words, the frequency shift is equal to N times the number of grating lines that pass a given point per second. For a grating, the frequency shift is independent of the wavelength. To obtain a large frequency shift in the range of several KHz, a circular grating is useful [7].

An acoustooptic Bragg cell shifts the frequency of diffracted light in the same way as a moving diffraction grating. In this case the traveling acoustic wave serves as the moving grating. The frequency shift of the first diffracted order is equal to the frequency used to drive the Bragg cell independent of the wavelength of light. A Bragg cell gives a frequency shift in the range of several MHz.

Some lasers can be designed to give two frequency outputs, for example, the Zeeman split He-Ne laser (see Chap. 1).

6.1.2 Heterodyne Interferometer for Optical Testing

Figure 6.2 shows an optical arrangement of a heterodyne interferometer [8]. The single-frequency output of a laser is split and each component directed to a Bragg cell B for frequency shifting. One cell is driven to give

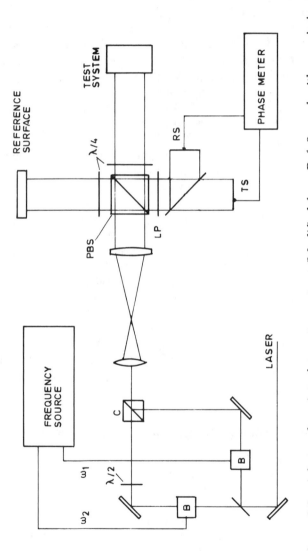

Figure 6.2 A heterodyne interferometer system. (Modified from Ref. 8, used with permission.)

an output of angular frequency w_1 and the other of w_2, with the difference frequency being a convenient video frequency including zero. The polarization of the beam exiting from one Bragg cell is rotated 90° by a $\lambda/2$ plate. The two beams are then combined by a polarization beam combiner C. The beams are now expanded.

The combined beam with two frequencies that are orthogonally polarized enters a Twyman-Green interferometer. A polarization beam splitter PBS splits it into two orthogonally polarized beams that travel to the two arms of the interferometer. A $\lambda/4$ plate in each arm converts the ongoing beams to circular polarization, and on return to linear polarization. The beams are recombined by the PBS and sent to the interference plane. A linear polarizer LP oriented at 45° to these polarizations causes the waves to interfere.

The reference signal RS for phase measurement is obtained from a small area detector located at a stationary point on the interference pattern. The test signal TS for the entire image is obtained by scanning the image with a single detector. The test and reference signals are supplied to a phase meter that converts the phase difference into an analog voltage. The phase information at all points can be obtained in real time by the use of an array of detectors, each followed by a phase detector providing parallel channels. This arrangement should give the fastest speed but will be complex and expensive. A system of high spatial resolution but of moderate speed has been made with an image dissector camera to scan the interference pattern [8]. Figure 6.3 shows the OPD map of a deformable mirror obtained using this interferometer.

As in a conventional inteferferometer, the test system is placed in one arm of the interferometer and a reference surface in the other. This arrangement is suitable for optical components or system testing. For surface profiling, different arrangements have been proposed.

6.1.3 Surface Profiling by Heterodyne Interferometry

Interferometers have long been used for surface profiling. In combination with heterodyne techniques these systems become much more sensitive [9,10]. Figure 6.4 shows a modified Twyman-Green interferometer for surface profiling using heterodyne interferometry. A focused laser spot in the test arm scans the object as it translates across the surface like a stylus but without the mechanical contact. The phase variation introduced in the beam as the object is scanned is given by $\phi = 4\pi h/\lambda$, where h is the variation in the profile height.

The input to the interferometer is a two-frequency laser beam, the frequency components f_1 and f_2 being orthogonally polarized. The passage

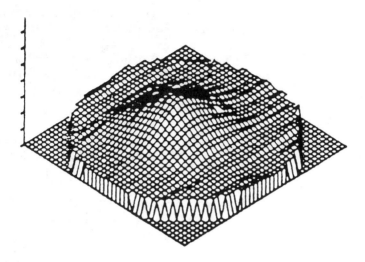

Figure 6.3 OPD map of a deformable mirror obtained using heterodyne interferometry. The vertical scale is 0.647 μm/tic. (From Ref. 8, used with permission.)

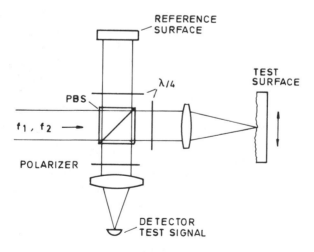

Figure 6.4 Heterodyne interferometer for surface profiling.

of light through the interferometer can be explained in the same way as for Fig. 6.2. The light reflected from the two arms interferes and is received on a photodetector which produces an electrical signal. The phase of this signal varies as the optical phase ϕ varies as a function of height variation. Since the object is being scanned point by point, the reference signal is not available here from the interferogram. The reference signal is obtained by causing the components of the input beam to interfere as will be explained in connection with Fig. 6.5.

The difficulty in using this interferometer for surface profiling is to perform a translation of the object without height fluctuations due to the carriage. Other problems are due to the mechanical or thermal instabilities in the two arms of the interferometer.

To overcome some of these difficulties a different optical arrangement has been used [11]. The surface is illuminated by two separated focused beams of light with slightly different frequencies whose reflected beams interfere. The phase of the sinusoidal intensity modulation is related to the height difference between the illuminated points on the surface. If one of the beams remains focused on a fixed point while the other beam is moved along the surface, height variations along the scanned line are measured. Fig. 6.5 shows an optical arrangement to realize this. The light source is a Zeeman split two-frequency He-Ne laser with a frequency difference f_2-f_1 of a few MHz. The two components are orthogonally linearly polarized. The light beam is split into two parts with the help of a polarization-insensitive beam splitter BS. A small fraction of the input is transmitted and received on a detector D_1. A polarizer P_1 placed before the detector at $45°$ to each polarization enables them to interfere, resulting in a signal with angular frequency w_2-w_1. This serves as the reference signal.

The reflected beams are directed to an interferometer consisting of a Wollaston prism and a microscope objective. The Wollaston prism introduces an angular divergence between the orthogonally polarized beams, so that the objective brings them to a focus at two distinct points on the surface being tested. Thus we have on the surface two spots of different angular frequencies w_1 and w_2 that are typically separated by about 100μm. The spot size depends on the numerical aperture of the microscope objective and is of the order of 2μm or less. The reflected beams are picked up by the microscope objective, are recombined by the Wollaston prism, and leave on a path parallel to the incident beam but displaced to eliminate the overlap. This beam passes through an adjustable neutral density filter ND to compensate for surface reflectivity and a polarizer P_2 at $45°$ before being incident on the measurement detector D_2.

In view of the discussion in Sec. 6.1, the signals in the reference and measurement photodetectors will be respectively given by

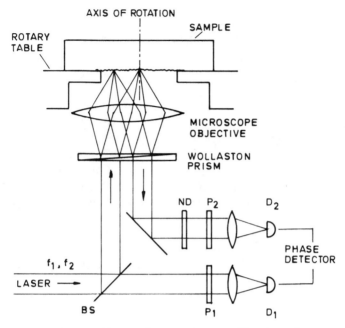

Figure 6.5   Differential heterodyne surface profiling interferometer. (Modified from Ref. 11, used with permission).

$$S_r = \cos (w_2 - w_1)t \qquad\qquad (6.6)$$

$$S_t = \cos [(w_2 - w_1)t + \phi] \qquad\qquad (6.7)$$

where ϕ is the phase difference introduced by the height variation h between the two spots ($\phi = 4\pi h/\lambda$). By monitoring the phase difference between the reference and the test signal with a phase meter, h can be determined as the surface is rotated and one of the spots scans a circle on the surface while the other spot remains fixed at the center of rotation.

The rotary table is usually an air-bearing type. As the table rotates, it will undergo some degree of translations and tilt with respect to the fixed frame. Axial motion does not affect the phase measurement beams, provided the motion is less than the depth of focus. Radial motion must be a small fraction of the focused beam size so that the axial beam remains at one point on the surface.

The tilt of the table that occurs when the normal to the table is not parallel to the axis of rotation does affect the phase measurement, as it introduces a differential path length change between the beams. A fixed tilt causes the normal to tilt, which results in an additive sinusoidal term in the phase measurement and can be removed in data analysis. Random tilting leads to nonrecoverable surface errors. Usually a random table tilt should be a fraction of a microradian.

A commercial instrument based on this principle has a vertical measurement range of 0.1 to 3000Å peak to valley with a sensitivity of less than 0.1Å. The scan length is 1 mm circumference, with a scan speed of 30 sec per 360° rotation.

Another heterodyne technique proposed for profiling uses two concentric focused laser beams on the sample surface [12]. As shown in Fig. 6.6, a small focused laser spot≈2 μm in diameter of angular frequency w_1 is overlapped on the same sample surface by a larger focused spot≈50 μm in diameter from the same laser source but at a slightly different angular frequency w_2. The different sizes of the focused spots are obtained by using two light beams of different apertures and focused by the same lens. The

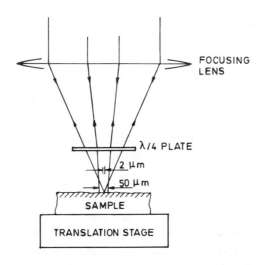

Figure 6.6 Generation of two concentric focused spots of different diameters on sample surface. (Modified from Ref. 12, used with permission.)

smaller spot on reflection carries the phase due to local height variations, whereas the larger spot takes a phase value that is averaged over the area of the spot. The phase difference between the two spots gives the profile height at the point of interest. Since the two spots have different frequencies, the phase difference can be determined by the heterodyne interference technique. By scanning the surface across the spots, the profile of the surface can be obtained. Since the sample vibrations impose almost the same phase changes on each individual laser beam, the error due to vibrations is eliminated. Figure 6.7 shows the schematic arrangement of the profilometer. Two laser beams, one of slightly shifted frequency, are produced by a Bragg cell. One is expanded in diameter and the other is reduced. After combining with a beam splitter BS, the two beams are focused with a (50 mm, f/0.95) lens on the sample. A λ/4 plate helps in rotating the original plane of polarization by 90° on reflection from the

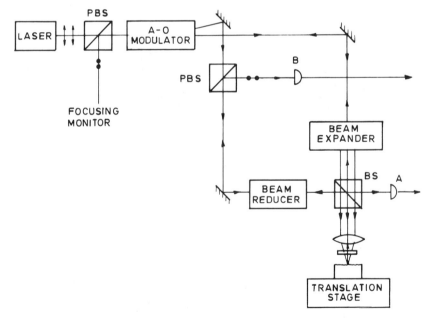

Figure 6.7   Concentric focused spots heterodyne interferometer for surface profiling. (Modified from Ref. 12, used with permission.)

sample. The reflected beams are recollimated by the lens and strike detector B through a polarizing beam splitter PBS after passage through the beam reducer. The detector B gives the heterodyne signal. The photodetector A provides the reference signal.

A phase variation greater than 2π radians due to surface height variation greater than $\lambda/2$ will result in ambiguous results. With a wavelength of about 0.6 μm, irregularities less than about 0.3 μm can be detected without any ambiguity. This range can be extended by operating the heterodyne interferometer at two closely spaced wavelengths λ_a and λ_b with two detectors, each one seeing one of the wavelengths [10]. The resultant intensities at the two detectors will be

$$I(\lambda_a) \propto \cos\left[(w_2 - w_1)t + \frac{4\pi h}{\lambda_a}\right] \tag{6.8}$$

$$I(\lambda_b) \propto \cos\left[(w_2 - w_1)t + \frac{4\pi h}{\lambda_b}\right] \tag{6.9}$$

where $w_2 - w_1$ is the angular frequency difference, which is the same for both wavelengths. The two detector signals have the same frequency, but there is a phase difference between them given by

$$\phi = 4\pi h \left(\frac{1}{\lambda_b} - \frac{1}{\lambda_a} \right)$$

$$= 4\pi h \left(\frac{\lambda_a - \lambda_b}{\lambda_a \lambda_b} \right)$$

$$= 4\pi h/\lambda_0 \tag{6.10}$$

where

$$\lambda_0 = \frac{\lambda_a \lambda_b}{\lambda_a - \lambda_b} \tag{6.11}$$

is the effective wavelength. Thus by measuring the phase difference between the two intensity variations, the range of the interferometer becomes $\lambda_0/4$. Any desired wavelength λ_0 can be generated by suitable choice of λ_a and λ_b. For example, two wavelengths 0.4765 μm and 0.4880 of an argon ion laser give $\lambda_0 = 20.22$ μm. Figure 6.8 shows an arrangement to imple-

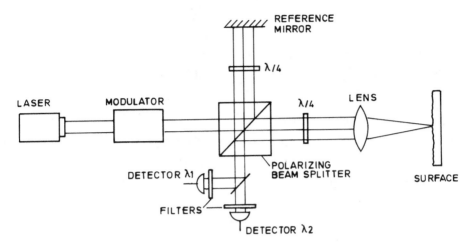

Figure 6.8   Dual wavelength heterodyne interferometer. Arrows and dots represent orthogonal linear polarizations. (Modified from Ref. 10, used with permission.)

ment this technique. The laser gives light linearly polarized at 45° to the plane of paper. This light can be resolved into two components in and perpendicular to the plane of the paper. The frequency of one of the components is shifted by an electrooptic modulator. The light enters an interferometer similar to that in Fig. 6.4. At the output, inteference filters placed before the detectors separate λ_a and λ_b.

6.1.4 Two-Frequency Interferometer for Small Displacement Measurement

In Sec. 6.1.2 and 6.1.3 we considered inteference between two beams of different frequencies. It is interesting to consider the interference of two beams, each beam having two frequencies that are slightly different [13-15]. Light from a two-frequency light source is introduced into a Michelson interferometer so that both frequencies travel the two paths. The amplitude of the beams leaving the two arms of the interferometer may be expressed as

$$A_1(t) = a_1[\exp(iw_1t) + \exp(iw_2t)] \qquad (6.12)$$

$$A_2(t) = a_2 \{exp[i(w_1t + \phi)] + exp[i(w_2t + \phi)]\} \tag{6.13}$$

where ϕ represents the phase due to excess path difference in one of the paths. The signal on the photodetector is proportional to the intensity given by $I(t) = A(t) \cdot A(t)^*$ where

$$A(t) = A_1(t) + A_2(t) \tag{6.14}$$

It can be shown that

$$I(t) = [I_1 + I_2 + 2\sqrt{I_1I_2} \cos \phi] +$$

$$[I_1 + I_2 + 2\sqrt{I_1I_2} \cos \phi] \cos 2(w_2 - w_1)t \tag{6.15}$$

This is an interesting result. The first term in brackets represents the usual expression for interference by two beams of the same frequency (homodyne intereference). The second term in brackets results from the heterodyne interference of two beams of different frequencies. This heterodyne component of angular frequency 2 $(w_2 - w_1)$ has no phase but is accompanied by fluctuation in the amplitude of the signal due to the phase difference ϕ. This means that the amplitude of the electrical signal of angular frequency $2(w_2 - w_1)$ will vary as ϕ varies. Hence the parameter causing the phase change can be measured as a function of the ac signal strength. A similar result is obtained in three-beam heterodyne interferometry [16].

The signal strength due to the first term in the above equation also varies with ϕ (homodyne detection). But the signal produced here is dc. Apart from the usual difficulties of operting with dc signal, it is difficult to detect small nonsinusoidal motion of a test body in a low frequency range in the presence of unwanted noise that is usually of low frequency. In contrast, the heterodyne component is shifted from null to $2(w_2 - w_1)$, which can be in the KHz and MHz region, so that the low-frequency noise has little effect on measurement. The time-varying amplitude of this term can be extracted by AM detection techniques because here, as in conventional heterodyne systems, the beat signal does not have the term in its phase.

This technique has been used to measure or detect small displacements, such as small vibrational amplitudes, in the range of a fraction of a nanometer. Figure 6.9 shows the optical arrangement to achieve this [14]. An ultrasonic light modulator is used to produce a two-frequency laser

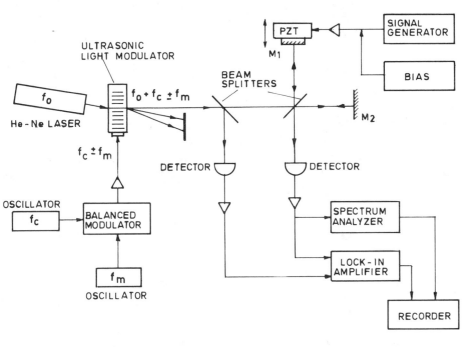

Figure 6.9 Two-frequency interferometer for small displacement measurement. (Modified from Ref. 14, used with permission.)

beam $f_o + f_c \pm f_m$. A Michelson interferometer is used and the vibrating mirror M_1 modulates the signal beam. The fixed mirror M_2 provides the reference beam.

6.2 PHASE SHIFTING INTERFEROMETRY

In heterodyne interferometry the phase difference between the two interfering beams is changed at a constant, continuous rate in time as a result of the frequency difference between the two beams. The phase of the heterodyne signal so produced is then related to optical phase. In the phase shifting technique, however, the phase between the interfering beams is changed by introducing discrete phase steps in the reference beam [17]. The intensity is then measured at a fixed grid of points with the help of a detector array for several phase steps. The intensity may be measured at the end of each phase step (this is the phase stepping technique) or it may be integrated while the phase is being shifted (the integrating bucket technique). In a two-beam interference pattern, the intensity distribution is given by

$$I(x, y) = I_0(x, y) \{1 + V \cos [\phi(x, y) + \alpha]\} \qquad (6.16)$$

where ϕ (x,y) is the phase difference distribution across the intereference pattern and V the modulation of the fringes. The equation has three unknowns, I_0, V, and ϕ. A minimum of three measurements is necessary to determine the phase ϕ.

The Three Step Method

In the three step method the intensity measurements are made by introducing three phase steps such as $-\alpha$, 0, and $+\alpha$ in the reference wave. Thus we may write the intensity equations as

$$I_1 = I_0[1 + V \cos(\phi - \alpha)]$$
$$= I_0[1 + V \cos \phi \cos \alpha + V \sin \phi \sin \alpha] \qquad (6.17)$$

$$I_2 = I_0[1 + V \cos \phi] \qquad (6.18)$$

$$I_3 = I_0[1 + V \cos (\phi + \alpha)]$$
$$= I_0[1 + V \cos \phi \cos \alpha - V \sin \phi \sin \alpha] \qquad (6.19)$$

From the eqs. (6.17) to (6.19) we may write

$$\tan \phi = \frac{1 - \cos \alpha}{\sin \alpha} \frac{I_1 - I_3}{2I_2 - I_1 - I_3} \qquad (6.20)$$

For a phase step of $\alpha = 120°$ we have

$$\tan \phi = \frac{\sqrt{3}(I_1 - I_3)}{2I_2 - I_1 - I_3} \qquad (6.21)$$

For a phase step of $\pi/2$ and phase values of $-\pi/4, \pi/4$, and $3\pi/4$ the measured intensities are

$$I_1 = I_0[1 + V \cos (\phi - \pi/4)] \tag{6.22}$$

$$I_2 = I_0[1 + V \cos (\phi + \pi/4)] \tag{6.23}$$

$$I_3 = I_0[1 + V \cos (\phi + 3\pi/4)] \tag{6.24}$$

These three equations can be manipulated to give

$$\tan \phi = \frac{I_2 - I_1}{I_3 - I_2} \tag{6.25}$$

The three-step method is not self calibrating, as the value α cannot be obtained from the measurements. Further, it is also the most sensitive to system errors.

The Four Step Method

The most common four-step method makes use of reference beam phase settings of $0, \pi/2, \pi$, and $3\pi/2$ to generate four sets of intensity measurements. The intensity equations are

$$I_1 = I_0[1 + V \cos \phi] \tag{6.26}$$

$$I_2 = I_0[1 + V \cos (\phi + \pi/2)] = I_0[1 - V \sin \phi] \tag{6.27}$$

$$I_3 = I_0[1 + V \cos (\phi + \pi)] = I_0[1 - V \cos \phi] \tag{6.28}$$

$$I_4 = I_0[1 + V \cos (\phi + 3\pi/2)] = I_0[1 + V \sin \phi] \tag{6.29}$$

The phase is ϕ given by

$$\tan \phi = \frac{I_4 - I_2}{I_1 - I_3} \tag{6.30}$$

The Carré Method

In the three- and four-step methods, the phase shift must be known by other means. A method developed by Carré [18] again makes use of four steps, but here the phase measurement is independent of the amount of the phase shift. The four phase steps are shifted by 2α and are given as -3α, $-\alpha$, α, and 3α. This yields the four intensity equation as

$$I_1 = I_0[1 + V \cos(\phi - 3\alpha)] \tag{6.31}$$

$$I_2 = I_0[1 + V \cos(\phi - \alpha)] \tag{6.32}$$

$$I_3 = I_0[1 + V \cos(\phi + \alpha)] \tag{6.33}$$

$$I_4 = I_0[1 + V \cos(\phi + 3\alpha)] \tag{6.34}$$

These equations give the phase shift and the phase at each point as

$$\tan^2 \alpha = \frac{3(I_2 - I_3) - (I_1 - I_4)}{(I_2 - I_3 + I_1 - I_4)} \tag{6.35}$$

$$\tan \phi = \tan \alpha \frac{(I_2 - I_3) + (I_1 - I_4)}{(I_2 + I_3) - (I_1 + I_4)} \tag{6.36}$$

These two equations can be combined to give

$$\tan \phi = \frac{\sqrt{[3(I_2 - I_3) - (I_1 - I_4)][(I_2 - I_3) + (I_1 - I_4)]}}{(I_2 + I_3) - (I_1 + I_4)} \tag{6.37}$$

The assumption made here is that the phase shift should be linear. Its actual value need not be known. This has the advantage of working when a linear phase shift is introduced in a converging or diverging beam where the amount of phase shift varies across the beam in contrast to the collimated beam.

The Five Step Method

The Carré algorithm for phase causes a problem when the phase difference is close to $m\pi$ where m is an integer [19]. In these conditions the numerator and the denominator of Eqs. (6.35) and (6.36) tend to zero so that there is increased uncertainty in the values of α and ϕ. Using the five-step technique with phase steps of -2α, $-\alpha$, 0, $+2\alpha$, this problem can be avoided. The corresponding intensity values are

$$I_1 = I_0[1 + V \cos(\phi - 2\alpha)] \tag{6.38}$$

$$I_2 = I_0[1 + V \cos(\phi - \alpha)] \tag{6.39}$$

$$I_3 = I_0[1 + V \cos \phi] \tag{6.40}$$

$$I_4 = I_0[1 + V \cos(\phi + \alpha)] \tag{6.41}$$

$$I_5 = I_0[1 + V \cos(\phi + 2\alpha)] \tag{6.42}$$

From these relations we obtain

$$\tan \phi = \frac{1 - \cos 2\alpha}{\sin \alpha} \frac{I_2 - I_4}{2I_3 - (I_1 + I_5)} \tag{6.43}$$

$$\cos \alpha = \frac{I_5 - I_1}{2(I_4 - I_2)} \tag{6.44}$$

Eq. (6.43) is least sensitive to errors in α for $\alpha = \pi/2$. In this case ϕ is given by

$$\tan \phi = \frac{2(I_2 - I_4)}{2I_3 - I_5 - I_1} \tag{6.45}$$

The numerator and denominator of this equation cannot go simultaneously to zero. It can therefore be used for all values of ϕ.

The Integrating Bucket Method

In the integrating bucket method the integrated intensity is recorded over a phase interval Δ while the phase is linearly varied with time [20]. The recorded intensity will be given by

$$I = \frac{1}{\Delta} \int_{\alpha - \Delta/2}^{\alpha + \Delta/2} I_0[1 + V \cos(\phi + \alpha)] \, d\alpha$$

$$= I_0[1 + \frac{\sin \Delta/2}{\Delta/2} V \cos(\phi + \alpha)] \tag{6.46}$$

Except for the factor $(\sin \Delta/2)/(\Delta/2)$, the equation remains the same, and various algorithms explained above can be used with the integrating bucket technique. The values of the factor for $\Delta = 120°$ and $90°$ are 0.83 and 0.90, respectively. It is therefore convenient to ramp the phase linearly and measure the integrated intensity and at the same time not lose much on modulation.

Multi-Step Method

In the multi-step method [21] the phase shifts are so chosen that N measurements are equally spaced over one period, so that $\alpha_n = n(2\pi/N)$. The phase is then given by

$$\tan \phi = \frac{\sum I_n \sin \alpha_n}{\sum I_n \cos \alpha_n} \tag{6.47}$$

Phase Ambiguities

Eqs. (6.20), (6.25), (6.30) for phase calculations calculate the phase modulo π. To determine the phase modulo 2π, the signs of the quantities in the numerators and denominators are examined, because they are proportional to $\sin \phi$ and $\cos \phi$. This, however, does not apply to Eq. (6.37). In this case we examine

$$I_2 - I_3 = (2I_0 \ V \sin \alpha) \sin \phi \tag{6.48}$$

$$(I_2 + I_3) - (I_1 + I_4) = (2I_0 \ V \cos \alpha \sin^2 \alpha) \cos \phi \tag{6.49}$$

and place the value of ϕ in the appropriate quadrant. The result still suffers from 2π ambiguities because the phase variation over the interference pattern may be several multiples of 2π, giving fringes of various orders. The 2π ambiguities are removed by adding or subtracting 2π from individual data points (pixels) until the phase difference between adjacent pixels is less than π. This means that fringes formed on the array detectors should have spacing so that the phase change between adjacent pixels is less than π. The fringe spacing in an interferometer can always be varied by suitable adjustments.

Phase Shifting Techniques

One of the most common methods of introducing phase shift is to shift a mirror of the interferometer with the help of a piezoelectric translator (PZT) as in Fig. 6.10, which shows a Twyman-Green interferometer for optical component or system testing. A step or ramp voltage input to the PZT gives the required phase shift for phase stepping or integrating bucket technique respectively. The phase shift between two interfering beams can also be introduced by polarization techniques [22] or gratings [23, 24].

Sources of Error in Phase Shifting Interferometry

The main sources of error in phase shifting interferometry (PSI) are incorrect phase shift (calibration error of the phase shifter), nonlinearity of

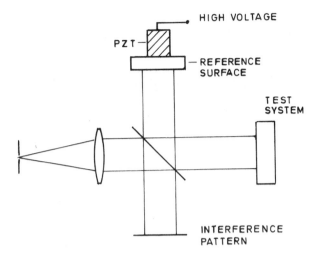

HIGH VOLTAGE

PZT

— REFERENCE
SURFACE

TEST
SYSTEM

INTERFERENCE
PATTERN

Figure 6.10 Twyman-Green interferometer for phase shifting interferometry.

the phase shift, and detection nonlinearities [17–19,25,26]. Phase shifter errors are significant in the calculation of phase by Eqs. (6.25) and (6.30); but Eqs. (6.37) and (6.45) are less sensitive to these errors. Error in the result of the four-step technique due to phase shifter calibration error can be minimized by calculating the phase by Eq. (6.25) using the first three intensity measurements and averaging this with the phase calculated by the same equation with the last three measurements of the four-step method. The influence of detector nonlinearity on the phase calculation reduces as we go from the three-bucket to the five-bucket method.

6.2.1 Phase Shifting Interferometer for Optical Component and System Testing

The conventional Twyman-Green or Fizeau interferometer can be easily modified for analyzing the interferogram obtained from a test component or system by phase shifting interferometry. The reference surface is mounted on a PZT, which is given a step or ramp voltage for phase shifting (Figs. 6.10, 6.11, and 6.12). Most of the conventional testing configurations can be adopted for PSI. For testing aspheric surfaces, both lateral and radial shear interferometers have been found suitable [23,27]. To extend the phase

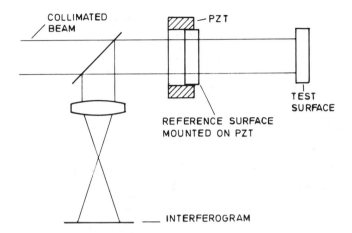

Figure 6.11   Fizeau interferometer for testing plane surfaces using the phase shifting technique.

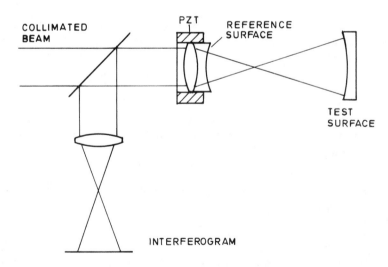

Figure 6.12   Fizeau interferometer for testing spherical surfaces using the phase shifting technique.

measurement range of PSI, two-wavelength PSI has been proposed [28]. The phase data obtained at the array of points is analyzed with the help of a microcomputer to give wavefront shape, peak-to-valley or rms surface errors, point-spread function, MTF, etc. as the case may be. Several commercial interferometers have been designed using the phase shifting technique.

6.2.2 Interference Microscopes Using PSI

Inteference microscopes may be used for surface profile measurement of rough surfaces. Figure 6.13 shows the optical schematic of a profile microscope based on the Mireau interferometer using PSI [29]. Interference is produced between a reference surface and the test surface. The reference

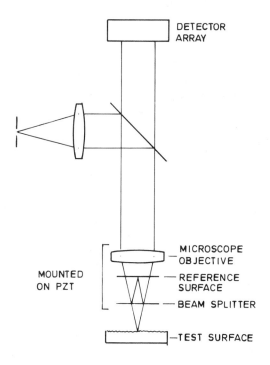

Figure 6.13   Interference microscope profilometer.

surface along with the beam splitter and the objective is moved with the help of a PZT transducer to produce phase shifts between the reference and test beams. The interference fringes are received on a linear detector array placed at the top of the microscope. For the purpose of adjustments, the fringes can also be seen via an eyepiece. The detector array is read out into computer memory several times, with the phase of the reference beam being changed between each measurement. The computer then calculates the height error at each of the detector points across the portion of the sample that is in the field of view. Height resolution of the order of 1Å or better has been obtained. The lateral resolution is determined by the diffraction spot of the objective.

6.2.3 Differential Interference Microscopes Using PSI

In the arrangement discussed in the previous section, the test surface is compared with a separately arranged reference surface. The size of the surface area that can be evaluated in one measurement pass is limited by the field of view of the optical system. A larger field of view reduces the lateral resolution.

An arrangement of two focused spots similar to that discussed in sec. 6.1.3, which does not require a separate reference surface, can be designed with PSI [9,30]. This arrangement is less sensitive to mechanical and thermal instabilities. The measurement is carried out by scanning the object across the focused spots. Thus it gives point-by-point measurement rather than a fixed length of the object at a time. Figure 6.14 shows the optical arrangement. A laser beam passing through an electrooptic phase modulator PM is resolved into two orthogonally polarized beams. A Wollaston prism angularly separates the two beams, and the microscope objective MO forms two focused spots on the object surface OS. The diameter of the laser spots on the surface is in the 1 to 10 μm range. The optical phase difference of the two beams reflected from the test object is proportional to the height difference h between the illuminated points on the surface. The phase difference is given by $\phi = (4\pi/\lambda)h$. The beams reflected from the surface are recombined by the Wollaston prism. On partial reflection by the beam splitter BS, these beams are brought to interference passing through a polarizer P before a photodetector D. The total amount of light reaching the detector depends on the object's reflectivity, which can be controlled by a suitable light attenuator placed in front of the laser source.

To implement PSI, the phase difference between the two interfering beams is shifted by applying voltage to the modulator in step or ramp form depending on whether the phase stepping or integrating bucket technique

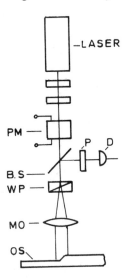

Figure 6.14   Differential interference microscope. (Modified from Ref. 30, used with permission.)

is to be used. The intensity values measured from the detector are stored and the phase value ϕ calculated by a computer.

Such an instrument can be used for step height measurement as well. The lateral resolution is given by the laser spot size. To minimize the errors due to tilt fluctuations of the microscope table while scanning the test surface, the separation of the two spots should be as close as possible. Some practical values are: spot sizes 10 μm to 1 μm and spot separation 56 μm to 3 μm, depending on the objective magnification (5× to 100×).

6.2.4 Hologram Interferometry Using PSI

In hologram interferometry we are interested in measuring the surface displacements of an object when it is stressed. These displacements show up as fringe patterns in real-time or double-exposure hologram inter-ferometry. Phase distribution related to surface displacements in the real-time intereference pattern can be evaluated by phase shifting interferom-etry [31]. Figure 6.15 shows an experimental arrangement for hologram interferometry using PSI. Initially a hologram of the object is recorded and replaced in its original position after processing. On loading the object, a set of holographic displacement fringes can be seen. The fringes are im-

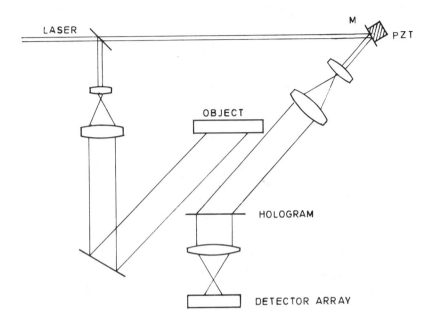

Figure 6.15 An optical scheme for phase shifting hologram interferometry.

aged on a photodiode array detector that reads intensity values at each point. The phase of the reference beam can be shifted with the help of the PZT transducer, which moves the mirror M. Several intensity records are made, and the phase is calculated at each pixel point, as explained earlier. The phase data is then converted into required information. The PSI technique has also been applied to speckle interferometry [32].

REFERENNCES

1. Crane, R. (1969). Interference phase measurement. *Appl. Opt., 8:* 536–542.
2. Sommargren, G. E. (1975). Up-down frequency shifter for optical heterodyne interferometry, *J. Opt. Soc. Am, 65:* 960–961.
3. Shagam, R. N., and Wyant, J. C. (1978), Optical frequency shifter for heterodyne interferometers using multiple rotating polarization retarders, *Appl. Opt., 17:* 3034–3035.
4. Kothiyal, M. P., and Delisle, C. (1984). Optical frequency shifter for heterodyne

interferometry using counterrotating wave plates,*Optics Letters, 9:* 319–321.

5. Li, Y., and Eichmann, G. (1986). Multipass counterrotating wave plate frequency shifters for heterodyne interferometry, *Optics Letters, 11:* 718–720.

6. Hu, H. Z. (1983). Polarization heterodyne interferometry using simple rotating polarizer 1: Theory and error analysis. *Appl. Opt., 22:* 2052–2056.

7. Stevenson, W. H. (1970). Optical frequency shifting by means of a rotating grating, *Appl. Opt., 9:* 649-652.

8. Massie, N. A. (1980), Real-time digital heterodyne interferometry: A system, *Appl. Opt., 19:* 154–160.

9. Makosch, G., and Solf, B. (1981). Surface profiling by electrooptical phase measurement. *SPIE, 316:* 42–53.

10. Pettigrew, R. M., and Hancock, F. J. (1979), An optical profilometer, *Precision Engineering, 1:*133–136.

11. Sommargren, G. E. (1981). Optical heterodyne profilometry, *Appl. Opt., 20:*610–618.

12. Huang, C. C. (1984). Optical heterodyne profilometer, *Opt. Eng., 23:*365-370.

13. Ohtsuka, Y., and Sasaki, I. (1977). Temporal interference effects by a pair of two-frequency laser beams—Application to extremely low velocity measurements, *Opt. Commun. 22:* 211–214.

14. Ohtsuka, Y. and Itoh, K. (1979). Two-frequency laser interferometer for small displacement measurements in low frequency range, *Appl. Opt., 18:* 219–224.

15. Ohtsuka, Y. and Tsubokawa, M. (1984). Dynamic two-frequency interferometry for small displacement measurements, *Opt. and Laser Tech., 16:* 25–29.

16. Kothiyal, M. P., and Delisle, C. (1985). Three beam heterodyne interferometer,*Opt. Commun. 56:* 145-149.

17. Creath, K. (1988). Phase measurement interferometry techniques, *Progress in Optics, 26:* 349–393 ed. E. Wolf.

18. Carré, P. (1966). Installation et utilisation du comparateur photoelectrique et interferential du Bureau International des Poids et Mesures, *Metrologia, 2:*13–23.

19. Hariharan, P., Oreb, B. F. and Eiju T. (1987). Digital phase shifting interferometry: A simple error compensating phase calculation algorithm, *Appl. Opt., 26:* 2504–2506.

20. Wyant, J. C. (1975). Use of ac heterodyne lateral shear interferometer with real-time wavefront correction system, *Appl. Opt. 14:* 2622–2626.

21. Brunning, J. H., Herriott, D. R., Gallagher, J. E., Rosenfeld, D. P., White, A. D., and Brangaccio, D. J. (1974). Digital wavefront measuring interferometer for testing optical surfaces and lenses, *Appl. Opt. 13:* 2693–2703.

22. Kothiyal, M. P., and Delisle, C. (1985). Shearing interferometer for phase shifting interferometry with polarization phase shifter, *Appl. Opt. 24:* 4439–4442.

23. Yatagai, T. (1984). Fringe scanning Ronchi test for aspherical surfaces, *Appl. Opt., 23:* 3676–3679.

24. Kwon, O. Y. (1984). Multichannel phase-shifted interferometer, *Optics Letters, 9:* 59–61.

25. Cheng, Y. Y., and Wyant, J. C. (1985). Phase shifter calibration in phase shifting interferometry, *Appl. Opt., 24:* 3049–3052.
26. Schwider, J., Burow, R., Elssner, K. E., Grzanna, J., Spolaczyk, R., and Merkel, K. (1983). Digital wavefront measuring interferometry: some systematic error sources, *Appl. Opt., 22:* 3421–3432.
27. Hariharan, P., Oreb, B. F., and Wanzhi, Z. (1984). Measurement of aspheric surfaces using a microcomputer controlled digital radial shear interferometer, *Opt. Acta, 31:* 989–999.
28. Cheng, Y. Y., and Wyant, J. C. (1984). Two wavelength phase shifting interferometry, *Appl. Opt. 23:* 4539-4543.
29. Wyant, J. C., Koliopoulous, C. L., Bhusan, B., and George, O. E. (1984). *Trans. Am. Soc. Lubrication Engs., 27:* 101.
30. Makosch, G., and Drollinger, B. (1984). Surface profile measurement with a scanning differential ac interferometer, *Appl. Opt., 23,* 4544-4553.
31. Hariharan, P. (1985). Quasi-heterodyne hologram interferometry, *Opt. Eng. 24:* 632–638.
32. Creath, J. (1985). Phase shifting speckle interferometry, *Appl. Opt., 24:* 3053–3058.

ADDITIONAL READING

Malacara, D. (1978), *Optical Shop Testing,* Wiley, New York.

CHAPTER 7

Hologram Interferometry and Speckle Metrology

7.1 HOLOGRAPHY

Holography is the process of recording a wave on a square-law medium and later releasing the same. A wave can carry information in the form of the modulation of amplitude, phase, and polarization. This information can be stored using holography. The basic idea was conceived by Gabor, who was attempting to improve the performance of the electron microscope. For more than a decade, holography remained practially dormant. Its revival came with the advent of the laser. Since then it has found applications in almost all branches of science, technology, and engineering.

Holography is a two-step process in which one first records and then reconstructs the image. Since both amplitude and phase information is to be recorded on media that respond to energy, we resort to interferometry. A coherent wave, called the reference wave, is added to the object wave at the plane of recording. Let $O(x, y) = O_0(x, y)e^{i\varnothing_0(x,y)}$ be the object wave. A reference wave $R(x, y) = R_0 e^{i\varnothing_R}$ is added to the object wave. Both waves are of the same frequency, so their time dependence is ignored in the analysis. The reference wave is usually either a spherical or a plane wave. The total amplitude at the recording plane is expressed as

$$a_T = R + O$$

Detectors in the optical region respond to energy; for example, a photo-multiplier responds to J/s, while a photographic plate responds to J/m². Therefore the irradiance distribution is recorded. The irradiance distribution at the recording plane is

$$I = |a_T|^2 = |R + O|^2$$

$$= R_0^2 + O_0^2(x, y) + 2R_0O_0(x, y) \cos [\phi_0(x, y) - \phi_R] \qquad (7.1)$$

The recording is done over a period T called the exposure time. The exposure E is thus IT. It may be seen from Eq. (7.1) that both amplitude and phase variations have been converted into irradiance variations to which the recording medium responds. We assume a photographic emulsion to be the recording medium, and after processing we assign to it an ampli-tude transmittance t(x, y). Further, we assume that the recording is linear, that is, that the amplitude transmittance is linearly mapped to the exposure variations [1]. The processed photographic record is called the hologram. It is characterized by the amplitude transmittance t(x, y) given by

$$t(x, y) = t_0 - \beta E (x, y)$$

where t_0 is the bias transmittance and β is a constant. To release the object wave from the hologram it is illuminated by a reference wave. The am-plitude of the wave just transmitted from the hologram is

$$R(x, y) \, t(x, y) = t_0 R(x, y) - \beta T[O_0^2(x, y) \, R(x, y)$$

$$+ O(x, y) \, R_0^2 + O^*(x, y) \, R^2(x, y)] \qquad (7.2)$$

where the constant $-\beta T R_0^2$ has been absorbed in t_0. It is seen that there are four waves of which one is an object wave multiplied by a constant. On viewing we see the object in its full glory although it has since been removed. Figure 7.1 (a) and (b) show a typical geometry of hologram recording and reconstruction. The hologram is therefore like a window with a memory. Different perspectives of the object are seen when it is viewed through different regions of the hologram. It may be mentioned that the object wave is isolated from the other waves provided the carrier frequency is larger than three times the maximum frequency present in the wave [1,2].

Hologram recording involves the phenomenon of interference, and hence all the conditions necessary for obtaining high contrast fringes in an

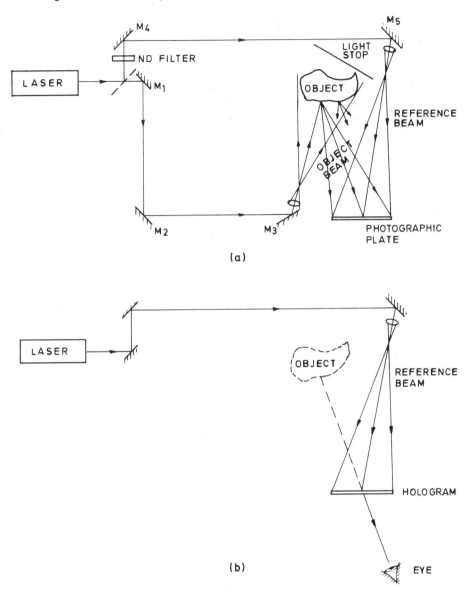

Figure 7.1   (a) A typical hologram recording geometry and (b) hologram reconstruction geometry.

interference patttern must be fulfilled. Further, the carrier frequency in holography is very high ($\approx$1000 l/mm), so that we must use high-resolution recording media, which require stability of the experimental setup.

On the other hand, reconstruction involves the phenomenon of diffraction. The fine structure on the hologram created by interference diffracts the incident wave. Therefore the conditions of monochromaticity and coherence are considerably relaxed. A large number of geometries have been developed for recording holograms [2,3,4].

7.2 HOLOGRAM TYPES

Holograms are classified as transmission type or reflection type depending on whether they can be reconstructed in transmission or reflection. Further, the exposure variation may cause either density variation or phase variation. The phase variation may be due to variation in refractive index, or in thickness or in both. Therefore the holograms are also called absorption or phase type. Then depending on the thickenss of the recording material with respect to the fringe spacing they are called thick (volume) or thin holograms. In a thick hologram the recording is done in the volume; the fringe planes lie inside the recording material. Usually the fringe planes are inclined with the surface, but in reflection holography they tend to run parallel to the surface.

Table 7.1 gives the various hologram types along with their theoretical diffraction efficiencies. The efficiency has been calculated for the plane waves recording; in no case would the practical diffraction efficiency exceed these values. The diffraction efficiency is defined as the ratio of irradiance in the useful order to that incident on the hologram; the Fresnel losses are ignored.

Detailed information on various holographic recording media is given in Chap. 1.

Table 7.1 Types of Holograms and Their Diffraction Efficiencies (%)

	Amplitude		Phase	
	Thin	Thick	Thin	Thick
Transmission	6.25	3.7	33.9	$\approx$100
Reflection	—	7.2	—	$\approx$100

A hologram may be reconstructed by radiation from spectral sources; even white light may be used. The use of a wavelength other than the one used in recording introduces wavelength-dependent aberrations [5]. A hologram is a diffraction-limited imaging system for point objects when used in a geometry identical to that used during recording and with the same wavelength. It can be achormatized for three wavelengths, but the net diffraction efficiency renders the system unsuitable for many imaging applications. On the other hand, good designs have been worked out when holooptical elements are to be used as Fourier transform lenses and scanners [6].

7.3 HOLOGRAM INTERFEROMETRY

Around 1965 a number of research groups almost simultaneously discovered that waves from diffuse objects can interfere. This gave birth to a new field now known as hologram interferometry (HI) or holometry [7,8]. HI permits interferometric comparison of a diffuse object with the master. Further, it became possible to observe interference between two time-delayed complex (diffuse) wavefronts. Here we shall discuss (1) real-time or single exposure or live-fringe HI (2) elapsed-time or double-exposure, frozen-fringe HI; and (3) time-average HI.

HI can be perfomed using the configuration of Fig. 7.1 (a) but incorporating a loading jig, and a fringe-control arrangement if need be.

7.3.1 Real-Time Hologram Interferometry

Consider that a recording is performed of an object, and the hologram is relocated exactly as it was during recording. When interrogated by the reference wave, an object wave identical to the one recorded will be release from the hologram. Further, the wave coming from the object will also be transmitted by the hologram due to its dc transmittance. Therefore two waves identical in their microstructure are available just after the hologram. These waves thus can interfere. The photographic process introduces a phase change of π. Thus the field will be dark due to destructive interference. Further, the wet development process lends some thickness variations to the emulsion, and hence invariably a few fringes are seen. To avoid the problems of relocation of the hologram and emulsion shrinkage, in-situ processing using a liquid gate is preferred. Fringes are formed when the object is deformed. These fringes are due to the interference between the wave coming from the deformed object and the stored wavefront from the hologram. The dark fringes are formed when

$$\phi'_0(x, y) - \phi_0(x, y) = 2m\pi \tag{7.3}$$

where $\phi'_0(x, y)$ is the phase of the wave from the deformed object. It is assumed that the microstructure of the object does not change on deformation. The fringes are formed in real time, and changes in the object can be continuously monitored until the fringes become too fine to be resolvable. The main disadvantages of real-time HI are that the fringe contrast is poor, that the hologram requires precise repositioning if in situ processing is not performed, and that any changes in the geometry make the hologram useless.

7.3.2 Double-Exposure Hologram Interferometry

Two records of the interference pattern, the object being deformed between the exposures, are made sequentially on the same plate. The doubly exposed hologram, on illumination with the reference wave, releases both object waves. The object therefore appears covered with high contrast fringes. The bright fringes are formed when

$$\phi'_0(x, y) - \phi_0(x, y) = 2m\pi \tag{7.4}$$

Double exposure HI gives fringes due to the change in the states of the object. This change is frozen and the holgram can be played later to obtain the fringe pattern. The fringes are of unit contrast.

7.3.3 Time Average Hologram Interferometry

A hologram recorded of a vibrating object over a period much longer than the period of vibration is a time-average hologram. The exposure recorded is given by

$$E = \int_0^T I(x, y; t)\, dt = \int_0^T (R_0^2 + O_0^2(x, y) + OR^* + O^*R)\, dt \tag{7.5}$$

where $O(x, y) = O_0(x, y)\, e^{i\delta(x,y;t)}$; the phase $\delta(x, y; t)$ is a function of both space coordinates and time. When the vibrating object is illuminated at an angle θ_1, and observed at an angle θ_2 with the normal to the object as shown in Fig. 7.2, the phase $\delta(x, y; t)$ is expressed as

$$\delta(x, y; t) = \frac{2\pi}{\lambda}\, A(x, y)\, (\cos\theta_1 + \cos\theta_2)\, \sin wt \tag{7.6}$$

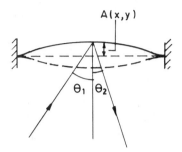

Figure 7.2   Calculation of phase change introduced by a vibrating body.

where $A(x, y)$ is the amplitude and w is the circular frequency of vibration. Since the desired wave is isolated during reconstruction, we write the amplitude of the reconstructed wave as

$$\propto O_0 \int_0^T e^{i\delta(x, y; t)} \, dt = O_0 J_0[kA(\cos \theta_1 + \cos \theta_2)], \text{ where } k = \frac{2\pi}{\lambda}$$

The irradiance distribution in the image is given by

$$I(x, y) = I_0 J_0^2[kA(\cos \theta_1 + \cos \theta_2)] \tag{7.7}$$

The object irradiance distribution is thus modulated by the J_0^2 function. The irradiance is zero at the zeros of the Bessel function J_0, and hence the amplitude of vibration can be calculated from the time-average interferogram. The decreasing irradiance of J_0 fringes makes them difficult to view and photograph when vibrational amplitudes are relatively high. Stroboscopic hologram interferometry may then be used.

7.3.4 The Characteristic Function

The time average integral

$$M_T = \frac{1}{T} \int_0^T e^{i\delta(x, y; t)} dt \tag{7.8}$$

is called the characteristic function. This is proportional to the amplitude of the reconstructed wave. The characteristic function for the sinusoidally vibrating object that is illuminated and viewed normally is, from Eq. (7.6)

$$M_T = \frac{1}{T} \int_0^T e^{(4\pi i/\lambda)\, A\, \sin\, wt}\, dt = J_0\!\left(\frac{4\pi}{\lambda}\, A\right) \tag{7.9}$$

The object irradiance is modulated by the square of the characteristic function. Table 7.2 gives the square of M_T for various kinds of motions and different hologram interferometric techniques. [9].

7.3.5 Calculation Of Phase Change On Loading

Consider a point P, on an object surface, that moves to a point P' when the object is deformed. Let $\bar{k}_1$ be the wave vector of the illumination beam and let $\bar{k}_2$ be that of the wave in the direction of observation—see Fig. 7.3 (a). It is easy to show that the phase difference ϕ_0 introduced by deformation is

$$\phi_0 = (\bar{k}_2 - \bar{k}_1) \cdot \bar{L}$$

$$= \bar{k} \cdot \bar{L} \tag{7.10}$$

where $\bar{L}(u\hat{\imath} + v\hat{\jmath} + w\hat{k})$ is the deformation vector and $\bar{k}$ is the sensitivity vector. A single observation of the interferogram determines a value of phase ϕ_0 that yields a measurement of the component of $\bar{L}$ in the direction of the sensitivity vector. Therefore the components of $\bar{L}$ can be calculated by setting up three equations, by making measurement on the same point either by changing the direction of observation through the hologram or by recording three holograms simultaneously [7].

Let us now consider that the illumination and observation beams lie in the x–z plane and enclose an angle θ as shown in Fig. 7.3 (b). Let the deformation vector $\bar{L}$ also lie in the same plane but make an angle ψ with the local normal at P. If the illumination and observation directions are symmetrical with respect to the normal, the magnitude of the sensitivity vector is 2k cos θ and its direction is along the normal. The component of $\bar{L}$ along this direction is L cosψ. Thus the phase change introduced by deformation is

$$\phi_0 = 2kL \cos\, \theta \cos\, \psi \tag{7.11}$$

The fringes are formed when L cos ψ = mλ/2 cos θ.

Table 7.2 Displacement and Characteristic Function $|M_T|^2$ of Types of Hologram Interferometry[a]

HI type	Displacement	$|M_T|^2$
Real-time	Static ($\bar{L}$)	$1 + c^2 - 2c \cos (\bar{k} \cdot \bar{L})$
Real-time	Harmonic of amplitude A(x, y)	$1 + c^2 - 2c J_0(\bar{k} \cdot \bar{A})$
Real-time time-average	Harmonic	$1 - J_0(\bar{k} \cdot \bar{A})$
Real-time with reference fringes	Harmonic	$1 + c^2 - 2c \cos(\bar{k} \cdot \bar{L}) \cdot J_0(\bar{k} \cdot \bar{A})$
Real-time stroboscopic	Harmonic: pulses at $t = \pi/2$ and $3\pi/2$	$1 + c^2 - 2c \cos(2\bar{k} \cdot \bar{A})$
Double-exposure	Static ($\bar{L}$)	$\cos^2 (\bar{k} \cdot \bar{L})/2$
Double-exposure stroboscopic	Harmonic: pulses at $t = \pi/2$ and $3\pi/2$	$\cos^2 (\bar{k} \cdot \bar{A})$
Time-average	Harmonic	$J_0^2 (\bar{k} \cdot \bar{A})$
	Constant velocity	$\text{sinc}^2 (\bar{k} \cdot \bar{L})/2$
	Constant acceleration from rest	$\dfrac{C^2 (\bar{k} \cdot \bar{L}) + S^2 (\bar{k} \cdot \bar{L})}{\bar{k} \cdot \bar{L}}$
Time-average	Irrationally related modes	$J_0^2(\bar{k} \cdot \bar{A}_1) J_0^2(\bar{k} \cdot \bar{A}_2)$
Temporally modulated	Harmonic motion frequency translated	$J_m^2(\bar{k} \cdot \bar{A})$
	Amplitude modulated reference wave $f_r(t) = e^{i(wt - \delta)}$	$J_m^2(\bar{k} \cdot \bar{A}) \cos^2 \delta$
	Phase modulated reference wave $f_r(t) = e^{iM_R \sin wt}$	$J_0^2 (\bar{k} \cdot \bar{A} - M_R)$

[a]c is the contrast and C and S are Fresnel cosine and sine integrals.

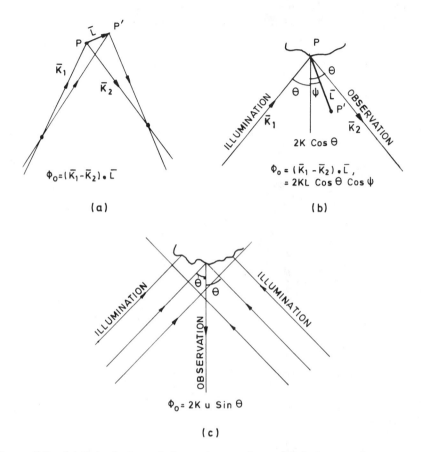

Figure 7.3   (a) Calculation of phase change due to deformation (b) phase change calculation for a two-dimensional case, and (c) phase change calculation for symmetrical illumination and observation directions with respect to the normal.

If the direction of $\bar{L}$ is known, its magnitude can be determined; otherwise the fringe pattern measures only the component along the sensitivity direction.

If the illumination and observation directions are symmetrical, as in the present case, only the normal component of deformation is measured. The sensitivity is a maximum when the illumination and observation directions are normal to the surface; each fringe then corresponds to 1/2 wavelength of displacement. It is possible to make the arrangement sensitive to the in-plane component alone. The object is illuminated symmetrically with respect to the normal, and the observation is made along the normal direction as shown in Fig. 7.3 (c). The phase difference ϕ_0 is now expressed as

$$\phi_0 = (\overline{k}_2 - \overline{k}_1) \cdot \overline{L} - (\overline{k}_2 - \overline{k}_1') \cdot \overline{L} = (\overline{k}_1' - \overline{k}_1) \cdot \overline{L}$$

$$= 2ku \sin \theta \qquad (7.12)$$

where u is the in-plane component. The fringe spacing will correspond to $\lambda / 2u \sin \theta$.

Figure 7.4 shows an interferogram of a fan blade impacted by a steel ball. The hologram was recorded with a double pulse ruby laster; the pulse separation was 10 μsec. The hologram was reconstructed using a He-Ne laser. Figure 7.5 shows a photograph showing the deformation of a human skull when the tooth is loaded.

Figure 7.4 Interferogram of a fan blade impacted by a steel ball. The double exposure hologram was recorded by a double pulse ruby laser.

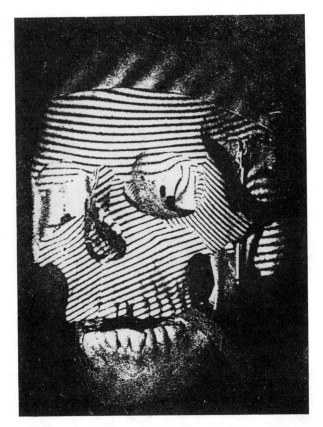

Figure 7.5   Interferogram showing deformation of a human skull if tooth is loaded. (Courtesy of Steinbichler H., Labor Dr. Steinbichler, F. R. G.)

7.3.6 Fringe Localization

Fringe localization in HI is a subject of great concern and has attracted the attraction of many researchers [7]. The fringes localize on a curve, and under special circumstances on the surface, which rarely coincides with the object. Therefore when the object is in focus no fringe may appear. The localization depends on the components of the deformation vector and the directions of illumination and observation. It could be very sharp and require precise focusing, or diffuse, wherein the fringes may be observed over a considerable distance. Fortunately, by stopping down the viewing aperture the fringes and the object can be photographed together. In fact, the aperture plays a central role in the theory of localization, and in its absence no fringe formation takes place [10]. Further, if the object undergoes a homogeneous deformation, the fringes will appear to be the intersection of the object surface with a set of equally spaced parallel planes, regardless of the shape of the object.

7.4 HOLOGRAPHIC NONDESTRUCTIVE TESTING

HI has been used to measure deformation and vibrational amplitudes in a variety of objects including biological. It provides interferometric sensitivity for these measurements that can be further increased using phase-detection techniques. An interferogram represents a pattern due to integrated phase changes arising out of deformation, rigid body displacements, and tilts. Often it is not easy to separate these components. Sandwich holography provides a means to compensate for tilts and out-of-plane motion by relative orientations of the two holograms [3].

A very interesting application of HI is for holographic nondestructive testing (HNDT) [11]. Except for tire testing, which is routinely used in industry, the application of HNDT is limited to laboratory experiments. In HNDT we look for flaws or anomalies in the fringe pattern obtained after the object is loaded. The success of HI for nondestructive testing depends on the proper choice of the method of excitation, which depending on the requirements, may be mechanical loading, pneumatic loading, thermal loading, acoustical or vibratory loading, or impact loading.

On the application of the load to the object, sometimes too many fringes are formed. This overcrowding masks the region of flaws or defects which then becomes impossible to locate precisely. Fringe control techniques are then employed. An optical arrangement in the reference or object beam paths capable of introducing linearly varying phase and quadratically varying phase is used for fringe compensation. Table 7.3 gives the kinds of flaws in a number of objects that have been detected using the various loading methods.[12].

Figure 7.6 gives a photograph of a helicopter rotor blade that was subjected to pneumatic loading between exposures. The regions of disbonds are clearly visible. Figure 7.7 shows an interferogram of a running engine.

Table 7.3   Kinds of Flaws Found with Various Loading Methods

	Mechanical Loading
Cracks	Bolt holes in steel channels, concrete, rocket nozzle liners, welded metal plates, turbine blades, glass
Debonds	Honeycomb panels, propellant liners
Fatigue damage	Composite materials
Deformation	Radial deformation in carbon cylinders and tubes

	Pneumatic Loading
Cracks	Welded joints, aluminium shells, bonded cylinders, ceramic heat exchanger tubes
Debonds or delamination	Composite panels and tubes, honeycomb structures, bonded cylinders, tires, rocket launch tubes, rocket nozzle liners
Weld defects	Welded plastic pipes, honeycomb panels
Weakness (thinness) and structural flaws	Aluminium cylinders, pressure vessels, tubes, composite tubes, composite domes
Solder joints	Medical implant devices
Bad brazes	Silicon pressure sensors

	Thermal Loading
Cracks	Glass tubes, ceramic tubes, train wheels, turbo fan blades
Debonds	Aircraft wing assemblies, honeycomb structures, laminate structures, rocket motor casing, rocket nozzle liners, rocket propellant liners, tyres, bonded structures
Delaminations	Antique paintings
Various defects	Circuit boards and electronic modules

	Vibratory Loading
Cracks	Train wheels, ceramic bars
Debonds	Turbine blades, rocket propellant liners, sheet metal sandwich structures, laminate structures, honeycomb structures, fiberglass-reinforced plastics, helicopter rotors, ceramics bonded to composite plates, brake discs, adhesive joints, metallic surfaces and elastomers

(Continued)

Table 7.3 (*Continued*)

Strength	CRT
	Impulse Loading
Cracks	Steel plates, wing plank splices turbine blades
Debonds	Foam insulation
Voids and thin areas	Aluminium plates

7.5 CONTOURING

Contouring refers to the sectioning of an object with fringe planes perpendicular to a specified direction, usually normal to the direction of viewing. The fringe map provides full three-dimensional analysis of the object. It is essential to have a contour map of the object for complete deformation studies. Real-world objects are complicated structures like automobiles and aircraft fins, having various surface finishes and orientations of surface normals with respect to the incident wavefront. Hologram interferometry provides a fringe pattern when such an object is loaded. It is not possible to directly lump from a two-dimensional map of displacement across the image to three-dimensional description of the object. In order to interpret the fringe data, the equilibrium surface contour must be known or measured, and two sets of data must be connected point by point across the image. Some other applications of contouring are: quality control in production lines, the transfer of contour information from prototypes or models to numerically controlled machines, bio-medical applications (particularly in vivo), and the study of mechanical wear. Contouring has been performed by hologram interferometry using dual wavelengths, dual refractive indices around the object and dual directions of illumination [13]. In each of these methods, two slightly altered surfaces are compared interferometrically.

7.5.1 The Dual Wavelength Method

Figure 7.8 (a) shows the schematic of the experimental setup. The object is illuminated by a collimated on-axis beam of wavelength λ_1. A hologram of the object is recorded by adding an off-axis collimated beam. The hologram is relocated precisely after processing, or in-situ processing is per-

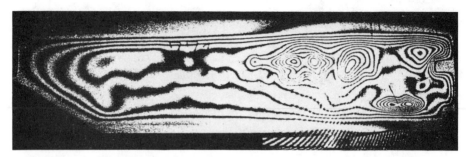

Figure 7.6   Interferogram of a helicopter rotor blade using double exposure hologram interferometry. (Courtesy of Steinbichler H., Labor Dr. Steinbichler, F.R.G.)

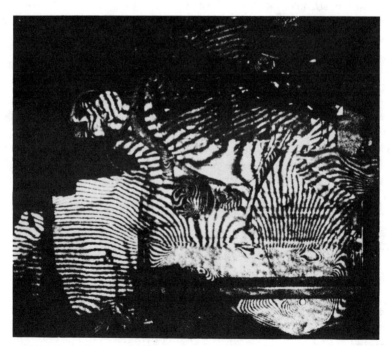

Figure 7.7   Interferogram of a running engine, (Courtesy of Steinbichler H., Labor Dr. Steinbichler, F.R.G.)

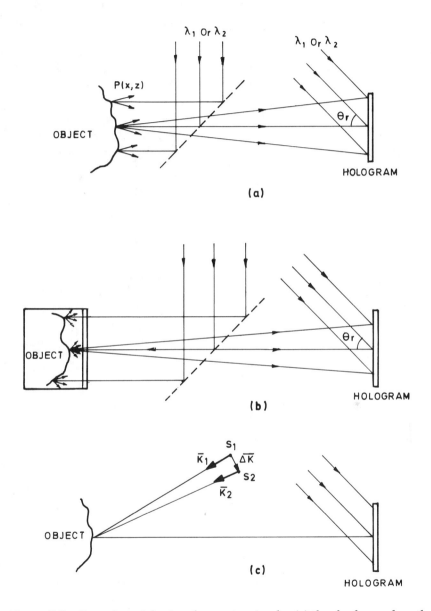

Figure 7.8 Experimental setup for contouring by (a) the dual wavelength method, (b) the dual refractive index method, and (c) the change of illumination direction method.

formed. The hologram is now illuminated by a collimated beam of wavelength λ_2, which reconstructs a virtual image of the surface. The object is also illuminated on-axis by the collimated beam of wavelength λ_2. Just behind the hologram, there are two waves of interest, one released from the hologram and the other transmitted through the hologram. These two waves produce an interference pattern on the surface or in its neighborhood. It may be observed that illumination of a hologram with a wavelength other than that used for recording results in both lateral and longitudinal displacement of the reconstructed image [2]. The lateral shift is removed if the collimated beam illuminates the hologram at an angle $\theta_r' = (\lambda_2/\lambda_1)\,\theta_r$, where θ_r is the angle of the reference beam with the hologram normal. Therefore the virtual image is displaced only longitudinally by Δz, given by

$$\Delta z = z_1 \frac{\lambda_1 - \lambda_2}{\lambda_2} \tag{7.13}$$

where z_1 is the z coordinate of any point on the object surface. Therefore the waves from the identical points on the virtual and actual surfaces will have a phase difference δ given by

$$\delta = \frac{2\pi}{\lambda_2}\, 2z_1\, \frac{\lambda_1 - \lambda_2}{\lambda_2} \tag{7.14}$$

The object is thus intersected by fringes with the incremental height $\delta z = z_{1(m+1)} - z_{1m}$ between two fringes of mth and $(m+1)^{th}$ order, where

$$\delta z = \frac{\lambda_2^2}{2(\lambda_1 - \lambda_1)} \tag{7.15a}$$

This can be rewritten as

$$\delta z = \frac{\lambda^2}{2\Delta\lambda} \tag{7.15b}$$

where λ_1 is very close to λ_2. Thus using a laser that emits in a number of discrete wavelengths, different contour intervals can be realized. If we take the wavelengths 488 nm and 476.5 nm of an argon ion laser, a contour interval of approximately 10 μm is obtained. Larger contour intervals are obtained with closely spaced wavelengths.

7.5.2 The Dual Refractive Index Method

The dual refractive index method is relatively simple and requires the change of refractive index of the medium surrounding the object. Both real-time and double-exposure hologram interferometric techniques , as in the dual wavelength method, can be employed. The object is immersed in a liquid of refractive index n_1 and a hologram is recorded with a collimated reference beam. The liquid is now changed to one of refractive index n_2 and another record is made on the same plate. The schematic of the experimental arrangement is shown in Fig. 7.8 (b). The double-exposure hologram on reconstruction gives an object image with fringes overlaid. The fringe spacing, for the illumination of the object by a collimated beam, is

$$\delta z = \frac{\lambda}{2\Delta n} \qquad (7.16)$$

where $\Delta n = |n_2 - n_1|$. For divergent illumination, a slightly different formula is used [2]. Using a set of liquids prepared by mixing, a range of contour intervals can be covered.

7.5.3 The Dual Illumination Method

In the dual illumination method, the contour fringe pattern on an object is generated by illuminating it from two different directions. Following Fig. 7.8 (c), the object is illumanted by spherical waves from sources S_1 and S_2. The illumination could be simultaneous or sequential, resulting in a single- or double-exposure recording of the hologram. Reconstruction from the hologram gives contour fringes superposed on the object. It is known that the interference pattern between two spherical waves is a family of hyperboloids of two sheets with foci at S_1 and S_2. In the far field of the sources, the fringe surfaces are approximately plane and parallel to the mean illumination direction. The fringe spacing is

$$\delta z = \frac{2\pi}{|\Delta \bar{k}|} \qquad (7.17)$$

where $\Delta \bar{k}$ is the difference between the two illumination vectors. The advantage of the method is its simplicity in principle and equipment. It can be used for contouring in real time. The drawback of the method is that the angle of illumination must be relatively large in order to get reasonable sensitivity. This may produce shadowing on certian parts of the object.

7.6 SPECKLE METROLOGY

A granular structure formed in space when a coherent wave is reflected from or transmitted through an optically rough surface is a speckle pattern. It arises from the self-interference of waves generated by the diffuse object. It can into prominence after the advent of the laser; a holographic reconstruction was always accompanied by a speckle pattern that was annoying to the viewer and that was therefore considered a bane of holographers. Around 1968 its information-carrying capacity was recognized. It has since been used for a number of optical measurements, including measurements of deformation vectors, strains, and surface roughness. Speckle methods are full-field methods and provide variable sensitivity between that of moiré methods and hologram interferometry. Further they are amenable to electronic detection, so that they are real-time inspection tools [14,15,16].

Let us consider a rough surface as in Fig. 7.9 (a) and (b) illuminated by a laser beam. Each point scatters the wave. At any point on the observation plane the light field arises due to the superposition of fields from a large number N of scatterers. it is given by [17]

$$u(x, y) = \frac{1}{\sqrt{N}} \sum_{j}^{N} a_j e^{i\phi_j(x,y)} \tag{7.18}$$

where $a_j/\sqrt{N}$ and ϕ_j are the random amplitude and phase of a wave scattered from the jth scatterer. We assume that the amplitudes a_j are statistically independent random variables and that the phases ϕ_j are also statistically independent random variables and distributed uniformly between $-\pi$ and π. It can be shown that the probability density function $P(I)$ of intensity, in a speckle pattern is given by $(1/\bar{I})e^{-(I/\bar{I})}$, where $\bar{I}$ is the mean intensity. The intensity in a speckle pattern thus obeys negative exponential statistics; this has been confirmed experimentally. The most probable intensity value is zero, but there are speckles in the pattern several times brighter than the mean value. The contrast in a linearly polarized and fully developed speckle pattern is 1. Speckle patterns can be added either by amplitude or by irradiance. The first-order statistics do not change except for a scaling constant when the speckle patterns are added by amplitude. On ther other hand, the statistics are completely changed when the speckle patterns are added by irradiance; they depend on the degree of correlation among the irradiance distributions.

A speckle pattern formed because of free propagation in space is called an objective speckle pattern. The intensity distribution in the image of a coherently illuminated diffuse surface also exhibits granular structure.

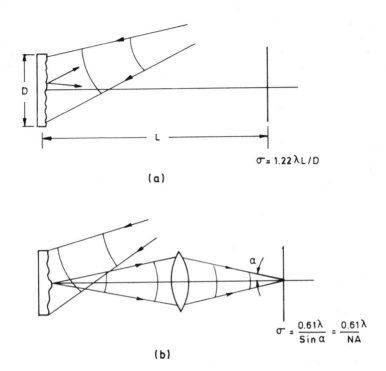

$\sigma = 1.22 \lambda L/D$

(a)

$$\sigma = \frac{0.61\lambda}{\text{Sin } \alpha} = \frac{0.61\lambda}{\text{NA}}$$

(b)

Figure 7.9 Speckle formation in (a) free-space propagation and (b) imaging geometry.

This speckle pattern is called a subjective speckle pattern. The pattern observed at planes other than the image plane will be called a defocused plane speckle pattern. A parameter of great importance to measurement is speckle size. It is obtained by the second-order speckle statistics [17]. It is tempting to look at the problem from the interference standpoint. Assume that a diffuser of diameter D is illuminated by a coherent beam and that the observation is made at a plane distant L from it. The smallest fringe width at this plane will be due to interference between waves from diametrically opposite scatterers. The fringe width will be $\lambda L/D$. This holds good for all diametrically opposite scatterers. Hence an average size to the objective speckle is assigned as $\sigma_0 = \lambda L/D$. It must be kept in mind that the speckles do not have well defined shapes or structure and hence only an average size is assigned. When an imaging optics is used, the speckle pattern is due to the superposition of a large number of dephased impulse functions. The average size is then given by the size of the Airy disc, i.e. $\sigma_s = 0.61 \, \lambda/NA = 1.2(1 + M) \, \lambda F$, where M is the magnification and F is the f number of the lens.

7.7 MEASUREMENT METHODS

Of the many applications of the speckle phenomenon we consider the measurement of the surface deformation vector and its derivatives, contour generation, and surface characterization. When an object is deformed, then the speckle pattern changes. The changes are essentially of two types, speckle pattern bodily shifts and irriadiance changes. In practice both these changes occur simultaneously and lead thus to decorrelation. Methods based on the measurement of the positional shift of the speckle pattern fall under speckle photography, while those exploiting irradiance variation fall under speckle interferometry. It may be remarked that the speckle phenomenon is basically an interference phenonmeon, but in speckle interferometry a reference beam is used to convert phase variations introduced by deformation into irradiance variation. It will be shown later that the fringes formed in speckle interferometry are correlation fringes.

7.8 SPECKLE PHOTOGRAPHY

An objective or subjective speckle pattern of an object is recorded on a high-resolution photographic plate. The object is deformed and a second exposure is made on the same plate. The plate on processing is called a doubly exposed specklegram. If the deformation does not lead to decorrelation, the specklegram contains two identical speckle patterns that are relatively shifted. The shift between speckles can be obtained by pointwise filtering or by whole-field filtering.

7.8.1 Pointwise Filtering

For pointwise filtering a narrow beam of laser light illuminates the specklegram at various grid points in succession to generate the displacement field. A screen placed at a distance receives the diffracted light. The beam is diffracted in a cone and the distribution in the halo is given by the autocorrelation of the aperture function. Assuming that the shift of the speckles is constant in the illuminated region, the halo is modulated by $\cos^2$ type fringes. These Young's fringes are always normal to the direction of displacement, and the fringe width $\bar{x}$ is related to the position shift Δx_H by

$$\Delta x_H = \frac{\lambda L}{x} \tag{7.19}$$

Thus both the direction and the magnitude of speckle movement can be obtained at each grid point. The sign ambiguity still remains. Figure 7.10 (a) shows an experimental arrangement of pointwise filtering and Figure 7.10 (b) shows the Young's fringes obtained.

7.8.2 The Wholefield Filtering Arrangement

Figure 7.11 (a) shows a schematic of the whole-field arrangement. The specklegram may be placed at the front focal plane of the lens L_1 or in the convergent beam. An aperture is placed at the Fourier plane and an image of the specklegram is made by lens L_2 by the light passing through the aperture. This arrangement is used to obtain a fringe pattern both in speckle photography and in speckle interferometry. The aperture allows through the light diffracted in that direction; the regions in which identical speckle pairs diffract in that direction will appear bright. In other words, $d_y \sin \theta$ mλ is the governing equation, where $\sin \theta = y/f$, y being the distance of the center of the aperture from the optical axis along the y direction and d_y is the y component of speckle shift in the specklegram. Since θ is fixed, the fringe formation is due to variation in d_y. The two adjacent fringes correspond to the change Δd_y in the y component:

$$\Delta d_y = \frac{f}{y} \tag{7.20}$$

Therefore the fringes represent the loci of incremental change in the displacement component. The fringe pattern corresponding to the x component can be obtained by locating the aperture on the x axis. It may be seen that the fringes get closer when the aperture is farther away from the optical axis. Figure 7.10 (b) shows the whole-field-filtered fringe pattern of a disc rotated between exposures.

When the object undergoes longitudinal displacement, the speckles are radially shifted at the recording plane. Such a doubly exposed specklegram is filtered by placing an aperture on the axis to calculate the magnitude of the longitudinal displacement. Such arrangement is sometimes called Fourier filtering set up.

Speckle photography has been used to measure in-plane displacement, out-of-plane displacement, and rigid body tilt. When the object moves bodily, the objective speckle pattern is used. In case of deformation mea-

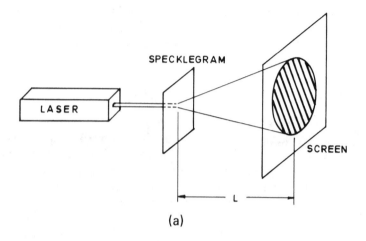

(a)

(b)

Figure 7.10   (a) Pointwise filtering of a specklegram and (b) Young's fringes obtained by pointwise filtering.

surement, a subjective speckle pattern is employed to maintain the correspondence between points on the object and the specklegram. The sensitivity is governed by the speckle size; measurement is possible only when the speckle shift is greater than the speckle size.

7.8.3 In-Plane Component Measurement

When a free-space geometry is used for recording, the speckle pattern shifts by an amount equal to the in-plane motion of the object. A double exposure specklegram therefore contains two identical but shifted speckle patterns. In imaging geometry the speckle shift is controlled by the magnification of the lens. Pointwise filtering at any point on the specklegram can be utilized for obtaining the in-plane component through the relation

$$u = \frac{\lambda z}{\bar{x}} \tag{7.21}$$

where u is the in-plane component, $\bar{x}$ is the fringe width, z is the separation between the hologram and the observation plane, and λ is the wavelength of the light used for filtering. Whole-field filtering does not yield any fringes when the displacement is uniform. Both pointwise filtering and whole-field filtering can be used when the object undergoes deformation.

7.8.4 Out-of-Plane Component Measurement

When the object translates longitudinally, the speckle pattern shifts radially. If the distance between the object and the recording plane is l_1 and the object shifts by $\in$ the radial shift Δr of the speckle at position r is [18]

$$\Delta r = \in \frac{r}{l_1} \tag{7.22}$$

The speckle pattern rapidly decorrelates, however, with increasing $\in$. Still it can remain correlated for large value of $\in$ if the object is illuminated by a spherical wave.

A doubly exposed specklegram is Fourier filtered to obtain the value of $\in$. It is illuminated by a spherical wave from a point source distant l_2 from the specklegram, and the filtering is performed by placing an aperture on the axis at a distance l_3 from the specklegram. The value of $\in$ is calculated from

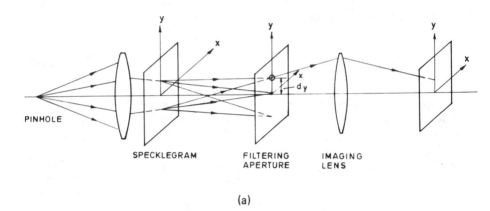

(a)

Figure 7.11 (a) A schematic of the whole-field filtering method and (b) fringe patterns obtained by filtering at various locations in the Fourier transform plane.

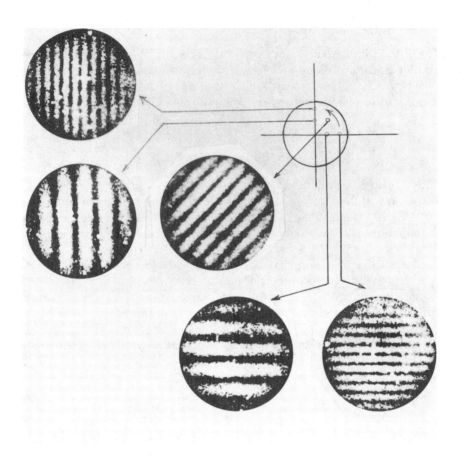

(b)

$$\in = \frac{\lambda l_1 l_2 l_3}{(l_2 + l_3)(r_m^2 - r_{m-1}^2)} \tag{7.23}$$

where r_m and r_{m-1} are the radii of the mth and $(m-1)^{th}$ circular fringes.

7.8.5 Tilt Measurement

The tilt of a diffuse object is measured by recording the specklegram in the Fourier transform plane of a lens. The double exposure specklegram is then pointwise filtered to obtain the magnitude of the tilt by measuring the fringe width. The tilt angle α is obtained from [19]

$$\alpha = \frac{f_1 \lambda}{f(1 + \cos \theta)\bar{x}} \tag{7.24}$$

where f and f_1 are the focal lengths of the lenses used for performing the Fourier transformation of the object and to display the fringes in the Fraunhofer plane, respectively: $\bar{x}$ is the fringe width, and θ is the angle of illumination. The illumination is by a collimated beam. The displacement of speckles is equal $f(1 + \cos \theta)\alpha$.

In practice an object may be illuminated by a spherical wave. The specklegram is then recorded not on the focal plane but at a plane conjugate to the source, as seen by attaching a small mirror on the rough object surface. This plane is called the tilt plane; it is insensitive to in-plane movement of the object. Therefore the speckle shift at this plane is due to the tilt alone. The speckle shift x_t at the tilt plane due to tilt of α of the object is [20]

$$x_t = \frac{f(1 + \cos \theta)\alpha}{1 + (b - f)/a} \tag{7.25}$$

where θ is the mean angle of illumination, a and b are the distances between source and object and object and lens, respectively. It is assumed that the lens axis is taken normal to the object surface. The tilt is then calculated by measuring the fringe width in a pattern obtained by point wise filtering.

7.8.6 Contouring

Speckle photography has been applied for contouring. Instead of the object surface, a plane in front of it is imaged on the recording plane. The object is tilted by a small angle α in between exposures. The double exposure specklegram is whole-field filtered to obtain contour fringes. The fringe formation is goverend by the the equation [21]

$$h_{n+1} - h_n = \frac{\lambda f}{dM} \tag{7.26}$$

where h_n corresponds to the depth at the nth fringe, f is the focal length of the transform lens or the distance between the specklegram plane and the aperture plane, d is the distance of the aperture from the axis, and M is the magnification of the imaging lens.

The tilt results in decorrelation of the speckle patterns; and hence the object is titled by a very small angle. Contour intervals are rather large.

7.9 SPECKLE INTERFEROMETRY

Speckle interferometry has been used for the measurement of out-of-plane and in-plane components. In earlier experiments, the Michelson inter-ferometer configuration with the mirror replaced by diffusers was used. The diffusers were imaged on the recording plane. The speckle pattern at the image plane was due to the coherent superposition of individual speckle patterns from each diffuser. A double exposure specklegram was recorded with one of the diffusers deformed between exposures. The correlation fringes obtained from the specklegram were of extremely low contrast. Nonlinearity of the photographic emulsion was exploited to im-prove the contrast.

The aperturing of the imaging lens provides two benefits: the contrast of the fringes dramatically increases, and multiplexing of the information is possible. It was mentioned earlier that the intensity distribution in the halo is given by the autocorrelation of the aperture function. Consider a square aperture as shown in Fig. 7.12 (a). Its autocorrleation function is shown in Fig. 7.12 (b). If instead of a single aperture, two apertures are used, as shown in Fig. 7.12 (c), the autocorrelation function is modified greatly; the autocorrelation function is shown in Fig. 7.12 (d). Therefore the halo now consists of three smaller halos; the filtering is done via one of the I-order halos. If the apertures are small, there is no need of aperturing the halo for filtering. It thus becomes obvious that the use of two apertures in front of the lens is equivalent to prefiltering. The obvious advantage is a large signal-to-noise ratio while filtering. The disadvantage may be longer exposure time while recording the speckle patterns. The aperturing has been found to be of great value when the photographic emulsion is used

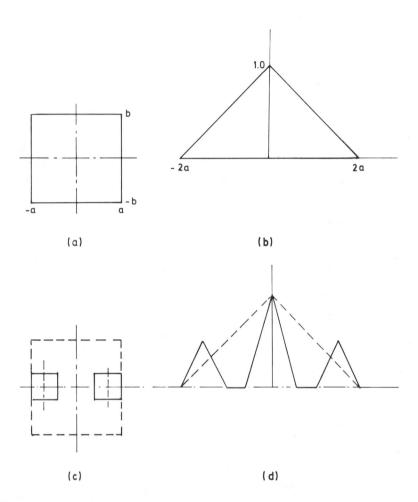

Figure 7.12   (a) a square aperture, (b) autocorrelation of the square aperture, (c) a pair of small square apertures, and (d) autocorrelation functions of the aperture in (c).

for recording. When electronic detection is used, the single but small aperture is used to match the speckle size with the resolution of the camera.

7.9.1 Out-of-Plane Component Measurement

Figure 7.13 (a) shows a schematic of the arangement for out-of-plane component measurement. A pair of apertures is placed in front of the lens. One of the apertures carries a ground glass plate while the other is open. A part of the laser beam directly falls on the plate; the diffuse beam issuing from it serves as a reference beam. The object is imaged via the other aperture. A double exposure record is made and the specklegram is whole-field filtered. Figure 7.13 (b) gives a photograph of the fringes obtained when a circular diaphragm clamped at the edge is centrally displaced. The fringes correspond to out-of-plane displacement $w(x,y) = m\lambda/2$, where m = 0,1,2,3, The fringe order is 0 at the clamped edge and increases inwardly.

Since both the reference and object beam propagate nearly along the same direction, normal to the object surface, the fringes thus obtained are due to the out-of-plane component.

7.9.2 In-Plane Component Measurement

The in-plane component can be measured using either dual illumination directions and a single direction of observation or single illumination direction and dual observation directions.

Let an object be illuminated by two plane waves lying in the x–z plane and making angles of θ and $-\theta$ with the normal to the surface. The observation is made along the direction of the normal to the surface. The phase difference δ when the object undergoes deformation $\bar{L} = u_i + v_j + w_k$ is

$$\delta = (\bar{k}_2 - \bar{k}_1) \cdot \bar{L} \qquad (7.27)$$

where $\bar{k}_2$ and $\bar{k}_1$ are propagation vectors along the observation and illumination directions. Thus for the geometry having symmetric illumination, the phase difference δ is given by

$$\delta = (\bar{k}_1 - \bar{k}_1') \cdot \bar{L}$$

$$= \frac{2\pi}{\lambda} \, 2u \sin \theta \qquad (7.28)$$

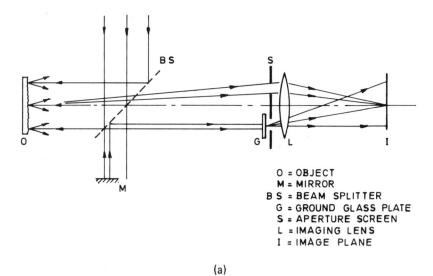

O = OBJECT
M = MIRROR
B S = BEAM SPLITTER
G = GROUND GLASS PLATE
S = APERTURE SCREEN
L = IMAGING LENS
I = IMAGE PLANE

(a)

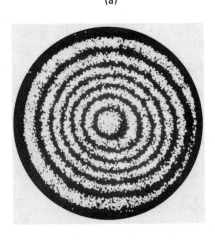

(b)

Figure 7.13   (a) A schematic of measuring out-of-plane displacement using speckle interferometry and (b) an interferogram of a circular aperture clamped at the edges and deflected in the center.

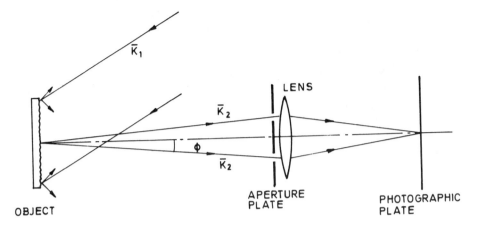

Figure 7.14 A two-aperture arrangement to measure in-plane displacement.

The bright fringes are formed when

$$u = \frac{m\lambda}{2 \sin \theta} \tag{7.29}$$

The u family of fringes are separated by $\lambda/2\sin \theta$. Fringes corresponding to the v component are obtained when the beams lie in the y–z plane.

Another convenient arrangement to measure the in-plane component is to use two apertures in front of the lens, thereby providing two observation directions, as shown in Fig. 7.14. The phase difference is given by

$$\delta = (\bar{k}_2 - \bar{k}_1) \cdot \bar{L} - (\bar{k}_2{}' - \bar{k}_1) \cdot \bar{L} = (\bar{k}_2 - \bar{k}_2{}') \cdot \bar{L}$$

$$= \frac{2\pi}{\lambda}\ 2u \sin \phi \tag{7.30}$$

Therefore the fringe formation is governed by

$$2u \sin \phi = m\lambda \tag{7.31}$$

Again the in-plane component is sensed. This arrangement was first used by Duffy [22]. Both methods described above automatically compensate for out-of-plane component and measure only the component determined by the geometry. The sensitivity of Duffy's arrangement is poor as $\phi < \theta$ and because ϕ is limited by the aperture of the lens.

The two illumination arrangements can be used to measure large out-of-plane displacement as well. In this case the object is illuminated by two waves that make angles of θ_1 and θ_2 with the normal to the surface. The phase difference δ is now expressed as [1]

$$\delta = (\bar{k}_1 - \bar{k}_1') \cdot \bar{L}$$

$$= \frac{2\pi}{\lambda} [u(\sin \theta_1 - \sin \theta_2) + w(\cos \theta_1 - \cos \theta_2)] \tag{7.32}$$

If we assume that $w > u$ and the angles θ_1 and θ_2 are large, approaching $\pi/2$, so that $|\cos \theta_1 - \cos \theta_2| > \sin \theta_1 - \sin \theta_2$, the first term can be neglected. Thus

$$w(x, y) = \frac{\lambda}{2(\cos \theta_1 - \cos \theta_2)} \tag{7.33}$$

Since $\cos \theta_1 \approx \cos \theta_2$, large out-of-plane deformation can be measured. Thus nonsymmetric illumination leads to desensitization of the speckle interferometry for measuring out-of-plane displacement. Speckle interferometry demands the same stability requirements as in hologram interferometry. The advantages are that the in-plane and out-of-plane components can be independently measured, that the fringes localize on the object surface, and that the resolution requirements are less severe.

7.10 SPECKLE SHEAR INTERFEROMETRY

Shear interferometry responds to the gradient of the path function. In speckle shear interferometry using imaging geometry, a point in the image plane receives contributions from two different points on the object, or a point on the object may be imaged as two points [23]. The separation between the points in the image plane or the object plane is known as image or object shear respectively. It is possible to introduce various kinds of shears, for example linear/lateral, radial, rotational, inversion, and folding. Table 7.4 gives the various shear types along with the optical elements used to introduce the shears.

We now look briefly at the process of fringe formation in speckle shear

Table 7.4   Shear Types and Optical Elements

Shear type	Definition (superposition of)	Optical elements
Linear/ lateral	f (x, y) and f (x + Δx, y + Δy)	Michelson interferometer, plane parallel plates, wedge plates, biprism, Ronchi gratings, Billets lens
Rotational	f (r, θ) and f(r, θ + Δθ)	Pair of Dove prisms
Radial	f (r, θ) and f(r + Δr, θ)	Billets lens
Inversion	f(x ,y) and f (−x, −y) or f(r, θ) and f(r, θ + π)	Pair of Dove prisms
Folding	f(x, y) and f(−x, y)	One Dove prism

interferometry. Let I_1 and I_2 be the irradiances recorded before and after the object is deformed. These are expressed as

$$I_1 = a^2(y) + a^2(y + \Delta y_0) + 2a(y)a(y + \Delta y_0) \cos \phi \qquad (7.34a)$$

$$I_2 = a^2(y) + a^2(y + \Delta y_0) + 2a(y)a(y + \Delta y_0) \cos(\phi + \delta) \qquad (7.34b)$$

where $a(y)$ and $a(y + \Delta y_0)$ are the amplitudes of the waves at any point on the image plane from two points at y and $y + \Delta y_0$; Δy_0 is the object shear along the y direction, ϕ is the random phase, and δ is the phase difference due to deformation. The phase difference δ can be expressed as

$$\delta = \delta(y + \Delta y_0) - \delta(y)$$

where

$$\delta(y + \Delta y_0) = (\bar{k}_2 - \bar{k}_1) \cdot \bar{L} (y + \Delta y_0) \qquad (7.35a)$$

$$\delta(y) = (\bar{k}_2 - \bar{k}_1) \cdot \bar{L} (y) \qquad (7.35b)$$

Assuming that the illumination beam lies in the y–z plane and makes an angle θ with the z axis, and that the camera axis is along the z axis, the phase difference δ can be expressed as

$$\delta = \frac{2\pi}{\lambda} \left[\frac{\partial v}{\partial y} \sin \theta + \frac{\partial w}{\partial y} (1 + \cos \theta) \right] \Delta y_0 \qquad (7.36)$$

The photographic plate records both the irradiance distributions in succession. Assuming linear recording, the amplitude transmittance of the specklegram is

$$t = t_0 - \beta(I_1 + I_2)T \qquad (7.37)$$

When the specklegram is whole-field filtered, bright fringes are formed when $\delta = 2m\pi$. This occurs where both the speckle patterns are well correlated. Therefore the fringe formation is governed by

$$\frac{\partial v}{\partial y} \sin \theta + \frac{\partial w}{\partial y} (1 + \cos \theta) = \frac{m\lambda}{\Delta y_0} = \frac{m\lambda M}{\Delta y_i} \qquad (7.38)$$

where Δy_i is the image plane shear and M is the magnification of the imaging lens. Both $\partial v/\partial y$ (strain) and $\partial w/\partial y$ (partial slope) can be obtained if the recordings are made for two different illumination angles. For normal illumination by collimated beam $\partial v/\partial y$ is not sensed. The fringe pattern then represents the partial y slope $\partial w/\partial y$, i.e.

$$\frac{\partial w}{\partial y} = \frac{m\lambda}{2\Delta y_0} \qquad (7.39)$$

Figure 7.15 (a) shows a fringe pattern depicting the partial y slope of an edge-clamped diaphragm centrally loaded. Table 7.5 gives a summary of various slope shear fringe patterns obtained using different shear types. It is seen that the radial shear gives fringes corresponding to $r\partial w/\partial r$, from which dw/dr may be found. Since radial shear increases linearly with r, it is more sensitive to loose clamping, etc. The rotation shear does not yield any fringe pattern for a circularly symmetric deflection profile and hence is best suited for NDT of objects that may undergo large out-of-plane deformation on loading. Figure 7.15(b) shows $\partial w/\partial \theta$ fringes for a cantilever.

(a)

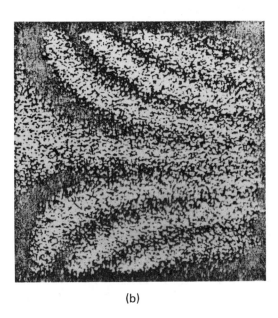

(b)

Figure 7.15 (a) Interferogram showing partial y-slope fringe pattern of an edge-clamped diaphragm centrally loaded and (b) interferogram showing $\partial w / \partial \theta$ fringe pattern of a cantilever.

Table 7.5 Shears and Slope Fringes

Shear	Equation governing slope fringes
Lateral shear Δy_0	$\dfrac{\partial w}{\partial y} = \dfrac{m\lambda}{2\Delta y_0}$
Radial shear [a] $\lvert \Delta r \rvert = (1 - M^2)r$	$r\dfrac{\partial w}{\partial r} = \dfrac{m\lambda}{2(1 - M^2)}$
Rotional shear $\Delta\theta$	$\dfrac{\partial w}{\partial \theta} = \dfrac{m\lambda}{2\Delta\theta}$
	$\dfrac{\partial w}{\partial \theta} = 0 \quad \text{when } w(r, \theta) = w(r)$

[a]M, magnification.

7.10.1 Tilt Measurement with Shear Interferometry

Tilt measurement can be done with radial shear, rotational shear, inversion shear, and folding shear configurations. They offer different sensitivities.

Radial Shear

The tilt β is given about the y axis. Therefore the path $w(r, \theta)$ introduced due to tilt is

$$w(r, \theta) = \beta x = \beta r \cos \theta \tag{7.40}$$

In radial shear the fringe formation is governed by

$$r\frac{\partial w}{\partial r} = \frac{m\lambda}{2(1 - M^2)} \tag{7.41}$$

This gives

$$\beta x = \frac{m\lambda}{2(1 - M^2)} \tag{7.42}$$

Eq. (7.42) shows that the fringes run parallel to the y axis. The tilt angle is calculated by

$$\beta = \frac{\lambda}{2(1 - M^2)\bar{x}} \tag{7.43}$$

where $\bar{x}$ is the fringe width. The largest tilt that can be measured depends on both the speckle size and decorrelation effects.

Rotational Shear

We assume again that the object is tilted about the y axis. Thus

$$w(r, \theta) = \beta x = \beta r \cos \theta.$$

The fringe information is governed by

$$\frac{\partial w}{\partial \theta} = \frac{m\lambda}{2\Delta\theta} \tag{7.44}$$

where $\Delta\theta$ is the rotational shear. From this equation we obtain

$$\beta = \frac{\lambda}{2\Delta\theta\bar{y}} \tag{7.45}$$

The fringe runs parallel to the x axis.

Inversion Shear

The phase difference δ for the rotational shear of θ' is given by

$$\delta = \frac{2\pi}{\lambda} 2[w(r, \theta + \theta') - w(r, \theta)]$$

$$= \frac{4\pi}{\lambda} \beta[r \cos(\theta + \theta') - r \cos \theta]$$

$$= \frac{4\pi}{\lambda} \beta[x (\cos\theta' - 1) - y \sin \theta'] \tag{7.46}$$

The fringe formation is governed by

$$2\beta[x(\cos\theta' - 1) - y\sin\theta'] = m\lambda. \tag{7.47}$$

Thus a straight-line fringe pattern is formed with the fringes inclined by $\pi - \theta'/2$ with the x axis. For inversion shear $\theta' = \pi$ and hence the fringes are perpendicular to the x axis or run parallel to the y axis. The angle of tilt is given by $\beta = \lambda/4\,\overline{x}$.

If the tilt is given about the x axis, the fringes now run parallel to it. The tilt angle is given by $\beta = \lambda/4\,\overline{y}$.

Folding Shear:

We again assume that the object is tilted about the y axis. Thus

$$w(x, y) = \beta x$$

When folding is about the x axis, the phase difference δ is expressed as

$$\delta = \frac{4\pi}{\lambda}\,[w(x, y) - w(-x, y)] \tag{7.48}$$

The equation governing the fringe formation is $4\beta x = m\lambda$. Therefore the tilt angle is calculated from

$$\beta = \frac{\lambda}{4\overline{x}}$$

The fringes run parallel to the y axis.

If the tilt is about the x axis, no fringe formation takes place. The folding shear configuration is sensitive to the component of the tilt about the axis perpendicular to the folding axis. If the angle between the folding axis and tilt axis is γ, then the tilt is obtained from the relation

$$\beta = \frac{\lambda}{4\overline{x}\sin\gamma} \tag{7.49}$$

7.10.2 Curvature Fringes

Shear interferometry gives the derivative of the displacment; both strain and slope fringes can be obtained. In some cases an experimentalist is interested in obtaining curvature fringes which are due to the second derivative of the out-of-plane component. This is obtained by the moiré of the slope fringes. We will now discuss a configuration that can yield out-of-plane component, slope, and curvature fringes from the same double exposure specklegram. The sensitivities of the three patterns are very different, and it is therefore doubtful if all three patterns obtained from the same specklegram are equally useful.

7.10.3 Generalized Configuration

Figure 7.16 (a) shows an experimental arrangement. The object is illuminated by a collimated beam and imaged by a lens on the recording plane. A screen containing four apertures is mounted in front of the lens. Two apertures carry wedge plates oppositely mounted, one aperture carries a ground glass plate, and one aperture carries a plane parallel plate to compensate for the extra path, as shown in Fig. 7.16(b). The wedge plates provide the linear shear. A double exposure specklegram is recorded and the object is deformed between exposures. On filtering, a number of halos are obtained; for example halo A_{01} is due to the interaction of light from apertures A_0 and A_1. The halo distribution is shown in Fig. 7.16 (c). If the field from halo A_{02} is filtered for image formation, only displacement fringes are obtained. Other halos provide fringe patterns belonging to slope and curvature. The experiments are conducted on a square diaphragm clamped at the periphery and displaced in the center. Figure 7.16 (d) is an interferogram showing displacement fringes obtained by filtering via A_{02} halo. The interferograms shown in Figs. 7.16 (e) and (f) are obtained by filtering via halos A_{13} and $A_{12} + A_{23}$, respectively. They represent slope fringes and curvature fringes as a result of moiré between the two slope patterns.

7.10.4 Multiplexing

It has been shown that aperturing a lens results in high-contrast fringes. Further, it is also possible to put more information on a single recording plate and retrieve it later. The amount of information that can be placed is essentially limited by the dynamic range. There are two ways to load more information, by theta modulation and by frequency modulation [24]. In theta modulation, a screen containing two apertures is placed in front of the imaging lens and a double exposure record is made with the object deformed between exposures. The screen is now rotated about the optical axis, and another double exposure record is made on the same plate with the object deformed to a different level. This way a large number of double exposure records can be made on the plate, provided its dynamic range

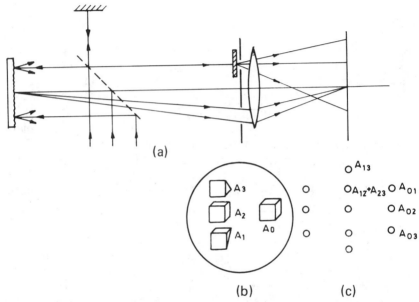

(a)

(b)   (c)

Figure 7.16   (a) Arrangement of a setup to measure displacement, slope, and curvature fringe patterns simultaneously, (b) aperture arrangement in front of the imaging lens, (c) halo distribution at the Fourier transform plane, (d) displacement fringe pattern on filtering from halo A_{02}, (e) slope fringe pattern on filtering from halo A_{13}, and (f) curvature fringe pattern on filtering from halo $A_{12} + A_{23}$.

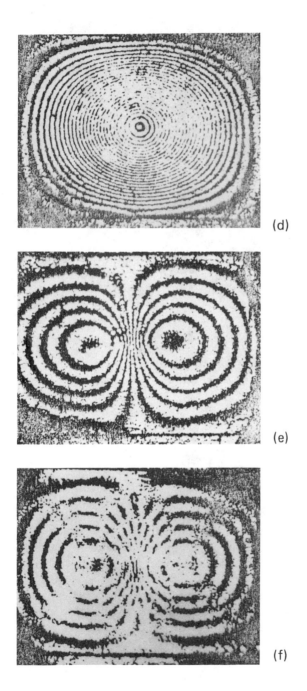

(d)

(e)

(f)

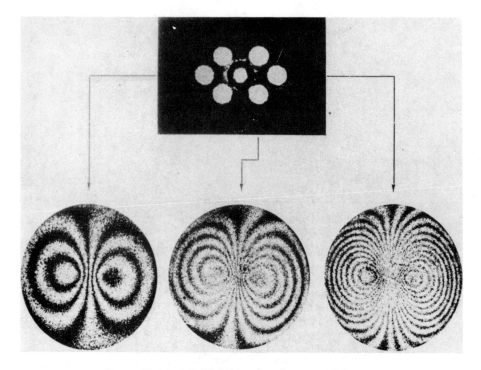

Figure 7.17   Multiplexing by theta modulation.

allows. The specklegram, on whole-field filtering, will display a large number of angularly displaced halos, these halos will be isolated from each other if the angle of rotation is properly chosen. Filtering via different halos will yield interferograms corresponding to different states of loading of the object. Figure 7.17 shows a series of interferograms extracted by filtering from their respective halos.

A similar objective is met if the separation between the pair of apertures is varied. This is known as frequency modulation, as the frequency of interference fringes due to the fields issuing from the two apertures changes when their separation is varied. During filtering, the halos are separated because the angles of diffraction are different.

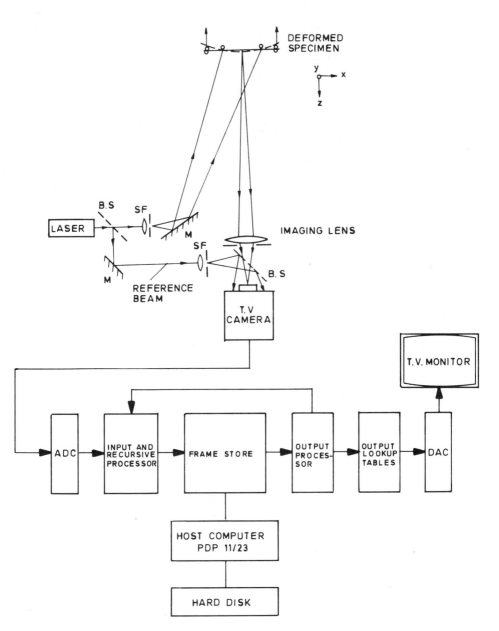

Figure 7.18   A typical configuration for ESPI/DSPI.

7.10.5 Electronic Speckle Pattern Interferometry

So far we have discussed methods that utilize photographic/holographic emulsions for recording. The resolution of the recording media can be relatively low; the minimum speckle size is typically in the range of 5 to 100 μm. Thus a standard television camera may be used to record the pattern. In other words, speckle interferometry can be performed with electronic detection. Video processing may be used to generate correlation fringes equivalent to those obtained photographically. The method is known as electronic speckle pattern interferometry (ESPI) [16,25]. The major feature of ESPI is its ability to present correlation fringes almost in real time. It can be done in a lit room and requires no messy photographic processing. The method has been demonstrated to work in adverse environments and even for long object distances. Irradiance correlation in ESPI is observed by a process of video signal subtraction or addition.

We shall discuss the subtraction process in detail. One of the common interferometric arrangements used for ESPI is shown in Fig. 7.18. The f number of the video camera is adjusted so that the speckle size is within the resolution limit. The object is imaged and the output of the camera with the object in its initial state is recorded on a video storage device (a video tape recorder, disc, or solid-state device). The object is then deformed and the live video signal is subtracted electronically from the stored signal. The correlated areas will give a zero resultant signal, while uncorrelated areas will give nonzero signals.

Let I_1 and I_2 be the irradiance distributions incident on the camera face plate before and after the object is deformed. Then

$$I_1 = a_1^2 + a_2^2 + 2a_1a_2 \cos \phi \tag{7.50a}$$

$$I_2 = a_1^2 + a_2^2 + 2a_1a_2 \cos (\phi + \delta) \tag{7.50b}$$

where δ is the phase difference introduced by deformation. The video signals are assumed to be proportional to the irradiances, and hence the subtracted signal V_s will be

$$V_s \propto (I_1 - I_2) = 2a_1a_2[\cos \phi - \cos(\phi + \delta)]$$
$$= 4a_1a_2 \sin(\phi + \frac{\delta}{2})\sin \frac{\delta}{2} \tag{7.51}$$

The subtracted signal has both negative and positive values. The monitor, however, will display negative signals as dark areas. To avoid this loss of signal, V_s is rectified before being displayed on the monitor. The brightness on the monitor is then proportional to $|V_s|$. Therefore the brightness B on the monitor is expressed as

$$b = 4K[a_1a_2 \sin^2 (\phi + \frac{\delta}{2})\sin^2 \frac{\delta}{2}]^{\frac{1}{2}} \qquad (7.52)$$

Where K is a proportionality constant. The brightness will be maximum when

$$\delta = (2m +1)\pi \qquad (7.53a)$$

and minimum when

$$\delta = 2m\pi \qquad (7.53b)$$

Thus the correlated regions appear dark on the monitor. The fringe contrast is improved by high-pass filtering. For speckle interferometry of vibrating objects or for an object illuminated with two pulses in succession, the video signal addition method is used, as the camera tube has a characteristic persistence time of the order of 0.1 sec. The two irradiances will be added if the time interval between the two illuminations is less than the persistence time. Another technique by which the speckle pattern data is processed digitally on a host comptuer is called digital speckle pattern interferometry (DSPI). The analog video signal from the camera is sent to a high-speed analog-to-digital converter (ADC). The ADC samples the incoming signal to yield a digital picture made up of an array of pixels, and the intensity at each pixel is quantized to a set of gray levels. The number of pixels and the number of gray levels depend on the frame store memory of the system used.

Two frames of this digital picture corresponding to the object in its initial and deformed states are stored in the host computer, which is interfaced to the frame store. These two frames are subtracted pixel by pixel and the data sent to a digital-to-analog converter (DAC). The processed data is displayed on the monitor. Real-time differencing can be achieved by means of a recursive processor incorporated in the digital image processing system. The difference signal is subjected to nonlinear processing such as level slicing or level windowing to enhance the fringe contrast.

Speckle Nondestructive Testing

ESPI/DSPI offers the advantages of real-time display of the fringe system in a lit room. Both these techniques have been applied for measurement of

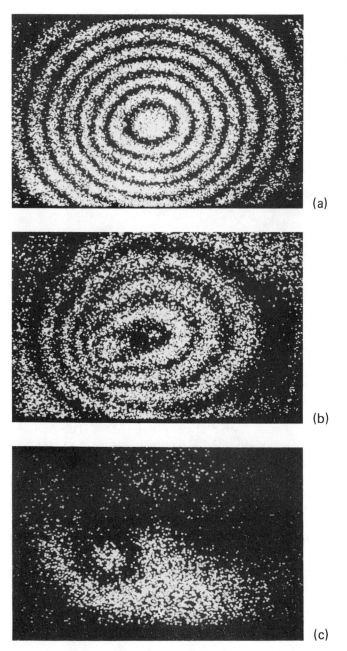

(a)

(b)

(c)

Figure 7.19 Comparative digital speckle pattern: (a) interferogram of a defective diaphragm, (b) interferogram of a defective diaphragm when partially compensated, and (c) interferogram of a defective diaphragm when fully compensated.

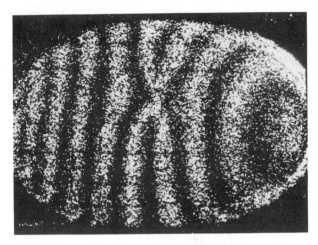

Figure 7.20   Interference pattern in the vicinity of a crack.

out-of-plane displacement, in-plane displacement, amplitude of vibration, etc. All shear configurations can be implemented on ESPI/DSPI. Thermal convection, diffusion, etc., have also been studied using these techniques. Here we report two applications, the comparison of two objects and crack detection. The two objects to be compared are imaged on the face plate of a vidicon tube. One of the objects is a master and other is a test object. Their responses to an external force field are to be compared, or if the test object has a flaw, its location and size are to be experimentally determined. Both objects are loaded simultaneously to the same extent. If their responses are identical, that is, if the deflection profiles are the same, no fringe pattern will be formed. Only in regions where the profiles differ do we obtain fringes. Figure 7.19 (a) shows an interferogram of a defective diaphragm when only it is loaded. The defective region is not visible because there are too many fringes. When the second diaphragm is also loaded, compensation takes place, and at full compensation only the defective region is seen. Figure 7.19 (b) is the partially compensated interferogram, while Figure 7.19 (c) is the fully compensated one, showing the defective region.

An in plane-sensitive configuration with out-of-plane loading is used for locating cracks. Figure 7.20 shows the interference pattern in the vicinity of a crack.

In all these experiments, a frame grabber with a 256 x 256 x 8 bit memory is used.

Contour Generation

All the methods used in hologram interferometry are easily adapted to speckle interferometry, and particularly so to ESPI/DSPI. But the fringe quality is poor, and not too many fringe planes can be made to intersect the object. Further, a method that always gives true depth contours, because the object is intersected by planes normal to the direction of viewing, can be easily performed on ESPI/DSPI. The method employs an in-plane-sensitive configuration. The first frame is stored and the object is now tilted about an axis perpendicular to the plane containing the illumination beams. The object covered with true depth contours can be seen in real time. The contour interval is $\Delta h = (\lambda/2 \sin \theta)\beta$, where β is the tilt of the object and 2θ is the interbeam angle. Because of the tilt of the object, the fringes decorrelate very fast, and hence only coarse contour intervals can be realized [26].

7.11 SURFACE CHARACTERIZATION

There are various methods for surface roughness measurement. Of these the stylus profilometer is the most popular. It can give true information on the surface profile provided the width of the stylus tip is small compared to the lateral size of the surface irregularities.

It suffers from the following drawbacks: i) it may damage the surface under examination because of direct contact of the diamond stylus with the surface: ii) it usually measures the average height deviations, independently of the correlation length of the surface roughness, unless some special procedure is adopted; and iii) it gives one-dimensional information only, and therefore it is impractical for surveying an entire two-dimensional surface. Because of these drawbacks, several optical techniques to measure surface roughness have been examined. These include interferometric methods and light scattering methods based on Beckmann theory. Interferometric methods can be employed on surfaces whose height variations are smaller than the wavelength of the light used. After the advent of the laser, speckle methods were developed to measure surface roughness [27]. These methods can be grouped under i) the coherent light speckle contrast method, ii) polychromatic speckle pattern methods, and iii) the speckle pattern correlation method. The coherent light speckle contrast method is useful for measuring fine-scale surface roughness with rms value less than 0.25 μm. The polychromatic speckle pattern methods

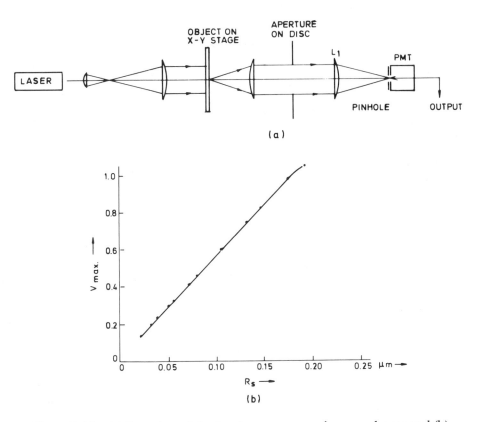

Figure 7.21 (a) Experimental setup to measure surface roughness and (b) plot of contrast V_{max} against rms roughness R_s. (From Ref. 27, used with permission.)

are employed for measuring moderate surface roughness of 0.2–5 μm. Speckle pattern correlation is applicable to comparatively rough surfaces of 1–30 μm. The coherent light speckle contrast method can be performed either at the image plane or at a defocused plane. We will describe only the method where the contrast of the speckle is measured at the image plane.

It is well known that a fully developed speckle pattern is formed when there are a large number of randomly distributed scatterers and when the random phase variations impressed on the scattered waves lie between $-\pi$ and π or beyond, The contrast V is defined as

$$V = \frac{(<I^2> - <I>^2)^{\frac{1}{2}}}{<I>} = \frac{<\Delta I^2>^{\frac{1}{2}}}{<I>}$$

where the angle brackets < . . . > denote the ensemble average. The contrast of a fully developed linearly polarized speckle pattern is 1. As the height variation of a rough surface is decreased, the phase changes are small and the contrast of the speckle pattern decreases. As an extreme case, a smooth surface does not produce any speckle pattern, or the contrast of the speckle pattern is 0. The contrast measurement can be done both at the image plane and at the diffraction plane. Certainly at the image plane the contrast will depend on the point-spread function of the imaging lens.

A beam from a He–Ne laser operating in a single mode is directed to illuminate a small area of the surface of an object placed on an x–y translation stage (fig. 7.21). An image of a small area is formed by lens L_1 at the sensitive surface of the photomultiplier tube. A tiny pinhole whose diameter is smaller than the smallest speckle in the image is placed in front of the photomultiplier. The point-spread function of the lens L_1 can be varied by bringing in any one of the apertures in a disc in the path of the beam.

The illumination beam scans the object surface when it is translated by the stage. The output from the photomultiplier is the current variation, which is proportional to the intensity falling on it. The analog current input, after amplification, is digitized by an ADC. The mean and mean square values are calculated from the sampled data and the average contrast plotted. It is found that the maximum contrast V_{max} varies linearly with rms roughness R_s up to 0.20 μm. This dependence is shown in Fig. 7.21 (b). This plot therefore serves as a calibration curve from which the roughness of any test surface is obtained by measuring the speckle contrast at the image plane.

7.12 WHITE LIGHT SPECKLE METROLOGY

Speckle photography is used to measure displacement of the object by measuring the positional shift of speckles; it does not matter how the speckles are generated. Laser speckle photography is restricted to limited range set by decorrelation. However, white light speckle photography can be applied over a larger range with attendant reduced sensitivity. Speckles for white light are generated artificially. One of the most common methods is to coat the surface of the object with a retroreflective paint that contains glass beads with diameters of 20 to 60 μm. The detailed structure of white light speckles is fixed on the surface of the object and does not exist in space as with laser speckles. White light photography has been used for measurement of rigid body translation, large displacements of engineering objects, vibrational studies, contouring, etc. It is a versatile tool for engineering metrology. It offers the following advantages: the experimental setup is simple, and stability requirements are not stringent; since the speckles are physically present on the surface, decorrelation effects due to excessive loading are minimal; film resolution is not a limiting factor; and the method can be applied to large engineering objects.

The essential requirement is a lens that images the whole object with-

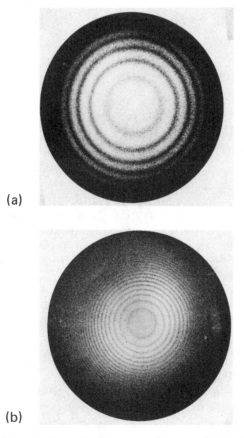

(a)

(b)

Figure 7.22   Out-of-plane displacement fringes obtained with white light for (a) $\in$ = 5 mm and (b) $\in$ = 20 mm.

out distortion and gives well resolved speckles. The sensitivity of this method is dependent on the size of the artificial speckles created on the object surface.

A white light specklegram can be interrogated with a narrow laser beam or can be Fourier filtered to extract the information. The methods applicable to laser speckles also apply to white light speckles. A white light specklegram can be recorded by placing multiple apertures at the lens plane. The filtering is then performed via one of the diffracted halos at the Fourier transform plane.

Again it is emphasized that white light speckle photography measures only the shift of speckles between exposures. An out-of-plane displacement or vibrational amplitude is measured because of the fictitious in-plane motion of the speckles. It is demonstrated for the case when the object is given a longitudinal translation $\in$ between the exposures. The in-plane shift Δr of the speckles is given by

$$\Delta r = \frac{\in r}{l}$$

where l is the distance between the object and the recording plane and r is the radial coordinate of a speckle as measured from the optical axis. The longitudinal displacement $\in$ is obtained from

$$\in = \frac{\Delta r}{r} l$$

The in-plane shift Δr can be measured by either pointwise filtering or Fourier filtering. With Fourier filtering, the white light specklegram is illuminated by a collimated beam, and a pinhole is placed on the axis at the distance l_f behind the specklegram. The bright fringes are formed whenever

$$\Delta r \sin \theta = m\lambda_s \quad \text{or} \quad \frac{\in r^2}{l l_f} = m\lambda_s$$

where λ_s is the wavelength of the light used for filtering. A set of circular fringes is formed. Figure 7.22 (a) shows a photograph obtained for $\in = 5$ mm and Fig. 7.22 (b) for $\in = 20$ mm. The sensitivity can be varied by using a divergent illumination wave instead of a collimated beam, as has been pointed out in the case of laser speckle photography.

REFERENCES

1. Goodman, J. W. (1968). *Introduction to Fourier Optics*, McGraw-Hill, New York.
2. Collier, R. J., Burckhardt, C. B., and Lin L. H. (1971). *Optical Holography*, Academic Press, New York.
3. Abramson, N. (1981). *The Making and Evaluation of Holograms*, Academic Press, New York.
4. Hariharan, P. (1984) *Optical Holography*, Cambridge Univ. Press, Cambridge.
5. Latta, J. N. (1971). Computer based analysis of holographic imagery and aberrations introduced due to a wavelength shift, *Appl. Opt., 10:* 609-618.
6. Amitai, Y. and Friesem, A. A. (1987). Recursive design technique for Fourier transform holographic lenses, *Opt. Eng., 26:* 1133–1139.
7. Vest, C. M. (1979). *Holographic Interferometry*, John Wiley, New York.
8. Schumann, W. and Dubas, M. (1979). *Holographic Interferometry*, Springer-Verlag, Berlin.
9 Sirohi, R. S. (1985). *Proceedings in Laser and Plasma Technology*, World Scientific Publishing Co., Singapore, pp. 484–501.
10. Stetson, K. A. (1969). A rigorous theory of the fringes of hologram interferometry, *Optik, 29:* 386–400.
11. Erf, R. K. ed. (1974). *Holographic Non-Destructive Testing*, Academic Press, New York.
12. Vest, C. M. (1982). Status and future of holographic non-destructive evaluation, *Proc. SPIE, 349:* 186–198.
13. Thalmann, R. and Dandliker, R. (1985). Holographic contouring using electronic phase measurement. *Opt. Eng., 24:* 930–935.
14. Erf, R. K., ed. (1978). *Speckle Metrology*. Academic Press, New York.
15. Ennos, A. E. (1978). Speckle interferometry, *Progress in Optics*, E. Wolf, ed., 16: 233–288, Elsevier, New York.
16. Jones, R., and Wyke, C. (1983) *Holographic and Speckle Interferometry*, Univ. Press, Cambridge
17. Dainty, J. C., ed. (1984). *Laser Speckle and Related Phenomena*, Springer-Verlag, Berlin.
18. Francon, M. (1979). *Laser Speckle and Applications in Optics*, Academic Press, New York.
19. Tiziani, H. J. (1972). A study of the use of laser speckles to measure small tilts of optically rough surfaces, *Opt. Commun., 5:* 271–276.
20. Gregory, D. A. (1976). Basic physical principles of defocused speckle photography: A tilt topology inspection technique. *Opt. Laser Technol., 8:* 201–213.
21. Jaisingh, G. K., and Chiang F. P. (1981). Contouring by laser speckle, *Appl. Opt., 20:* 3885–3887.
22. Duffy, D. E. (1972). Moiré gauging of in-plane displacement using double aperture imaging, *Appl. Opt., 11:* 1778-1781.
23. Sirohi, R. S. (1984). Speckle shear interferometry—A review, *J. Opt.* (India),*13:* 95–113

24. Joenathan, C., Mohanty, R. K., and Sirohi, R. S. (1984). Multiplexing in speckle shear interferometry, *Opt. Acta, 31:* 681–692.
25. Løkberg, O. J. and Slettemoen, G. A. (1987). Basic electronic speckle pattern interferometry. *Applied Optics and Optical Engineering X:* 455–504.
26. Ganesan, A. R. and Sirohi, R. S. (1988). A new method of contouring using Digital Speckle Pattern Interferometry, *Proc. SPIE, 954:* 327–332.
27. Asakura, T. (1978). Surface roughness measurement, *Speckle Metrology,* R. K. Erf., ed. Academic Press, New York.

CHAPTER 8

Optical Data Processing

Certain operations like spatial filtering, edge enhancement, correlation, convolution, matrix multiplication, addition, subtraction, etc. are in the domain of optical data processing [1-4]. The signal in optics is two dimensional. It can be a real-valued irradiance variation or a complex-valued amplitude variation. In coherent optical data processing the input to the processor is a two-dimensional transmittance variation that may be proportional to the irradiance variation or the amplitude variation of the signal. The processor is usually a Fourier transform processor; the Fourier transform of an input is physically available and can be manipulated to bring out desired features. The manipulation is performed by a variety of filters. It will be shown that the Fourier transform of the input is displayed by diffraction in the far-field, and the lens merely brings the transform to its back focal plane. The merits of an optical processor are its speed and its massive parallel processing capability.

8.1 FOURIER TRANSFORMATION

For Fourier transformation [5-9] we consider the arrangement shown in Fig. 8.1. The transparency is illuminated by a spherical wave emanating from a point source S. A lens is placed at a distance d from the transparency. We would like to find out the field distribution at a plane distant d_0 from the lens.

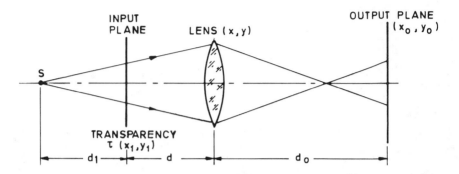

Figure 8.1   Calculation of the field at the output plane when the transparency is illuminated by a divergent spherical wave.

A lens is characterized by a transmittance function t(x, y) defined as [1]

$$t(x, y) = \exp\left[\pm \frac{ik}{2f}(x^2 + y^2)\right]$$
(8.1)

This is valid under parabolic approximation, and the constant phase factor
is ignored. The + sign applies to a negative lens and the − sign to a positive
lens. The input transparency of transmittance $\tau(x_1, y_1)$ is illuminated by a
spherical wave issuing from the point source S. The field at the plane of
transparency just in front of it is given, under paraxial approximation, as

$$u_f(x_1, y_1) = \frac{A}{d_1} \exp(ikd_1) \exp\left(\frac{ik}{2d_1}(x_1^2 + y_1^2)\right)$$

where A is the amplitude at a unit distance. The field just behind the
transparency is

$$u_b(x_1, y_1) = \frac{A}{d_1} \exp(ikd_1) \exp\left(\frac{ik}{2d_1}(x_1^2 + y_1^2)\right)\tau(x_1, y_1)$$
(8.2)

The field at the lens plane is obtained using Fresnel approximation. It is
given by

$$u_f(x, y) = \frac{\exp(ikd)}{i\lambda d} \iint u_b(x_1, y_1) \exp\left\{ \frac{ik}{2d} [(x_1 - x)^2 + (y_1, y)] \right\} dx_1\, dy_1$$

The field behind the positive lens is

$$u_b(x, y) = u_f(x, y)\exp\left[\frac{-ik}{2f} (x^2 + y^2) \right]$$

Again we use the Fresnel approximation to obtain the field at any point on the observation plane. This is obtained as

$$u(x_0, y_0) = \frac{\exp(ikd_0)}{i\lambda d_0} \iint u_b(x, y) \exp\left\{ \frac{ik}{2d_0} [(x_0 - x)^2 + (y_0 - y)^2] \right\} dx\, dy$$

Substituting for $u_b(x, y)$ and then $u_f(x, y)$ in this equation, the field $u(x_0, y_0)$ is given by

$$u(x_0, y_0) = \frac{A}{d_1} \frac{\exp[ik(d_1 + d + d_0)]}{(i\lambda)^2 dd_0} \iiiint \tau(x_1, y_1)$$

$$\exp\left[ik(x_1^2 + y_1^2)\left(\frac{1}{d_1} + \frac{1}{d} \right) \right]\exp\left[i\frac{k}{2d_0} (x_0^2 + y_0^2) \right]$$

$$\exp\left[i\frac{k}{2} (x^2 + y^2) \left(\frac{1}{d} + \frac{1}{d_0} - \frac{1}{f} \right) \right]$$

$$\exp\left[-ik\left\{ x\left(\frac{x_1}{d} + \frac{x_0}{d_0} \right) + y\left(\frac{y_1}{d} + \frac{y_0}{d_0} \right) \right\} \right] dx_1\, dy_1\, dx\, dy \qquad (8.3)$$

The integral on x, y can be evaluated if we assume that all the field diffracted by the transparency is received by the lens, that is, that there is no diffraction at the lens plane. The limits of integration are then taken from $-\infty$ to ∞. The field now is given as

$$u(x_0, y_0) = \frac{A}{d_1} \frac{\exp[ik(d_1 + d + d_0)]}{(i\lambda)^2 dd_0} \frac{i}{2\alpha}$$

$$\exp\left[i\frac{k}{2} (x_0^2 + y_0^2) \left(\frac{1}{d_0} - \frac{1}{2\alpha\lambda d_0^2} \right) \right]$$

$$\iint \tau(x_1, y_1) \exp\left[i\frac{k}{2} (x_1^2 + y_1^2) \left(\frac{1}{d_1} + \frac{1}{d} - \frac{1}{2\alpha\lambda d^2}\right)\right]$$

$$\exp\left(-2\pi i \frac{x_1 x_0 + y_1 y_0}{2\alpha\lambda^2 dd_0}\right) dx_1 \, dy_1 \qquad\qquad (8.4)$$

where $2\alpha\lambda = \dfrac{1}{d} + \dfrac{1}{d_0} - \dfrac{1}{f}$

The integral will represent the Fourier transform of $\tau(x_1, y_1)$, provided the quadratic phase term is 1. This happens when

$$\frac{1}{d_1} + \frac{1}{d} = \frac{1}{2\alpha\lambda d^2}$$

This condition can be simplified to

$$\frac{1}{d_1 + d} + \frac{1}{d_0} = \frac{1}{f}$$

Therefore the Fourier transform of the input is obtained at the observation plane, provided it is conjugate to the source plane.

Eq. (8.4) can now be written as

$$u(x_0, y_0) = \frac{A}{d_1} \frac{\exp[ik(d_1 + d + d_0)]}{i\lambda} \frac{d + d_1}{d_1 d_0}$$

$$\exp\left[i\frac{k}{2} (x_0^2 + y_0^2) \frac{f - d}{d_0 f + df - dd_0}\right]$$

$$\iint \tau(x_1, y_1) \exp\left[-2\frac{\pi}{\lambda} i \frac{(x_1 x_0 + y_1 y_0)f}{d_0 f + df - dd_0}\right] dx_1 \, dy_1 \qquad (8.5)$$

We now define the phase term as

$$\phi(x_0, y_0) = \frac{k(x_0^2 + y_0^2)(f - d)}{2(d_0 f + df - dd_0)}$$

and the scale factor as

$$S_f = \frac{d_0 f + df - dd_0}{f}$$

The phase term gives the phase of the transform, and the scale factor gives its size. Both are independent of d_1. They depend only on the lens focal length and the relative positions of input and output planes with respect to the lens plane. Eq. (8.5) represents the Fourier transform of the transparency $\tau(x_1, y_1)$ multiplied by a qaudratic phase factor. The condition for vanishing of the quadratic phase term inside the integral as given by eq. (8.5) is an imaging condition. Thus the Fourier transform of an input is available at a plane conjugate to the point source plane. Table 8.1 gives the various Fourier transform configurations along with their phase terms and scale factors.

Out of seven configurations to obtain the Fourier transform of an input, Fig. 8.2 shows only five; the remaining two can be realized by locating the input between the lens and the front focal plane, and at the lens plane itself. It can be seen that only two provide a Fourier transform of the input that is free from a quadratic phase term. This so called pure Fourier transform is obtained both in collimated illumination and in divergent illumination, provided the input is placed at the front focal plane of the lens [3]. Although divergent illumination requires fewer optical components, we shall restrict ourselves to the Fourier transform configuration in which the transparency is illuminated by a collimated beam.

8.2 THE FOURIER TRANSFORM PROCESSOR

Figure 8.3 shows a schematic of a Fourier transform processor. A beam from a He-Ne laser is expanded and collimated. Collimation may be checked by any one of the methods based on shear interferometry or Talbot interferometry. A pair of Fourier transform lenses L_1 and L_2 of focal lengths f_1 and f_2 is positioned with the focal planes coincident. The transparency $\tau(x_1, y_1)$ is placed at the input plane, which is the front focal plane of lens L_1. Its Fourier transform is displayed at the back focal plane of lens L_1. This plane is also called the frequency plane or filter plane. A filter is placed at this plane that modifies both the amplitude and the phase of the transform. The lens L_2 takes the Fourier transform of the field exiting at the filter plane, which is also its front focal plane. The Fourier transform is displayed at its back focal plane, which is called the output plane. The input and output planes are conjugate planes. The field distribution at the filter plane can be expressed as

$$T(f_x, f_y) = c \iint \tau(x_1, y_1) \exp[-2\pi i(f_x x_1 + f_y y_1)] \, dx_1 \, dy_1 \qquad (8.6)$$

where c is a complex constant, and spatial frequencies f_x and f_y are defined in terms of coordinates x_0 and y_0 at the filter plane as $f_x = x_0/\lambda f$, and $f_y =$

Table 8.1 Fourier Transform Configurations, Phase Terms, and Scale Factors

Source position (from lens)	Object position (relative to lens)	Output plane	Phase term (x_0, y_0)	Scale factor (S_f)
(a) ∞	f in front	f behind	—	f
∞	d in front	f behind	$\dfrac{k}{2f^2}(f-d)(x_0^2+y_0^2)$	f
∞	at the lens	f behind	$k(x_0^2+y_0^2)$	f
(b) ∞	behind lens distance d′ from the output plane	f behind	$\dfrac{k}{2d'}(x_0^2-y_0^2)$	d′
(c) $p=d_1+f$	f in front	$d_0=\dfrac{f(d_1+f)}{d_1}$	—	f
(d) $p=d_1+f$	z in front	$d_0=\dfrac{f(d_1+f)}{d_1}$	$\dfrac{k(f-z)(x_0^2+y_0^2)}{2[d_0(f-z)+fz]}$	$\dfrac{d_0(f-z)+fz}{f}$
(e) $p=d_1+f$	at the lens	$d_0=\dfrac{f(d_1+f)}{d_1}$	$\dfrac{k(x_0^2+y_0^2)}{2d_0}$	d_0

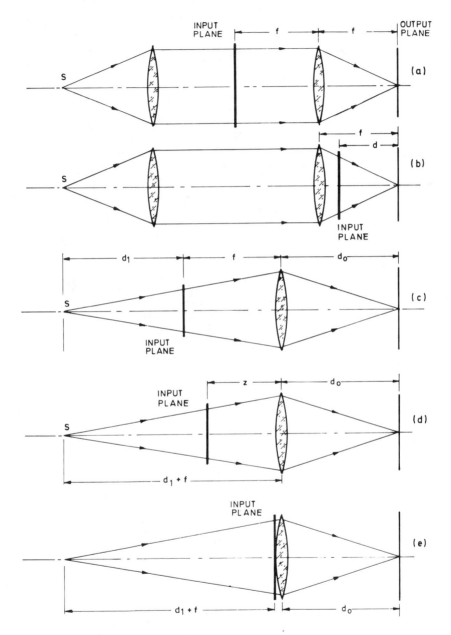

Figure 8.2 Various configurations to obtain a Fourier transform of an input.

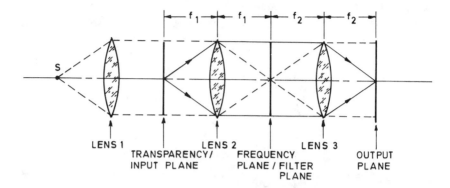

Figure 8.3   A Fourier transform processor.

$y_0/\lambda f$ and denoted in lines/mm. The frequency and space coordinates are linearly related. $T(f_x, f_y)$ is the Fourier transformer of $\tau(x_1, y_1)$. Let the filter be denoted by $H(f_x, f_y)$; this represents the transmittance of the filter. The field at the Fourier transform plane just after the filter is $T(f_x, f_y) H(f_x, f_y)$. The Fourier transform of this is displayed at the output plane, e.g. the back focal plane of the lens L_2. Thus

$$u(x_2, y_2) = c' \iint T(f_x, f_y) H(f_x, f_y) \exp[-2\pi i(f_x x_2 + f_y y_2)] \, df_x \, df_y$$

$$= c' \iint \tau(-\xi-x_2,-\eta-y_2) h(\xi,\eta) \, d\xi \, d\eta \tag{8.7}$$

where h (x, y) is the impulse response of the filter i.e.

$$H(f_x, f_y) = \iint h(x, y) \exp[-2\pi i(f_x x + f_y y)] \, dx \, dy.$$

If no filter is replaced at the frequency plane, that is if the spectrum of $\tau(x_1, y_1)$ is allowed without any modification, then the impulse response of the filter can be represented by a δ function. The field distribution at the output plane will be

$$u(x_2, y_2) = c''\tau(-x_2, -y_2)$$

The output consists of the input function referred to the reflected co-ordinates. The presence of the filter modifies the input. Indeed, the output is obtained as a convolution of the input function with the impulse response function of the filter, again referred to the reflected coordinates. A number of operations can be performed using variety of filters. Instead of describing the variations in detail, we discuss some applications under the classification of filter types.

8.3 FILTERS

Filters [8] may modify both amplitude and phase. In some applications only a portion of the spectrum is allowed for image formation. Depending on the way filters operate on the spectrum, they are classified as blocking filters, amplitude filters, phase filters, or complex filters.

8.3.1 Blocking Filters

Blocking filters [10, 11] are binary in nature; they remove a certain portion of the spectrum completely and allow the rest for image formation. They are of three types: low-pass filters, band-pass filters, and high-pass filters. Low-pass filters allow only low frequencies to pass through without attenuation. An example of a low-pass filter is a pin-hole placed on axis. If the diameter of the pinhole is 2d, it will pass frequencies up to $d/\lambda f$ when placed on axis at the back focal plane of the transforming lens. The frequency $d/\lambda f$ is the cut-off frequency. A band-pass filter allows a range of intermediate frequencies. An annular aperture is an example of a band-pass filter. If it is contained by a pair of diameters $2d_1$ and $2d_2$ ($d_2 > d_1$), the range of frequencies that is allowed is $d_1/\lambda f$ to $d_2/\lambda f$. A high-pass filter permits only high frequencies to pass through. An opaque dot of diameter 2d will allow all frequencies above $d/\lambda f$. Since the information about the edge of an object is contained in the higher frequencies, high-pass filtering is obviously used for edge enhancement. Figure 8.4 (a) gives an original image of a resolution test chart, and Fig. 8.4 (b) is its high-pass filtered image. Periodicity in a pattern can be removed by blocking filters. The removal of the raster scan in a television picture or of half-tone dots in a newspaper or periodical photograph can be also done by a blocking filter.

8.3.2 Amplitude Filters

Amplitude filters modify only the amplitudes of all frequencies in the spectrum. Interesting applications are in contrast enhancement, differentiation, integration, etc. For differentiation, the transmittance of the filter must increase linearly with frequency [12]. Taking the one-dimensional case, the filter may be represented as

$$H(f_x, f_y) = af_x$$

where a is a constant. The field distribution in the image now can be expressed as

$$u(x_2, y_2) = c'a \iint f_x\, T(f_x, f_y)\, \exp[-2\pi i(f_x x_2 + f_y y_2)]\, df_x\, df_y \qquad (8.8)$$

It is now shown that $u(x_2, y_2)$ is the partial derivature of $\tau(x_1, y_1)$ with respect to x_1. We write

$$\tau(x_1, y_1) = c_1 \iint T(f_x, f_y)\, \exp[2\pi i(f_x x_1 + f_y y_1)]\, df_x\, df_y$$

as the inverse Fourier transform of $T\,(f_x, f_y)$, which gives the input function. Differentiating this equation with respect to x_1 leads to

$$\frac{\partial \tau(x_1, y_1)}{\partial x_1} = 2\pi i c_1 \iint f_x\, T(f_x, f_y)\, \exp[2\pi i(f_x x_1 + f_y y_1)]\, df_x\, df_y \qquad (8.9)$$

The integrals in Eqs. (8.8) and (8.9) give the Fourier transforms of $f_x\, T(f_x, f_y)$; in one case in reflected coordinates. Therefore Eq. (8.9) gives the partial x derivative of the input function.

Similarly, it can be shown that an integrated value of the function is

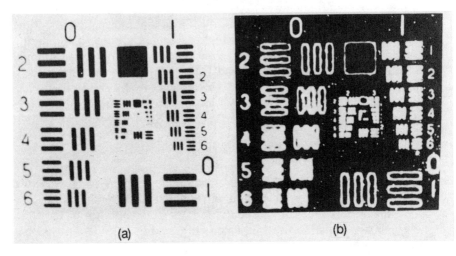

(a)   (b)

Figure 8.4 (a) Original object and (b) its high pass filtered image. (From Ref. 3, courtesy Springer-Verlag.)

obtained if the filter transmittance varies linearly as $1/f_x$. Due to singularity at $f_x = 0$, a small region around $f_x = 0$ is ascribed a constant value. It is possible to obtain a differentiated image of a phase object using a filter whose transmittance varies quadratically with f_x.

8.3.3 Phase Filters

Phase filters modify only the values of the phases of different frequency components either continuously or discretely. It is very difficult to obtain continuous variation of phase in the desired way. Using bleaching, and control over processing, a phase distribution may be obtained that is related linearly to the irradiance distribution recorded on the photographic plate. It is relatively easier to vary phase discretely. One of the well-known examples of a phase filter is the phase plate used to see phase objects with a phase contrast microscope. The phase plate retards or advances the phase of the zero order by $\pi/2$. In order to show how phase variations can be converted into irradiance variations to which detectors in the optical region respond, we consider a weak phase object characterized by the transmittance [13]

$$\exp[i\phi(x_1)] = 1 + i\phi(x_1) \tag{8.10}$$

The zero order is now retarded or advanced by $\pi/2$; therefore the field at the back focal plane of the objective of the microscope will be

$$= \delta(f_x) \exp\left(\pm \frac{i\pi}{2}\right) + i\ \Phi(f_x) \tag{8.11}$$

where Φ is the Fourier transform of $\phi(x)$.

Eq. (8.11) when further Fourier transformed gives

$$= \pm i + i\phi(x_2) = i[\pm 1 + \phi(x_2)]$$

The irradiance distribution in the image then becomes

$$i(x_2) = 1 \mp 2\phi(x_2)$$

Thus the phase variations have been converted into irradiance variations; in one case increased phase results in decreased brightness, while in the other case the opposite is true. The example of a weak phase object is a convenient choice; phase contrast microscopy is applicable to the general class of phase objects. The use of phase filters in optical correlators (instead of matched filters) has resulted in high irradiance and narrow correlation peaks.

8.3.4 Complex Filters

Complex filters are usually composite filters; one filter is for the control of amplitude and the other is for the phase of the spectrum. They are difficult to realize. Possible applications include deblurring a motion blurred picture, removal of defocus blur, and correction of aberrated images. As an example, electon micrographs are optically processed to remove the degrading influence of the spherical aberration of the electron lens. Let us consider the deblurring of a linear motion blurred picture. An object $o(x, y)$ is imaged through an ideal camera moving with a constant velocity normal to the direction of sight. We can attribute a blurring function $h(x, y)$ to this process; instead of a point, a point object is imaged as a line of length vT, where v is the uniform velocity and T is the exposure time. The irradiance distribution in the image will be

$$i(x_1, y_1) = \iint o(x, y)\, h(x - x_1, y - y_1)\, dx\, dy \tag{8.12}$$

where $o(x, y)$ is the irradiance distribution in the object. Assuming that the amplitude transmittance of the transparency is linearly related to the irradiance distribution, the Fourier transform is expressed as

$$I(f_x, f_y) = O(f_x, f_y)\, H(f_x, f_y) \tag{8.13}$$

where $O(f_x, f_y)$, $I(f_x, f_y)$, and $H(f_x, f_y)$ are the Fourier transforms of $o(x, y)$, $i(x, y)$, and $h(x, y)$, respectively. In fact we want to isolate $O(f_x, f_y)$, which on further Fourier transformation will give us the object distribution $o(x, y)$. Thus

$$O(f_x, f_y) = \frac{I(f_x, f_y)}{H(f_x, f_y)} \tag{8.14}$$

Thus the image restoration is performed by multiplying the spectrum of the blurred image with a filter function $1/H$. Thus a filter that is the inverse of H performs this deblurring operation. In practice, $h(x, y)$ may be known, or it may be obtained from a photograph. In the example of linear blur, the function h(x, y) is a slit of length vT, and its Fourier transform is

$$H(f_x) = vT \ sinc(vTf_x) \qquad (8.15)$$

The inverse of $H(f_x)$ has singularities at zeros of $H(f_x)$ and is impossible to realize as such. Instead we realize the inverse filter as the composite of two

$$\frac{1}{H} = \frac{H^*}{|H|^2} \qquad (8.16)$$

The photographic part $|H|^2$ is prepared separately, and H^* is prepared holographically and is known as a Vander Lugt filter. The composite filter, when placed at the Fourier transform plane, performs the deblurring operation. Various methods of deblurring have been described by Tsu-jiuchi and Stroke [14]. It may be noted that no restoration is possible at the zeros of H, where the information is completely lost. The deblurring works remarkably well.

Figures 8.5 (a) and (b) illustrates an example where a linear-motion-blurred picture is restored using a clipped inverse sine filter. In another example, shown in Fig. 8.6, a scene blurred by a complex motion, here an S type, has been restored. It may be mentioned that complex filters can be realized using synthetic holography [15].

(a)   (b)

Figure 8.5   (a) A linear blur image and (b) corrected image with an inverse filter. (From Ref. 3, courtesy Springer-Verlag.)

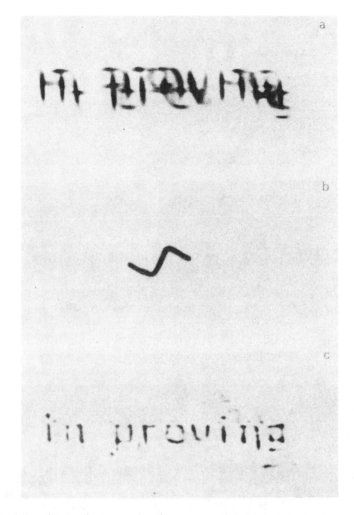

Figure 8.6 (a) A photograph of a smeared image, (b) the impulse response of the filter, and (c) image blur removal with an inverse holographic filter. (From Ref. 3, courtesy Springer-Verlag.)

8.3.5 Matched Filtering

Matched filtering [3] is used to detect the presence of a signal buried in white noise. A filter with an impulse response h(x) is called a matched filter if

$$h(x) = u^* (-x)$$

where u(x) is the signal in the input to be detected. The filter $H(f_x)$ is placed at the frequency plane. The correlation peaks appear at positions of the signal u(x) at the output plane. The matched filtering system is also called a frequency plane correlator.

8.4 THE JOINT FOURIER TRANSFORM CORRELATOR

Figure 8.7 shows a schematic of a joint Fourier transform correlator [3]. The inputs $g_1(x, y)$ and $g_2(x, y)$ to be correlated are placed at the input plane. The amplitude transmittance of the input is given by

$$u(x_1, y_1) = g_1(x_1 - a, y_1) + g_2(x_1 + a, y_1)$$

where 2a is the separation between the two inputs. The lens L_1 takes the Fourier transform, which is photographically recorded. Assuming a linear recording, the amplitude transmittance of the transparency is given by

$$t(x, y) = t_0 - \beta |u(x, y)|^2$$

where $x = \lambda f f_x$ and $y = \lambda f f_y$. This transparency is illuminated by a collimated beam, and a further Fourier transform is taken by the lens L_2. We obtain a field distribution at the output plane of

$$u(x_0, y_0) = \iint t(x, y) \exp\left[- \frac{2\pi i}{\lambda f} (xx_0 + yy_0)\right] dx\, dy$$

$$= t_0 \delta(x_0, y_0)$$

$$-\beta[g_1 \otimes g_1 + g_2 \otimes g_2 + g_1 \otimes g_2 * \delta(x_0 - 2a, y_0)$$

$$+ g_1 \otimes g_2 * \delta(x_0 + 2a, y_0)] \tag{8.17}$$

where $\otimes$ represents cross-correlation and $*$ convolution of g_1 and g_2. The output plane contains correlated outputs separated by 4a. If the correlation of g_1 and g_2 is to be isolated, certain conditions on their physical sizes and separation must be imposed.

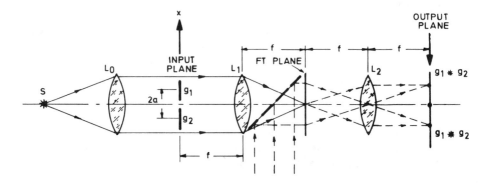

Figure 8.7   A schematic of a joint Fourier transform processor.

8.5 IMAGE SUBTRACTION

Image subtraction [16] is an interesting application of optical information processing. It is a process of difference detection between time-delayed images of a scene or between a master photograph of an object and its replicas. The applications are in urban growth studies, earth resource studies, meterology, automatic surveillance, forestry, strategic defense, pattern recognition, etc. Both optical and electronic techniques are employed for image subtraction. The optical techniques are relatively easy to implement. Several methods have been invented that can be grouped under the categories of coherent holographic or spatial filtering methods, periodic or random carrier modulation methods, and interferometric methods. In some methods the inputs are presented sequentially, while in others they are presented simultaneously. In these methods, a phase change of π is introduced between the inputs to be compared and common information content is discarded. The methods of image subtraction basically differ in the way in which the phase change of π is introduced. Most of the optical methods of image subtraction deal with photographic imagery such as photographs of the scenes recorded on the film transparency. The differences between the scenes can be obtained either as differences between the amplitudes or as differences between the irradiances of the two input scenes. It may be remarked that irradiance subtraction is more sensitive to difference discrimination than is amplitude subtraction.

Let $I_1(x, y)$ and $I_2(x, y)$ be the irradiances of the two scenes. The aim is to obtain $I_1(x, y) - I_2(x, y)$. An input transparency is prepared in the following way: The irradiance $I_1(x, y)$ is modulated by a grating. The irradiance $I_2(x, y)$ modulated by the same grating is properly registered but shifted by half the period (a phase shift of π). Assuming amplitude transmittance proportional to irradiance, the transmittance of this transparency is given by

$$t(x, y) = I_1(x, y)\left(1 + \cos\frac{2\pi x}{d}\right) + I_2(x, y)\left(1 + \cos\frac{2\pi(x + d/2)}{d}\right)$$

$$= I_1(x, y) + I_2(x, y) + [I_1(x, y) - I_2(x, y)]\cos\frac{2\pi x}{d} \qquad (8.18)$$

where d is the pitch of the sinusoidal grating. In practice a Ronchi grating may be used. This input is placed at the input plane of the Fourier transform processor. At the frequency plane are observed three orders; the direct order carries information about the sum of the irradiances, while the diffracted orders carry the information about the difference of the irradiances. An aperture placed at the Fourier transform plane isolates one of the diffracted orders, which is used for image formation. The field distribution at the output plane is proportional to $I_1(x, y) - I_2(x, y)$. The method requires the mechanical shift of the grating by half a period and proper registration of the two irradiance distributions. A method based on polarization and a custom polarizer filter was developed in which the mechanical shift of the grating is not involved [17]. A lightweight processor using holographic lenses is found to perform better.

8.6 THETA MODULATION

Selective regions of a scene can be filtered out by placing gratings at different orientations. Figure 8.8 (a) shows a simple example of a scene modulated by gratings, and its frequency spectrum is shown in Fig. 8.8 (b). Information about the different regions is now spatially separated and can be extracted using a blocking filter. This theta modulation [18] offers a high potential when applied to the scene on a pixel-by-pixel basis. Indeed, it is defined as a technique that converts each pixel of the original scene $u(x, y)$ into a grating cell whose grating angle θ is proportional to the amplitude transmittance of the pixel:

$$\theta(x, y) = ku(x, y)$$

where k is a constant that sets a limit on the grating angle, which can not exceed π, i.e.

$$k = \frac{\pi}{u(x, y)|_{max}}$$

where $u(x, y)|_{max}$ represents the maximum amplitude in the scene. The object in the theta modulated form consists of grating cells at all the pixels with orientations proportional to the amplitudes [4]. The encoding of the object may be done as follows: The scene using a CCD camera is stored in the frame store where the different grey levels are stored in different locations. For example, in a frame store of 256 x 256 x 6 bit, the object scene is stored in 256 x 256 pixels, and the 64 grey levels are stored accordingly. Using a computer-controlled interferometer, so that the fringe orientation can be varied proportionally to the grey level at that pixel, and computer-controlled x–y scanners, the 256 x 256 x 64 grey level picture is transformed to 256 x 256 grating cells with 64 orientations on a transparency. This transparency is placed at the input plane of a Fourier transform processor. At the frequency plane are obtained diffracted spots whose size is governed by the pixel dimensions. All those pixels having the same grating orientation (the same amplitude) will diffract in a particular direction, generating two diffraction orders. All the diffracted spots would lie on a circle due to the constant grating pitch over the whole picture. The orientation step is so chosen as to separate spatially these diffraction orders, and hence only a limited number of grey levels can be coded. The range, however, can be increased, by combining both frequency and theta modulation. In that case a range of grating orientations will generate diffraction orders lying on circles of different radii of curvature.

A binary filter, say a hole, is placed at the frequency plane. This allows only one of the orders of interest to pass through. Thus all those pixels that have contributed to this order will now be imaged at the output plane. In other words, sharp lines appear in the image that represent the contour of one amplitude value of $u(x, y)$

One can also choose different kinds of filters and achieve nonlinear processing as well. For example, the filter $h(\theta)$ may have exponential transmittance dependent on θ. The output $u_0(x, y)$ will be exponentially related to the original object $u(x, y)$. Thresholding may be achieved if we choose a filter whose transmittance $h(\theta)$ is

$$h(\theta) = 0 \qquad 0 < \theta < \theta_0$$

$$= 1 \qquad \theta_0 < \theta < \pi$$

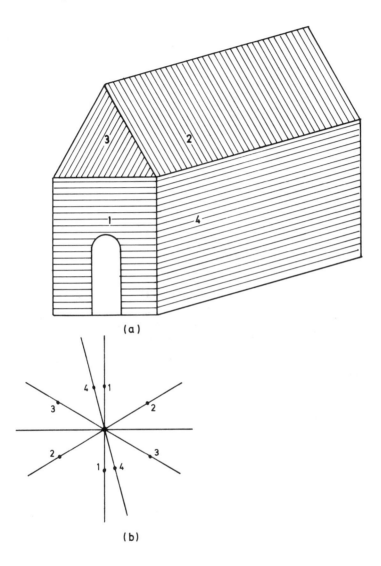

Figure 8.8 Theta modulation. (a) An object modulated by gratings of different constants and orientations and (b) a spectrum of the object.

This filter eliminates regions of amplitude transmittance below a certain value in the output image. But varying θ_0, the threshold level can be changed.

A disadvantage with coherent processing is that use of television displays and/or LED arrays as inputs to the processor are precluded. Further, it tends to be troubled by noise produced by dust, scratches, and other imperfections in the optical components. Also, when the output from such systems is presented, it is usually in the form of irradiances; the phase information is lost.

8.7 INCOHERENT PROCESSING

Incoherent processing [19,20] can be discussed under two different headings: diffraction-based processing systems and geometrical-optics-based systems.

We first consider an energy correlator. This was first discussed by Lohmann and Armitage for correlating the energy spectra of alphanumeric characters in a character recognition system [21]. Figure 8.9 shows the schematic arrangement of the correlator. A transparency $t_1(x_1, y_1)$ is coherently illuminated by a point source of quasi-monochromatic light of wavelength λ. Its Fourier transform is displayed at the frequency plane P_2, where a moving diffuser destroys the coherence of the Fourier transform and produces effectively a self-luminous irradiance distribution proportional to $|\tilde{t}_1(x_2/\lambda f, y_2/\lambda f)|^2$ where $\tilde{t}_1(x_2/\lambda f, y_2/\lambda f)$ is the Fourier transform of $t_1(x_1, y_1)$. Another transparency $t_2(x, y)$ is located at the plane P_3. The output at P_4 is the convolution of the energy spectrum of $t_1(x, y)$ with that of $t_2(x, y)$ and is expressed as

$$I_{OUT}(x, y) = \iint \left|\tilde{t}_1\left(\frac{\xi}{\lambda f}, \frac{\eta}{\lambda f}\right)\right|^2 \left|\tilde{t}_2\left(\frac{x - \xi}{\lambda f}, \frac{y - \eta}{\lambda f}\right)\right|^2 d\xi\, d\eta \quad (8.19)$$

If either $t_1(x, y)$ or $t_2(x, y)$ is inverted, the output is a cross-correlation of the two energy spectra. Normally t_1 and t_2 are real valued, and hence $|\tilde{t}_1|^2$ and $|\tilde{t}_2|^2$ are not only real but symmetric, and the convolution and correlation are equivalent. The success of the technique as an alphanumeric character recognition machine depends on the fact that the energy spectra associated with different characters differ sufficiently. In practice a transparency of known or required pattern is used for t_2, and the unknowns to be searched are located at t_1. The output of the detector placed on axis gives the cross-correlation coefficient defined as

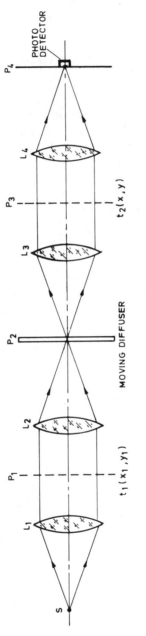

Figure 8.9   A schematic diagram of an energy correlator.

$$C_{m,n} = \iint |f_m\left(\frac{\xi}{\lambda f}, \frac{\eta}{\lambda f}\right)|^2 \, |f_n\left(\frac{\xi}{\lambda f}, \frac{\eta}{\lambda f}\right)|^2 \, d\xi \, d\eta \qquad (8.20)$$

It can be shown using the Schwartz inequality that the correlation coefficient is maximum when the character in t_1 matches with the desired character in t_2.

The advantage of the energy spectrum correlator lies in the position invariance of the unknown character. The disadvantage is that 180° symmetric objects like the forms of the numerals 6 and 9 have identical energy spectra. Further, some characters, like O and Q, have only subtle differences in energy spectra, so that their recognition is made difficult.

There are several incoherent processing systems that can be analyzed purely from a geometrical-optics approach. They can be grouped under two heads: those involving imaging, and shadow casting methods:

Let us consider the schematic shown in Fig. 8.10. This could be classified as a shadow casting method.

The transparency $t_1(x, y)$ is placed in the beam. The light passed through $t_1(x, y)$ is now transmitted through $t_2(x, y)$. The transparency $t_2(x, y)$ can be translated in its own plane. Both t_1 and t_2 are assumed to lie on the same plane.

The intensity transmitted at any point (x, y) will be given by

$$t_1(x, y) \, t_2(x + \xi, y + \eta)$$

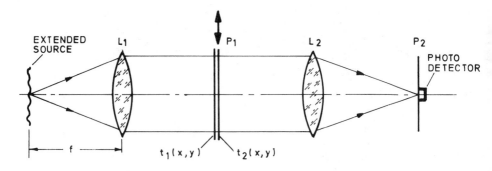

Figure 8.10   A geometrical-optics-based cross-correlator.

where ξ and η are the translations given to the transparency t_2. The lens gathers all the light transmitted by the transparency, and hence the output of the detector is proportional to

$$P(\xi, \eta) = \iint t_1(x, y)\, t_2(x + \xi, y + \eta)\, dx\, dy \qquad (8.21)$$

This represents the cross-correlation between t_1 and t_2. The output will be maximum when the characters in t_1 and t_2 are the same and $\xi = \eta = 0$. The two transparencies can be separated by a considerable distance, resulting in shadow casting of t_1 on t_2. Alternatively, the transparency t_1 may be imaged on t_2 using an additional lens.

REFERENCES

1. Goodman, J. W. (1968). *Introduction to Fourier Optics,* McGraw-Hill, New York.
2. Shulman, A. R. (1970). *Optical Data Processing,* John Wiley, New York.
3. Casasent, D., ed. (1978). *Optical Data Processing,* Springer-Verlag, Berlin.
4. Lee, S. H. (1981). *Optical Information Processing-Fundamentals,* Springer-Verlag, Berlin.
5. Bracewell, R. N. (1979). *The Fourier Transform and Its Applications,* 2nd ed., McGraw-Hill, New York.
6. Duffieux, P. M. (1983). *The Fourier Transform and Its Applications to Optics,* 2nd. ed., John Wiley, New York
7. Francon, M. (1979). *Optical Image Formation and Processing,* eng. ed. translation B. M. Jeffe, Academic Press, New York.
8. Gaskill, J. D. (1978). *Linear Systems, Fourier Transforms and Optics,* John Wiley, New York.
9. Steward, E. G. (1983). *Fourier Optics - An Introduction,* Ellis Horwood, Chichester, England.
10. Lipson, H. ed. (1972). *Optical Transforms,* Academic Press, London.
11. Cathey, W. T. (1974). *Optical Information Processing and Holography,* John Wiley, New York.
12. Sirohi, R. S., and Ram Mohan, V. (1977). Differentiation by spatial filtering, *Opt. Acta., 24:* 110–113.
13. Born, M., and Wolf, E. (1971). *Principles of Optics,* Pergamon Press, Oxford.
14. Tsujiuchi, J. and Stroke, G. W. (1971). Optical image deblurring methods, *Applications of Holography,* Barekette, E.S. et. al, ed., Plenum, pp. 259–308.
15. Lohmann, A. W. and Paris D. P. (1968). Computer generated spatial filters for Coherent Optical Data Processing, *Appl. Opt. 7:* 651–655
16. Ebersole, J. F. (1975). Optical image subtraction, *Opt. Eng.* 14: 436–447.
17. Dashiell, S. R., and Lohmann, A. W. (1973). Image subtraction by polarization-shifted periodic carrier, *Opt. Commun., 8:* 100–104.
18. Armitage, J. D., and Lohmann, A. W. (1965). Theta modulation in optics, *Appl.*

Opt. 4: 399–403.
19. Rogers, G. L. (1977). *Non-coherent Optical Processing,* John Wiley, New York.
20. Rhodes, W. T. and Sawchuk, A. A. (1981). Incoherent optical processing, *Optical Information Processing,* Lee, S. H., ed., Springer-Verlag, pp. 69–110
21. Armitage, J. D. and Lohmann, A. W. (1965). Character recognition by incoherent spatial filtering, *Appl. Opt., 4:* 461–467.

CHAPTER 9

Photoelasticity

9.1 INTRODUCTION

Photoelasticity is an experimental technique for obtaining stress distribution on a body subjected to an external load. The technique is carried out on scaled down models. It is based on the fact that certain materials when subjected to stress become birefringent; the refractive index of the material becomes direction dependent. The state of polarization of the wave propagating through a birefringent medium changes. The transmitted wave is analyzed for its state of polarization, and the stress distribution is then computed. The birefringence of the material is related to the stress through the stress–optic law, which forms the basis of photoelasticity.

9.2 THE STRESS–OPTIC LAW

The optical properties of a stressed material at any point can be represented by an index ellipsoid, the principal axes of which coincide with the principal axes of stress at that point. Let n_1, n_2, and n_3 be the principal refractive indices of the material at a point, and let σ_1, σ_2, and σ_3 be the principal stresses. The following equations express the relationship between principal refractive indices and principal stresses [1–7,12]:

$$n_1 - n_0 = a\sigma_1 + b(\sigma_2 + \sigma_3) \tag{9.1a}$$

$$n_2 - n_0 = a\sigma_2 + b(\sigma_1 + \sigma_3) \tag{9.1b}$$

$$n_3 - n_0 = a\sigma_3 + b(\sigma_1 + \sigma_2) \tag{9.1c}$$

where n_0 is the refractive index of the unstressed material and a and b are constants depending on the material. These equations can be written in the form

$$n_1 - n_2 = c(\sigma_1 - \sigma_2) \tag{9.2a}$$

$$n_1 - n_3 = c(\sigma_1 - \sigma_3) \tag{9.2b}$$

$$n_2 - n_3 = c(\sigma_2 - \sigma_3) \tag{9.2c}$$

where c = a − b is known as the stress–optic coefficient. This is the most general case of a triaxial state of stress.

In the case of a two-dimensional stress analysis, the above equations reduce to a single equation,

$$n_1 - n_2 = c(\sigma_1 - \sigma_2) \tag{9.3}$$

Consider a plane parallel plate of thickness d of a stressed material. A beam of linearly polarized light is incident normally on the plate. Inside the plate two linearly polarized waves that have orthogonal polarizations will propagate with different velocities. At the exit face they emerge with a relative phase difference δ which is given by

$$\delta = \frac{2\pi}{\lambda} d \, (n_1 - n_2) \tag{9.4}$$

where λ is the wavelength in a vacuum. These two waves will emerge in phase whenever

$$\delta = 2m\pi \qquad m = 0, \pm1, \pm2, \ldots \tag{9.5}$$

Thus

$$n_1 - n_2 = \frac{m\lambda}{d} \tag{9.6}$$

Substituting for $(n_1 - n_2)$ in Eq. (9.3), we obtain

$$\sigma_1 - \sigma_2 = m\frac{\lambda}{dc} = m \, \frac{F}{d} = mf \tag{9.7}$$

where $F = \lambda/c$ is a constant called the material fringe value and f is another constant called the model fringe value. F represents the principal stress difference necessary to produce unit change in the fringe order in a model of unit thickness. It may be seen that F is a constant for the material and light used, and is independent of material thickness, while f depends on material thickness as well. F is usually employed for comparing different photoelastic materials, while f is convenient for stress conversion in individual tests. As is obvious, the photoelasticity provides information regarding differences of principal stresses. Therefore, to affect separation of stresses, a knowledge of σ_1 or $\sigma_1 + \sigma_2$ is necessary. The measurement of $\sigma_1 - \sigma_2$ and the directions of principal stresses is done with the help of a polariscope.

9.3 THE POLARISCOPE

Two experimental arrangements, the plane polariscope and the circular polariscope, are used in photoelasticity.

9.3.1 The Plane Polariscope

The plane polariscope consists of a suitable light source and two polarizers, as shown in Fig. 9.1 (a). The first polarizer provides a linearly polarized beam that is incident on the model. The second polarizer, which

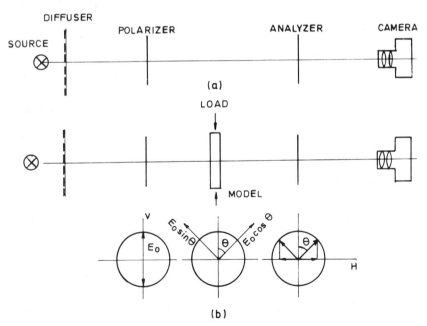

Figure 9.1   (a) A schematic of a plane polariscope and (b) the effect of a stressed model.

is often called an analyzer, resolves the component waves emerging from the model into one plane, so that the phase difference introduced by the model can be measured from the interference pattern.

9.3.2 Effects of a Stressed Plate in the Plane Polariscope

Let us consider that the beam emerging from the polarizer is vertically polarized, i.e. the $\bar{E}$ vector is vertical; see Fig. 9.1 (b). This is represented as $\bar{E} = \hat{j} E_0 e^{i(kx-wt)}$ at the entrance face of the model. Let the principal axis in the model be inclined by an angle θ with the vertical. The amplitudes of the waves transmitted along the principal axes at the exit face are given by

$$E_1 = E_0 \cos \theta \ e^{i(kx-wt)} \tag{9.8a}$$

$$E_2 = E_0 \sin \theta \ e^{i(kx-wt+\delta)} \tag{9.8b}$$

where the phase difference δ between the two waves is given by

$$\delta = \frac{2\pi}{\lambda}(n_1 - n_2)d \tag{9.9}$$

An analyzer is placed behind the model. Only that component of the field that is along the transmission direction of the analyzer is allowed through. Assuming the transmission axis of the analyzer to be perpendicular to that of the polarizer, the amplitude of light transmitted is given by

$$E_t = E_0 \cos \theta \sin \theta e^{i(kx-wt)} - E_0 \sin \theta \cos \theta \ e^{i(kx-wt+\delta)}$$

$$= \frac{E_0}{2} \sin 2\theta \ e^{i(kx-wt)} (1 - e^{i\delta})$$

The intensity transmitted is therefore given by

$$I_t = I_0 \sin^2 2\theta \sin^2 \frac{\delta}{2} \tag{9.10}$$

An equation similar to this can also be derived when the transmission axes of analyzer and polarizer are parallel.

The intensity in the transmitted beam is zero under the following two conditions:

Condition 1 (Isoclinics)

The transmitted intensity I_t is zero when

$$\theta = \frac{m\pi}{2} \qquad m = 0, 1, 2, \ldots \tag{9.11}$$

This is satisfied by all those points on the plate where the principal stress directions are parallel to the transmission axes of polarizer and analyzer. In general, these points lie on continuous curves forming a system of dark bands known as isoclinics. The angle θ made by the transmission axis of the polarizer or analyzer with the reference direction, usually taken vertical or horizontal, is called the parameter of the isoclinic concerned. As the cross polarizers are rotated, isoclinics of different parameters are obtained. It may be noted that the isoclinic is independent of the wavelength used and the magnitude of the load used for stressing the object.

Condition 2 (Isochromatics)

The intensity of transmitted light will be zero for any value of θ provided the phase difference δ is

$$\delta = 2m\pi \qquad m = 0, \pm1, \pm2, \ldots \tag{9.12a}$$

Alternately,

$$(n_1 - n_2)d = m\lambda \tag{9.12b}$$

Using the stress-optic law (Eq. 9.3), we have

$$m = \frac{\sigma_1 - \sigma_2}{f} \tag{9.13}$$

Therefore, in the stressed member, whereever $\sigma_1 - \sigma_2$ is such that m takes 0 or integer values, the transmitted intensity will be 0. In monochromatic light, it corresponds to continuous dark fringes. If white light is used, the appearance will be colored except in zero order. Each band corresponds to a certain value of $\sigma_1 - \sigma_2$ and hence is called isochromatic. The fringes corresponding to m = 0, 1, 2, . . . are called fringes of zero, first, second, etc. order respectively in monochromatic light. Further, the maximum in-plane shear stress τ_{max} is $\frac{1}{2}(\sigma_1 - \sigma_2)$. Thus

$$\tau_{max} = \frac{mF}{2d}$$

The isochromatics are, therefore, lines of equal in-plane maximum shear stress.

Distinction Between Isoclinics and Isochromatics

It is seen that two systems of fringes, isoclinics and isochromatics, appear when a stressed medium is placed in the field of a plane polariscope. In white light illumination, isoclinics appear black, while isochromatics appear colored, except in zero order. In monochromatic light, both of these are black, but isochromatics are more sharply defined. If the polarizers are rotated in the crossed position while keeping the load on the specimen constant, the isoclinic pattern varies while the isochromatic pattern remains unchanged. Conversely, when the load is varied for a fixed setting of a polarizer, the isochromatic pattern varies while the isoclinic pattern remains unchanged.

9.3.3 The Circular Polariscope

Both isoclinics and isochromatics appear together in the plane polariscope; the isoclinics are broad and dark. They usually hide the isochromatics. Sometimes it is desirable to eliminate the isoclinics altogether. This is achieved in the circular polariscope. A circular polariscope is obtained by inserting two quarter-wave plates between the polarizer and the analyzer with their principal axes at 45° with those of the polarizer. The axes of the polarizer and the analyzer may be parallel or crossed, and so also those of the quarter-wave plates. There are, therefore, four ways in which a circular polariscope can be set up. Figure 9.2 shows a schematic diagram of a circular polariscope, where we assume that the polarizer and the analyzer are crossed, with the polarizer's transmission axis vertical, and the fast and slow axes of quarter-wave plates QP_1 and QP_2 orthogonal to each other and inclined at an angle of 45° with the transmission direction of the polarizer. The propagation of a wave through the polariscope will now be described.

A beam of collimated light is incident on the polarizer. The wave amplitude just after the polarizer can be written as

$$E_v = E_0 e^{iwt}$$

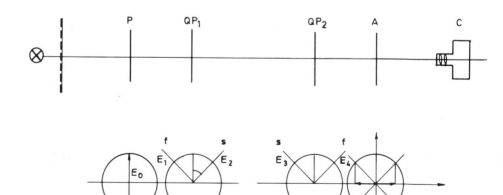

Figure 9.2   A schematic of a dark field circular polariscope.

where w is the frequency of the wave and E_v represents the instantaneous E field in the wave, the $\bar{E}$ vector being vertical. The wave propagates to the first quarter-wave plate. Therefore the wave incident on the plate is given by this equation except for the constant plase that the wave acquires because of propagation from polarizer to quarter-wave plate. This constant phase and others to be acquired later are ignored. In the quarter-wave plate, the wave splits into two waves that acquire a phase change of $\pi/2$ just before emergence. Let the fields transmitted along the fast and slow axes of the quarter-wave plate QP_1 be E_1 and E_2 respectively; then

$$E_1 = \frac{E_0}{\sqrt{2}} e^{iwt} \tag{9.14a}$$

$$E_2 = \frac{E_0}{\sqrt{2}} e^{i(wt+\pi/2)} \tag{9.14b}$$

These fields are incident on the quarter-wave plane QP_2 that is crossed to QP_1. Therefore the field E_1 will acquire a phase change of $\pi/2$ as it propagates along the slow axis of QP_1. The fields E_3 and E_4 transmitted by the quarter-wave plate QP_2 along its slow and fast axes are

$$E_3 = \frac{E_0}{\sqrt{2}} e^{i(wt+\pi/2)} \tag{9.15a}$$

$$E_4 = \frac{E_0}{\sqrt{2}} e^{i(wt+\pi/2)} \tag{9.15b}$$

The field transmitted by the analyzer is the vectorial summation of the components parallel to its transmission axis. For the analyzer with transmission axis horizontal (crossed with the polarizer), the field transmitted through the analyzer is

$$E_t = \frac{E_3}{\sqrt{2}} - \frac{E_4}{\sqrt{2}} = 0 \tag{9.16}$$

Therefore the field of view appears dark. If the analyzer's transmission axis is parallel to that of the polarizer, however, the field of view will be bright. Table 9.1 gives the details of all four arrangements.

9.3.4 Intensity Distribution with the Model

The model or the stressed member is placed between the quarter-wave plates QP_1 and QP_2 as shown in Fig. 9.3. It is illuminated by the beam transmitted by a polarizer and QP_1 in combination. The field in the beam,

Table 9.1 Characteristics of Four Arrangements of a Circular
Polariscope

S. No.	Arrange-ment	Quarter-wave plates	Polarizer and analyzer	Field of view
1	A	Crossed	Crossed	Dark
2	B	Crossed	Parallel	Bright
3	C	Parallel	Crossed	Bright
4	D	Parallel	Parallel	Dark

in general, can be expressed with respect to the reference axes x and y as

$$E_i = E_x \hat{\imath} + E_y \hat{\jmath} \tag{9.17}$$

This in general represents an elliptically polarized beam. Let the principal
axes make an angle α with the x axis. The field transmitted along the σ_1 and
σ_2 axes will be

$$E_{m1} = E_1 e^{i\delta_1} \tag{9.18a}$$

$$E_{m2} = E_2 e^{i\delta_2} \tag{9.18b}$$

where δ_1 and δ_2 are the absolute phases introduced by the model and $\delta = \delta_1 - \delta_2$ is the phase difference. Futher, $E_1 = E_x \cos \alpha + E_y \sin \alpha$ and $E_2 = E_y \cos \alpha - E_x \sin \alpha$ are the fields incident on the model along the principal
axes.

The transmitted fields as referred to the reference axes are

$$E_x' = E_{m1} \cos \alpha - E_{m2} \sin \alpha \tag{9.19a}$$

$$E_y' = E_{m2} \cos \alpha + E_{m1} \sin \alpha \tag{9.19b}$$

Substituting for E_{m1} and E_{m2} in these equations in terms of E_x and E_y, we
obtain

$$E_x' = e^{i\delta_2}[e^{i\delta} \cos \alpha(E_x \cos \alpha + E_y \sin \alpha) - (E_y \cos \alpha - E_x \sin \alpha)\sin \alpha]$$

$$= \frac{e^{i\delta_2}}{2} [E_x(e^{i\delta} + 1) + (e^{i\delta} - 1)(E_x \cos 2\alpha + E_y \sin 2\alpha)] \tag{9.20a}$$

Similarly,

$$E_y' = \frac{e^{i\delta_2}}{2} [E_y(e^{i\delta} + 1) - (e^{i\delta} - 1)(E_y \cos 2\alpha - E_x \sin 2\alpha)] \tag{9.20b}$$

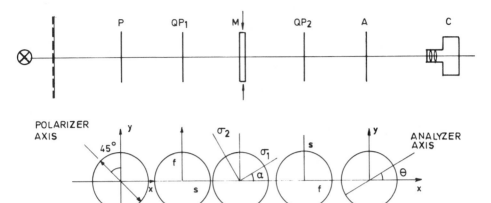

Figure 9.3 A circular polariscope with a model placed between quarter-wave plates.

This field is now incident on the quarter-wave plate. We, however, assume a phase plate that introduces absolute phases of ϕ_1 and ϕ_2 and the relative phase of $\phi = \phi_1 - \phi_2$. The fast axis is assumed to be along the x axis. Thus the fields transmitted by the phase plate are

$$E_x'' = E_x' e^{i\phi_1} \tag{9.21a}$$

$$E_y'' = E_y' e^{i\phi_2} \tag{9.21b}$$

The analyzer's transmission axis makes an angle θ with the x axis. The field transmitted by the analyzer will be the sum of the components along the transmission axis,

$$E_\theta = E_x'' \cos\theta + E_y'' \sin\theta$$

Substituting for E_x' and E_y', we obtain

$$E_\theta = \frac{e^{i\delta_2}e^{i\phi_2}}{2} [\{E_x(e^{i\delta} + 1) + (e^{i\delta} - 1)(E_x \cos 2\alpha + E_y \sin 2\alpha)\}e^{i\phi}\cos\theta$$

$$+ \{E_y(e^{i\delta} + 1) - (e^{i\delta} - 1)(E_y\cos 2\alpha - E_x \sin 2\alpha)\}\sin\theta] \tag{9.22}$$

We will now study this for the case of a circular polariscope where $\phi = \pi/2$ and $\theta = 45°$. Thus

$$E_{45} = \frac{e^{i(\delta_2+\phi_2)}}{2\sqrt{2}} \; [(E_y + iE_x)(e^{i\delta} + 1)$$

$$+ \; (e^{i\delta} - 1)\{(\cos 2\alpha - i \sin 2\alpha)(E_x i - E_y)\}] \tag{9.23}$$

When the circular polariscope is used in a dark field setting, both the polarizer and the analyzer are crossed. Therefore

$$-E_y = iE_x \tag{9.24}$$

Substituting for E_y in the above equation gives

$$E_{45} = \frac{e^{i(\delta_2+\phi_2)}}{\sqrt{2}} \; E_x(e^{i\delta} - 1)(i \cos 2\alpha + \sin 2\alpha) \tag{9.25}$$

The transmitted intensity I is

$$I = E_{45} \, E_{45}{}^* = E_x{}^2 \, (1 - \cos \delta)$$

$$= 2E_x{}^2 \sin^2 \frac{\delta}{2} = I_0 \sin^2 \frac{\delta}{2} \tag{9.26}$$

This equation shows that when the circular polariscope is set for a dark field, the transmitted intensity is zero when

$$\sin^2 \frac{\delta}{2} = 0 \quad \text{or} \quad \delta = 2m\pi \quad m = 0, 1, 2, \ldots \tag{9.27}$$

This condition results in the appearance of an isochromatic fringe pattern.

On the other hand, when the polariscope is set for a bright field, the transmitted intensity is given by

$$I \; \alpha \; \cos^2 \frac{\delta}{2}$$

Thus the extinction is obtained when

$$\cos^2 \frac{\delta}{2} = 0 \quad \text{or} \quad \delta = \frac{(2m + 1)\pi}{2} \quad m = 0, 1, 2, \ldots \tag{9.28}$$

This corresponds to half-order isochromatics. Thus both full-order and half-order isochromatics are obtained by the use of a circular polariscope in dark and bright field settings. However, the accuracy achieved in determining the values of the stresses depends considerably on the accuracy with which fringe orders are measured at various points on the photoelastic model. It is, therefore, imperative that the fringe orders be determined with higher accuracy than half. Various compensation techniques to determine fractional fringe orders are available [1,2,7].

9.3.5 Compensation Techniques

The Babinet-Soleil Method

A Soleil compensator [1,2,7] is introduced in the polariscope; it is kept either before or after the model. It introduces a constant phase difference (Chap. 2). The optical effect caused by the insertion of the compensator is shown in Fig. 9.4. When the compensator is placed in the field of the polariscope with its axis parallel to the σ_2 direction, the optical response of

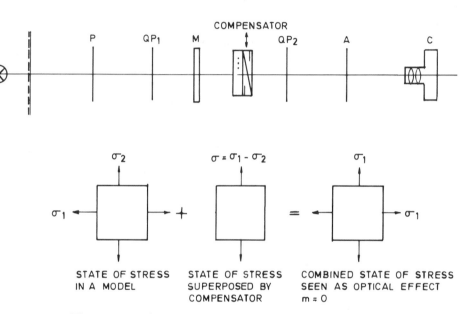

Figure 9.4   The operation of Babinet-Soleil compensator.

the combined system of model and compensator can be varied by controlling the effective birefringence of the compensator. If the birefringence of the compensator is set equal to that of the model (mF) but of opposite sign, the fringe order becomes zero. The compensator has to be oriented along either one of the principal stress directions and adjusted to cancel the optical response of the model.

In practice, at a given point in the model, where the fringe order is to be determined accurately, the isoclinic parameter is first determined by observing the minor principal stress σ_2 direction, and then the compensator is adjusted to cancel out the relative retardation of the model. The reading of the micrometer on the compensator is proportional to the fringe order m at the point of observation.

The Tardy Method

The Tardy Method [2,7] is commonly employed to measure the fringe order at any point of the model, since no additional equipment is necessary and the polarizer and analyzer themselves serve as compensating devices. This is a point-by-point method. It can be best understood by considering the effects of a stressed model in a circular polariscope with the local stress direction σ_1 aligned to the transmission axis of the polarizer. The analyzer position is arbitrary. The easiest procedure is to start with a correctly set plane polariscope. The polarizer is then aligned so that the isoclinic falls directly across the desired point. Then the quarter-wave plate QP_1 is inserted at 45° to the polarizer. This procedure is equivalent to setting the principal stress direction at 45° with respect to the fast axis of the QP_1 and illuminating the model by a circularly polarized beam. Thus the transmitted field E_θ can be obtained from Eq. (9.22) by setting $\alpha = 45°$ and $E_y = -iE_x$. Then

$$E_\theta = \frac{e^{i\delta_2} - e^{i\phi_2}}{2} E_x [\{(e^{i\delta} + 1) - i(e^{i\delta} - 1)\}i \cos \theta$$

$$+ \{(e^{i\delta} - 1) - i(e^{i\delta} + 1)\} \sin \theta]$$

$$= \frac{e^{i\delta_2} - e^{i\phi_2}}{2} E_x [\{(e^{i\delta} - 1) + i(e^{i\delta} + 1)\} \cos \theta$$

$$+ \{(e^{i\delta} - 1) - i(e^{i\delta} + 1)\} \sin \theta] \tag{9.29}$$

The transmitted intensity $I_0 = E_\theta E_\theta^*$ is given by

$$I_\theta = E_x^2 [1 + \sin(\delta - 2\theta] \tag{9.30}$$

At the dark field setting, $\theta_0 = 45°$. If a fringe does not pass through the desired point for this position of the analyzer, the partial fringe order can be obtained by rotating the analyzer an additional angle $\gamma = \theta - 45°$ until one of the adjacent fringes has moved to the desired point. The condition for dark fringes is that

$$\sin(\delta - 2\theta) = -1$$

or

$$\delta - 2\theta = (2n - \tfrac{1}{2})\pi \qquad n = 0, 1, 2, \ldots$$

or

$$\delta - 2\gamma = 2n\pi \qquad (9.31)$$

In terms of isochromatic fringe value, the phase difference δ introduced by the model is $\delta = 2m\pi$. Therefore the extinction occurs when

$$m = n \pm \frac{\gamma}{\pi} \qquad (9.32)$$

where γ is in radians. The γ term is additive when the next lower fringe is moved to pass through the point of interest, and it is subtractive when the next higher fringe is brought into coincidence with the desired point.

Other Methods of Compensation

The Sénarmont method is based on rotating the quarter-wave plate in the standard polariscope to obtain extinction at the desired point on the model under study. The photometric method is used both with plane and with circular polariscopes.

9.3.6 Errors in Polariscopes

The main source of errors in a polariscope is the quarter-wave plate. This plate introduces a phase change of $\pi/2$ only for a particular wavelength. Thus if a quarter-wave plate is not properly matched to the source wavelength, the intensity of the transmitted light in a dark field circular polariscope is given by

$$I' = I_0(1 - \sin^2 2\alpha \sin^2 \varepsilon) \sin^2 \frac{\delta}{2} \qquad (9.33)$$

where it is assumed that the phase difference introduced by the quarter-wave plate is $\phi = \pi/2+\varepsilon$ instead of $\pi/2$. It is seen that the dark fringes occur whenever $\sin^2(\delta/2) = 0$, giving the well-known isochromatics. The maxima of intensity occur wherever $\sin(\delta/2) = 1$, but the intensity of maxima is $(1 - \sin^2 2\alpha \sin^2\varepsilon)I_0$ instead of I_0. The reduction in intensity is zero at places where $\sin 2\alpha = 0$, i.e., where the principal stress directions coincide with the axes of the quarter-wave plates. This occurs only along the 45° isoclinic. For other isoclinic angles, there is a reduction in intensity, the reduction being largest when $2\alpha = 90°$, i.e., when the principal stresses coincide with the transmission directions of polarizer and analyzer. The intensity variation across the isochromatics may result in errors of measurement. The other sources of error are the variation of intensity across the field of the polariscope due to variation in transmittance, and averaging through the thickness of the stressed model in a diffuse field polariscope.

9.4 POLARISCOPE CONSTRUCTION

There are three types of polariscopes generally used for observations on photoelastic models. They are diffuse light polariscopes, lens polariscopes, and reflection polariscopes.

9.4.1 The Diffuse Light Polariscope

The diffuse light polariscope is one of the simplest and least expensive polariscopes available, yet it can be employed to produce quality photo-elastic results. The polariscope field can be made as large as possible, since it depends on the size of the available linear polarizers and quarter-wave plates. The Schematic of a diffuse light polariscope is shown in Fig. 9.3. It has a provision for illumination by either a white light source or a mono-chromatic light source. The polariscope can be set up for both dark field and bright field observations. Fractional fringe orders of isochromatics can be obtained using either the Babinet-Soleil compensator or the Tardy method. Another advantage of this polariscope is that the surface finish of the model need not be very good. Further, it is easier to operate than a lens polariscope.

9.4.2 The Lens Polariscope

The lens polariscope employs a collimated beam for the illumination of the model. The polarizer, analyzer and quarter-wave plates are placed in the collimated beam as shown in Fig. 9.5. A point source of light, either white or monochromatic, is placed at the focal point of the lens L_1. The lens L_2focuses the beam, which is then accepted by the camera. The camera lens images the model on the film plane.

The lens polariscope is required for fringe sharpening and fringe mul-tiplication. Further, when oblique incidence is used for say separation of stresses, the lens polariscope is used. The field dimensions are usually governed by collimating optics and are much smaller than those of the diffuse light polariscope.

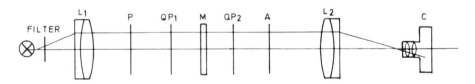

Figure 9.5   A schematic of a lens polariscope.

9.4.3 The Reflection Polariscope

The reflection polariscope consists of two assemblies of polarizer and quarter-wave plate attached to a common frame. These assemblies are connected mechanically so that they rotate in unison. A schematic diagram of a reflection polariscope is shown in Fig. 9.6. Instead of a photoelastic model, a thin layer of birefringent material is bonded to the surface of the

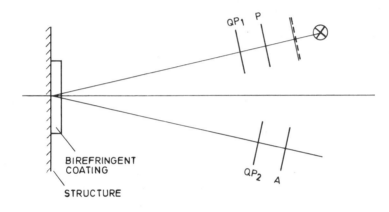

Figure 9.6   A schematic of a reflection polariscope.

member with a reflective backing. The deformation of the member under load is transferred to the coating, and the resultant photoelastic effect is studied with a reflection polariscope. It is assumed that the strains in the member and the coating are the same. Therefore the thickness of the coating is kept as small as possible.

The reflection polariscope can be equipped with a Babinet-Soleil compensator and other accessories to make it more versatile.

9.5 PHOTOELASTIC MATERIALS

One of the most important factors in photoelastic analysis is the selection of the proper material to be used for the model. An ideal photoelastic material does not exist, and the experimenter has to make a selection from the available materials that suits his purpose best.

9.5.1 The Requirements for a Good Photoelastic Material

The requirements for a good photoelastic material may be summarized as follows:

It should be transparent and free from any initial birefringence in the unstressed state.

It should exhibit linear characteristics with respect to stress-strain properties, stress-fringe order properties, and strain-fringe order properties.

It should possess both mechanical and optical isotropy and homogeneity.

It should exhibit a high modulus of elasticity and a high ultimate strength.

It should not creep excessively.

Its sensitivity should not change markedly with small variations in temperature.

It should be easy to machine without chipping or inducing machining stresses.

It should not exhibit time edge defects. The tendency of photoelastic materials to develop fringes along the free edges exposed to the environment for a long time is called the time edge effect. This effect may be induced by the humidity in the air.

Some of the materials that exhibit the photoelastic effect are glasses, celluloid, gelatin, rubber, cellulose, nitrates, vinyls, several phenolformaldehydes, Columbia Resin CR-39, perspex, Plexiglas, epoxy resins, etc. Of these the most commonly used materials are catalin-61-893, castolite, Columbia Resin CR-39., epoxy resin, and urethane rubber.

9.5.2 Choice of Model Materials

The properties desirable in a photoelastic material differ according to the nature of the data required and the experimental procedure to be employed. For isochromatic observation, the optical sensitivity should be as high as possible, i.e., the material fringe value should be as low as possible. To retain shape and form of the model under load, especially in the stress-freezing method, the elastic modulus should be as high as possible. The photoelastic materials are compared using a quantity Q known as the figure of merit. This is defined as the ratio of the elastic modulus to the material fringe value, i.e., $Q = E/F$. The figure of merit should be as high as possible. It should remain constant with load and also with time under load. Table 9.2 gives the properties of photoelastic materials at room temperature. Considerable variation in the values of the properties of photoelastic materials is possible between different samples of the same material. Therefore the values to be used in an investigation should be obtained from calibration tests on the mateial.

Table 9.2   Properties of Some Photoelastic Materials

Material	C (m^2/N) $\times 10^{-10}$	F (N/m fr) $(=\lambda/C) \times 10^3$	E (N/m^2) $\times 10^7$	Q (fr/m) $\times 10^2$
Glass	-196	-300	7000	2330
Plexiglas	-453	-130	300	230
Perspex	-471	-125	280	220
CR-39	34	17.2	200	1160
Celluloid	20–196	30–300	200–250	670–680
Epoxy	57	10.4	317	3050
Makrolon	78	7.5	260	3500
Hysol 4485	0.3	0.17	0.3	180
Gelatin	0.6	0.09	0.03	33

9.5.3 Fringe Multiplication and Fringe Sharpening

In the region of low fringe order or low fringe gradient, the fringes are sparsely distributed. The isochromatic pattern may be inadequate to allow accurate evaluation of the stresses. The number density can be increased by multipassing the model; the technique is similar to phase amplification in hologram interferometry and multipassed interferometers.

The model is placed in an index matching fluid between two partially transmitting mirrors and is illuminated by a collimated beam. The multiply reflected beams pass through the model a number of times. In order to isolate the multiply reflected beams, one of the mirrors is slightly tilted. Thus the multiply reflected beams are angularly separated, and hence they can be filtered out by placing a pinhole at the focal plane of a lens L_1 as shown in Fig. 9.7. The effect of the inclined mirror is shown alongside. All those rays that pass through the model without any reflection will be focused by the lens L_2 on the axis; filtering by placing a pinhole on axis will yield an isochromatic pattern having the intensity distribution $\sin^2(\delta/2)$. The rays, which suffer two reflections, one on each mirror, will pass through the specimen three times. On filtering, these rays generate an isochromatic fringe pattern with intensity distribution proportional to $\sin^2(3\delta/2)$; a multiplication of 3 has taken place. It may be noticed that the multiply reflected rays do not travel in the same region of the model and hence an averaging takes place. If we consider the rays with four reflections, the model is traversed 5 times, and hence a multiplication of 5 occurs. More elaborate calculations will indicate that higher multiplicity images of the model have very low intensities; good results can be obtained up to a multiciplicity of 5 tor 7 by optimizing the reflectivity and

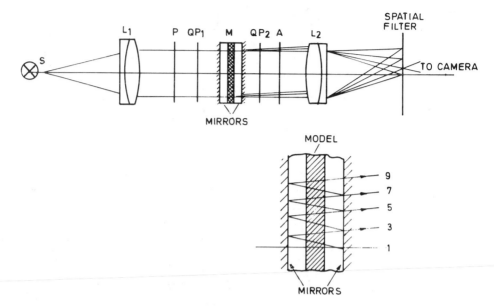

Figure 9.7   Fringe multiplication to enhance sensitivity.

transmissivity of the mirrors. If the tilt angle is about 0.005 rad, errors introduced by averaging are very small. Spatial isolation of multiplied images requires a long focal length of the lens L_2.

The intensity distribution in an isochromatic pattern obtained from a dark field polariscope is sinusoidal. It is difficult to locate the center of the fringes accurately. This is valid for the bright field case as well. The accuracy with which the centers of the fringes can be determined, and consequently the accuracy of measurement, can be increased by sharpening the fringes. The model is kept in an index matching fluid between two parallel mirrors. The beam bounces back and forth, and a large number of multiply reflected beams participate in image formation. Unlike in the Fabry-Perot interferometer, here the intensities of the multiply reflected beams fall off much more rapidly. Therefore only a few beams really contribute to image formation. The fringes are sharpened but are not as narrow as expected from the Fabry-Perot interferometer.

It may be noted that fringe multiplication enhances the sensitivity while fringe sharpening increases the accuracy of measurement. Both these techniques are attributed to Post [8].

9.5.4 Photoelastic Data

The fundamental photoelastic data are the isochromatic fringe order m and the isoclinic parameter θ. The observation can be made either point-wise or as a whole field with predecided mesh pattern overlaid on the model surface.

For the determination of fractional fringe order at each mesh point, the Tardy method or the Babinet-Soleil compensator method can be used. For improving the sensitivity of the isochromatic measurements for the whole field of the model, the fringe multiplication technique followed by the interpolation technique are adopted to arrive at the fringe values at each mesh point. The material fringe value for a given material is normally determined by testing an identically cast calibration specimen.

Since a three-dimensional photoelastic model is reduced to a number of two-dimensional slices, the nature and type of photoelastic data obtained are basically the same. The interpretation of the observations, however, will be entirely different for two and three dimensional photoelastic models. We first consider two-dimensional models.

9.6 THE TWO-DIMENSIONAL PROBLEM

Transmission photoelasticity as applied to two-dimensional models deals with an elastic continuum having plane stress distribution. Sometimes a three-dimensional model having plane strain type behavior can also be simplified to a plane stress problem, for example in pipes or tunnels. The significant aspect of the plane stress behavior of a model is that the stress σ_3 normal to its faces is zero. This means that both the principal stresses σ_1 and σ_2 are contained in the plane of the model.

9.6.1 Stresses in a Model

Difference of principal stresses at any point in a stressed photoelastic material is given in terms of fringe order and material fringe value as

$$\sigma_1 - \sigma_2 = \frac{mF}{d} \tag{9.7}$$

The maximum in plane shear stress τ_{max} is given by

$$\tau_{max} = \frac{1}{2}(\sigma_1 - \sigma_2) = \frac{mF}{2d} = \frac{mf}{2} \tag{9.34}$$

The shear stress at any point can be expressed as

$$\tau = \frac{1}{2} (\sigma_1 - \sigma_2) \sin \theta = \frac{mF}{2d} \sin \theta \qquad (9.35)$$

where θ is the isoclinic parameter at that point, indicating the principal stress direction with respect to the reference axis.

In any model, the general state of stress can be defined in terms of σ_1, σ_2 and τ_{xy}. The photoelastic observations however, give us only $\sigma_1 - \sigma_2$ and the orientations of principal stresses. For determining the particular values of σ_1 and σ_2; further analytical or experimental methods must be employed. For particular points on the model, the separation of σ_1 and σ_2 can be simplified by utilizing the following special conditions:

At free boundary one of the principal stresses is zero. Therefore the tangential stress is given by $\sigma_1 = mf$ if $\sigma_2 = 0$ and $\sigma_2 = mf$ if $\sigma_1 = 0$. The nature of this tangential stress can be determined using Babinet-Soleil compensator.

At a free corner, both σ_1 and σ_2 are zero. Hence the fringe value will always be zero. This fact helps in numbering the fringes in complex fringe patterns.

Along the boundary that is not free but subjected to a normal uniformly distributed load p, the tangential stress can be calculated. In the case of a pressure vessel subjected to an internal pressure p, one can write $\sigma_1 - \sigma_2 = \sigma_1 + p = mf$, since $\sigma_2 = -p$, or $\sigma_1 = mf - p$.

In any general case of photoelastic observations the directions associated with σ_1 and σ_2 can be established from the isoclinic parameter θ. Compensation methods can be effectively used to determine the nature and magnitude of the principal stresses in the boundary.

9.6.2 Separation of Principal Stresses

It has been emphasized earlier that photoelastic data give the difference of principal stresses. For defining the complete state of stress at any object point, one should either determine σ_1 and σ_2 and their directions or find out $\sigma_x, \sigma_y,$ and τ_{xy} with reference to a given system of coordinates. There are a number of methods by which the individual values of the principal stresses can be evaluated. Some of these utilize complementary information in addition to the photoelastic data. We will describe here only the oblique incidence method and the lateral extensometer. The holophotoelasticity that provides both isochromatics and isopachics will be described in Sec. 9.7.

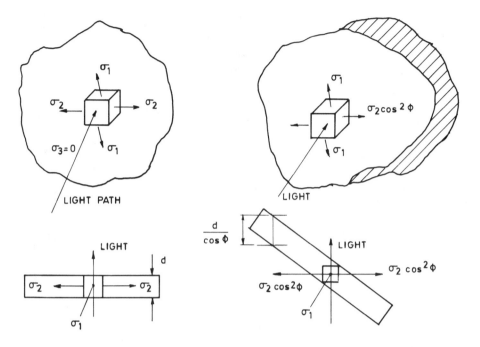

Figure 9.8   The oblique incidence method for the separation of principal stresses.

9.6.3 Oblique Incidence Method

In the frequently used oblique incidence method, additional photoelastic observations are made when the light beam passes through the model at an oblique angle, see Fig.9.8. These observations, in combination with the observations made at normal incidence, enable one to separate the σ_1 and σ_2 values.

When the incidence is not normal, the influence of the following three additional factors should be considered: change in the light path in the model, change in the secondary principal stresses, and the refraction of light at the entrance and exit faces of the model. This last factor is not very important, as its effect can be eliminated by immersing the model in an index matching liquid.

Let us consider a model of thickness d. When the angle of incidence is ϕ, the effective thickness in the model is

$$d_e = \frac{d}{\cos \phi} \tag{9.36}$$

Further, the secondary principal stresses in the plane normal to the observation are σ_1 and $\sigma_2 \cos^2\phi$. We now write for the stress-optic law at normal and oblique incidences as

$$\sigma_1 - \sigma_2 = \frac{mF}{d} \tag{9.37a}$$

$$\sigma_1 - \sigma_2 \cos^2 \phi = \frac{m_\phi F}{d \cos \phi} \tag{9.37b}$$

where m_ϕ is the fringe order for oblique incidence. Solving these equations for σ_1 and σ_2, we obtain

$$\sigma_1 = \frac{F}{d} \cos \phi \; \frac{(m_\phi - m \cos \phi)}{\sin^2 \phi} \tag{9.38a}$$

$$\sigma_2 = \frac{F}{d} \; \frac{(m_\phi \cos \phi - m)}{\sin^2 \phi} \tag{9.38b}$$

This method is used for separating stresses along an axis symmetry where one rotation of the model about the axis provides sufficient data to separate the stresses along the entire line. If the directions of the principal stresses are not known, three fringe patterns, one at normal incidence and two at oblique incidences, are needed for the separation of stresses [12].

9.6.4 Lateral Extension Measurement

Lateral extension measurement method is based on the measurement of ε_{zz}, the strain normal to the plane of the model. The strain is related to the sum of principal stresses as

$$\varepsilon_{zz} = - \frac{\mu}{E} (\sigma_1 + \sigma_2) \tag{9.39}$$

where μ is the Poisson ratio of the material. The strain is a linear function of the change in the thickness Δd of the model, i.e.

$$\varepsilon_{zz} = \frac{\Delta d}{d} \tag{9.40}$$

Thus we have

$$\sigma_1 + \sigma_2 = - \frac{E}{\mu} \frac{\Delta d}{d} \tag{9.41}$$

The measurement of Δd can be made by a mechanical device or by an interferometer. The mechanical device, or extensometer measures Δd on a mesh of points, while the interferometer provides a whole-field measurement. The fringe pattern obtained by the interferometer is called isopachics.

9.7 PHOTOELASTIC HOLOGRAPHY

Holography has been applied to photoelasticity, as it provides both isochromatic and isopachic fringe patterns. There are other advantages of holography, particularly to dynamic photoelasticity [10]. The subject of holography has been dealt with in Chap. 7. Here we extend the applications of holography to photoelasticity. We will obtain the intensity distributions in the reconstructed virtual image for single-exposure holography of a stressed model and for double-exposure holography with the model stressed between exposures.

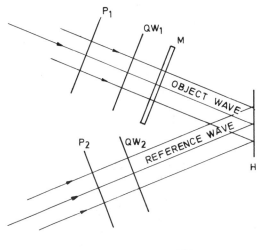

P_1, P_2	POLARIZERS
QW_1, QW_2	QUARTER WAVE PLATES
M	MODEL
H	RECORDING PLATE / FILM

Figure 9.9   A schematic of an experimental setup for holophotoelasticity.

9.7.1 Single Exposure Holography of a Stressed Model

The hologram recording geometry is shown in Fig. 9. 9. The beam issuing from the laser is usually linearly polarized with the $\bar{E}$ vector oscillating in the vertical plane. The stressed model is birefringent and supports two orthogonally linearly polarized waves. Therefore the model is illuminated with a circularly polarized wave, and the reference wave is also circularly polarized and of the same handedness. The fields just after the model can be expressed as

$$O_1 = O_0 e^{i\delta_1} \tag{9.42a}$$

$$O_2 = O_0 e^{i\delta_2} \tag{9.42b}$$

These waves are orthogonally polarized. The field at the recording plane will be proporational to O_1 and O_2. We will absorb this proportionality constant in the amplitudes. The phases δ_1 and δ_2 were defined earlier.

The addition of a circularly polarized reference wave at the recording plane results in simultaneous recording of two holograms. The intensity recorded is given by

$$I_H \propto |(O_0 e^{i\delta_1} + a_r e^{i\delta_r})|^2 + |(O_0 e^{i\delta_2} + a_r e^{i\delta_r})|^2 \tag{9.43}$$

where $a_r e^{i\delta_r}$ is the reference wave. The amplitudes of the interfering waves can be easily equalized. Further assuming linear recording, the amplitude of the reconstructed wave is

$$a \propto (e^{i\delta_1} + e^{i\delta_2}) = e^{i\delta_2}(1 + e^{i(\delta_1 - \delta_2)})$$

The intensity distribution in the reconstructed virtual image is

$$I_i \propto (1 + \cos \delta) \quad \text{where} \quad \delta = \delta_1 - \delta_2 \tag{9.44}$$

This expression is similar to that obtained with a bright field circular polariscope. Figure 9.10 shows a photograph of the reconstructed virtual image of a disc under diametral compression obtained from a single exposure hologram. It may be seen that no assembly of quarter-wave plate and analyzer is placed behind the model during recording; the state of polarization in the reference beam serves the function of this assembly. If the state of polarization in the reference beam is orthogonal to that in the object beam, i.e. the reference beam is of opposite handedness, the isochromatics corresponding to the dark field circular polariscope are obtained.

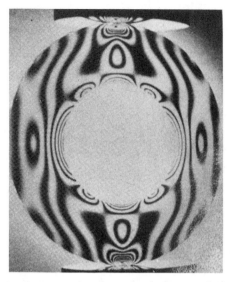

Figure 9.10   A photograph of an isochromatic fringe pattern of a disc under diametral compression obtained from a single exposure hologram of a stressed model. (From Ref. 11.)

9.7.2 Double Exposure Holography of a Model Stressed Between Exposures

The object is illuminated with a circularly polarized wave, and a reference wave that is also circularly polarized with the same handedness is added at the recording plane. The object wave is described as, $O_0 e^{i\delta_0}$, where δ_0 is the phase introduced by the model. The first exposure records the inital state. The object is now loaded and the second exposure is made on the same recording plate. Initially, illumination with circularly polarized light is done so that no changes in the experimental arrangement are needed when the object is loaded and the second exposure is made. Further, a bright field arrangement is chosen so that a bright virtual image of the object is obtained during reconstruction. The total intensity recorded, ignoring the amplitude factors associated with various waves as they can be equalized, is given by

$$I_H \propto (2|e^{i\delta_0} + e^{i\delta_r}|^2 + |e^{i\delta_1} + e^{i\delta_r}|^2 + |e^{i\delta_2} + e^{i\delta_r}|^2) \tag{9.45}$$

The factor 2 is introduced to make the exposure same for both recordings. Again, assuming linewar recording and writing terms that give the virtual image, the amplitude of the reconstructed wave is

$$A_H \propto (2e^{i\delta_0} + e^{i\delta_1} + e^{i\delta_2}) \tag{9.46}$$

The intensity distribution in the reconstructed virtual image is

$$I = I_0[6 + 4\cos(\delta_1 - \delta_0) + 4\cos(\delta_2 - \delta_0) + 2\cos(\delta_1 - \delta_2)]$$

$$= I_0\left[1 + 2\cos\left(\frac{\delta_1 + \delta_2 - 2\delta_0}{2}\right)\cos\left(\frac{\delta_1 - \delta_2}{2}\right) + \cos^2\left(\frac{\delta_1 - \delta_2}{2}\right)\right] \tag{9.47}$$

It is thus seen that the second term in Eq. (9.47) contains information about the isopachics while the second and third terms contain information about the isochromatics. There has been a controversy on the nature of these fringe patterns. Figure 9.11 shows an interferogram depicting both iso-

Figure 9.11 A photograph of isochromatic and isopachic fringe patterns of a disc under diametral compression obtained from a double exposure hologram (From Ref. 11.)

chromatics and isopachics obtained from a double exposure hologram recorded with circularly polarized illumination. It may be seen from the equation that dark isochromatics appear when

$$\delta = \delta_1 - \delta_2 = (2m + 1)\pi \qquad m = 0, 1, 2, \ldots$$

The intensity of the dark isochromatics is not zero. The bright isochromatics appear when $\delta = 2m\pi$. The intensity of the bright isochromatics is given by

$$I = 2I_0\left[1 + (-1)^m \cos\left(\frac{\delta_1 + \delta_2 - 2\delta_0}{2}\right)\right] \qquad (9.48)$$

To analyze the formation of the fringe pattern we assume m to be even, i.e., m = n where n is an even number. The intensity in the bright isochromatic is then given by

$$I_n = 2I_0\left(1 + \cos\frac{\delta_1 + \delta_2 - 2\delta_0}{2}\right) \qquad (9.49)$$

The dark isopachic will occur simultaneously, provided

$$\delta_1 + \delta_2 - 2\delta_0 = (K + \frac{1}{2})2\pi \qquad K = 0, 1, 2, \ldots \qquad (9.49)$$

The dark isopachic will have zero intensity. When we go to the next bright isochromatic, however, the intensity is given by

$$I_{n+1} = 2I_0\left(1 - \cos\frac{\delta_1 + \delta_2 - 2\delta_0}{2}\right) \qquad (9.51)$$

At the condition of the dark K^{th} isopachic, the intensity will be $4I_0$. In fact this implies that the isopachic has changed from a dark fringe to a bright fringe in crossing the isochromatic. The condition of a dark isopachic, from Eq. 9.51, is

$$\delta_1 + \delta_2 - 2\delta_0 = 2K\pi \qquad (9.52)$$

This shows that the isopachic has changed by one half order in going from one bright isochromatic to the next bright isochromatic. This simple interpretation is valid when the two families of fringes are nearly perpendicular, as shown in Fig. 9.12. In the other extreme case, when isochromatics and isopachcis run parallel to each other, this interpretation breaks down [9,11]. It is therefore proper to use methods by which the two families of fringes can be separated. The influence of birefringence, and hence the isochromatic pattern, can be eliminated by twice passing the beam through the model and a Faraday rotator. One can also us a model of a material with little or no response to birefringence, such as PMMA. For such a

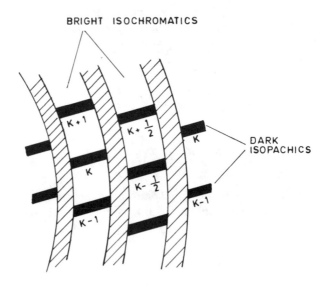

Figure 9.12   Isochromatic and isopachic fringe patterns.

model, only the isopachic pattern will be observed.

Finally, it may be worthwhile to mention that holophotoelasticity can be performed in real time. Further some advanced techniques used in photoelasticity are reported by Stanford [11]

9.8 THREE-DIMENSIONAL PHOTOELASTICITY

So far the analysis has been restricted to the evaluation of a two-dimensional stress field; the stress along the third dimension is considered constant. When the stress varies also along this direction, we have the three-dimensional stress state. A number of methods have been developed for the study of this state. Of these we describe only two experimental methods, namely the frozen stress method and the scattered light method.

9.8.1 The Frozen Stress Method

The stress freezing method is possibly the most powerful method of experimental stress analysis. In comparison to numerical methods it is considered neither cost nor time-effective, and hence its application in industry is declining. The procedure for stress freezing consists of heating the model to a temperature slightly above the critical temperature and then cooling it slowly, typically at less than 2°C per hour under the desired loading to room temperature. The load may be applied to model either before or after reaching the critical temperature. Extreme care is required to ensure that the model is subjected to the correct loading, as spurious stresses due to bending and gravitational loads may be induced because of low rigidity at the critical temperature.

After the model is cooled to room temperature, the elastic deformation responsible for optical anisotropy is permanently locked. The optical anisotropy is not disturbed even when the model is cut into thin slices for examination under the polariscope. The slices should be cut with sharp tools to avoid chipping the edges and corners; this mechanical operation should not induce additional stress in the slices. Two methods for data collection and interpretation are generally adopted. These methods are sub slicing and oblique incidence. The oblique incidence method is more practical. The details of these methods can be found in several books [2,3,7].

9.8.2 Scattered Light Photoelasticity

Scattered Light and Polarization

When a beam of light passes through a medium containing fine particles dispersed in the volume, a part of the beam is scattered. The intensity of scattered light, when the scatterers are much smaller than the wavelength of light, varies as λ^{-4}. This phenomenon was investigated by Rayleigh in detail and is called Rayleigh scattering. The most beautiful phenomena of

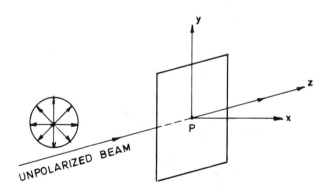

Figure 9.13   Scattering of unpolarized light by a scatterer at P.

the red sunset and the blue sky are due to scattering from the gaseous molecules in the atmosphere. Further, the light from the blue sky is partially linearly polarized. In some observation directions the scattered light is linearly polarized. Consider a scattering center located at P as shown in Fig. 9.13. Let the incident light be unpolarized light that can be resolved into two orthogonal linearly polarized components with random phases. The incident component vibrating in the y–z plane, when absorbed, will set the particle (or rather the electrons in the particle) vibrating along the y direction. The reradiated wave will have zero amplitude along the y direction. On the other hand, if the particle is oscillating along the x direction, the reradiated wave will have zero amplitude in that direction. Thus when the observation direction lies in the x–y plane passing through P, the scattered wave will be plane polarized. The particle acts as a linear polarizer.

Let us now consider that the incident ray is linearly polarized with the y–z plane being the plane of vibration. The electrons in the particle will oscillate along the y direction. There will be zero amplitude of the reradiated wave when the observation direction is along the y axis. The particle thus acts as an analyzer.

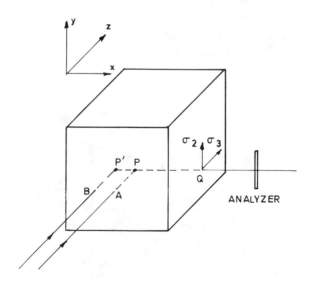

Figure 9.14   An unpolarized beam incident on a stressed model. Observation through an analyzer.

These observations on scattered light form the basis of scattered light photoelasticity.

Scattering from the model

Consider a stressed model in the path of a narrow beam of unpolarized light. We assume that the model has numerous scatterers. Let us now consider the light scattered by the point P inside the model when the observation direction is perpendicular to the incident direction. The light scattered by this point is resolved into components along the stressed directions σ_2 and σ_3 as shown in Fig. 9.14. In traversing a distance PQ in the stressed medium, these two orthogonally polarized components acquire a phase difference. If an analyzer is placed in the observation direction, the transmitted intensity will depend on the phase difference acquired. Since the incident beam is unpolarized there is no influence of traverse AP in the model. Consider that the transmitted intensity is zero for a certain location P of the scatterer; this occurs when the phase difference is a multiple of 2π. As the beam moves to illuminate another scatterer P' in the same plane along the line of sight, the transmitted intensity will undergo cyclic variations between minima and maxima depending on the additional phase acquired due to traverse over the distance PP'. It is, however, assumed that the directions of principle stresses σ_2 and σ_3 do not change over PP'.

Let m_1 and m_2 be the fringe orders when the light scattered from point P and P' respectively is analyzed, Then

$$m_1 F = (\sigma_3 - \sigma_2)x_1$$

and

$$m_2 F = (\sigma_3 - \sigma_2)x_2$$

where $x_1 = PQ$ and $x_2 = P'Q$. Therefore

$$(\sigma_3 - \sigma_2) = F \frac{dm}{dx}$$

The principal stress difference at any point along the observation direction is proportional to the gradient of the fringe order.

We now consider another situation where the incident beam is linearly polarized. We assume, for the sake of simplicity, that the principal stress directions are along x and y axes. The transmission axis of the polarizer makes an angle α with the x axis. The incident wave of amplitude E_0 is resolved along the x and y directions: these represent two orthogonal linearly polarized components that travel with different velocities and pick

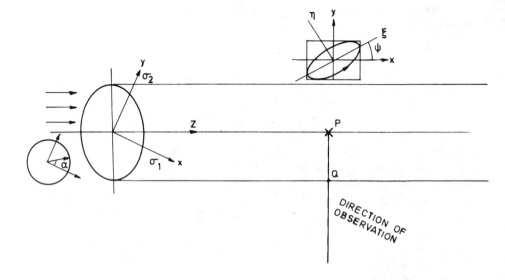

Figure 9.15 A polarized beam incident on a stressed model.

up a phase difference δ. Thus at any plane normal to the direction of propagation, the state of polarization of the wave in general will be elliptical. It is expressed (Fig. 9.15) as

$$\frac{E_x^{\,2}}{E_0^2\cos^2\alpha} + \frac{E_y^{\,2}}{E_0^2\sin^2\alpha} - \frac{2E_xE_y}{E_0^2\sin\alpha\cos\alpha}\cos\delta = \sin^2\delta$$

The major axis of the ellipse makes an angle ψ with the x axis, where

$$\tan 2\psi = \tan 2\alpha\,\cos\delta$$

When $\delta = 2K\pi$, K taking both negative and positive integer values, the state of polarization of the wave is linear with orientation $\psi = \pm\,\alpha$. For positive values of K the state of polarization is the same as that of the incident light. The stress-optic law for this situation can be expressed as

$$\sigma_1 - \sigma_2 = F\frac{dm}{dz}$$

The scatterer at P in any plane is therefore excited by an elliptically po-

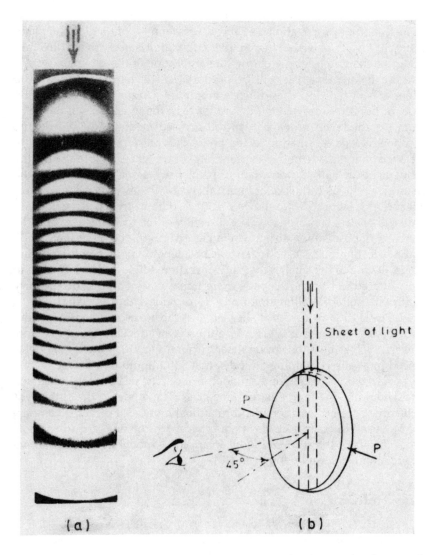

Figure 9.16 (a) A scattered light stress pattern of a disc under diametral compression. (From Ref. 13, used with permission.) (b) A schematic showing direction of illumination and observation.

larized light. In scattered light photoelasticity, we are looking normal to the direction of propagation, i.e., the observation is confined to the plane of the elliptically polarized light. If the observation is made in the direction along the major axis of the ellipse, the amplitude of the reradiated wave received by the observer will be proportional to the magnitude of the minor axis of the ellipse, and hence minimum. On the other hand, if the observation direction is along the minor axis, the intensity will be maximum. As the beam propagates in the stressed medium, the ellipse just described keeps on rotating as the phase difference increases. Therefore the observation made in the scatttered light normal to the direction of the incident beam will show variation in intensity along the length of the beam. There is no influence of birefringence in the model on the traverse distance PQ. Figure 9.16(a) shows a scattereed light stress pattern of a disc under diametral compression. The direction of the incident light and the direction of observation are shown in Figure 9.16(b). It may be mentioned that this technnique is used to find the beat length in polarization holding fibers. Owing to the smaller intensity of scattered light, the source used in the scattered light polariscope is more intense; a high-pressure mercury lamp with suitable collimating optics to provide a thin pencil of light is used. The laser is an attractive alternative as the beam can be used without optics. A He-Ne laser of 10–15 mW output is an adequate source, as more powerful lasers give too strong a background illumination. The model is placed in a tank containing an index matching liquid to avoid refraction and polarization changes at the model surface. The walls of the immersion tank should be made from stress-free glass. The model in the tank needs to be given both translational and rotational motions. Further, a compensator is used in a separate stage equipped with rotation and translation. More details of the scattered light polariscope, model materials, data interpretation, etc. can be found in [13].

REFERENCES

1. Aben, H., (1979). *Integrated Photo-Elasticity*, McGraw-Hill, New York.
2. Dally, J. W., and Riley, W. F. (1978). *Experimental Stress Analysis*, McGraw-Hill, New York.
3. Durelli, A. J., and Riley, W. F. (1965). *Introduction to Photomechanics*, Prentice Hall, Englewood Cliffs, New Jersey.
4. Frocht, M. M. (1966). *Photoelasticity*, 2 vols., John Wiley, New York.
5. Jessop, H. T., and Harris, F. C. (1960). *Photoelasticity: Principles and Methods*, Dover, New York.
6. Holister, A. S. (1967). *Experimental Stress Analysis: Principles and Methods*, Cambridge Univ. Press, Cambridge.
7. Kuske, A. and Robertson, G. (1974). *Photoelastic Stress Analysis*, John Wiley, New

York.

8. Post, D. (1970). Photoelastic fringe multiplication for tenfold increase in sensitivity, *Exp. Mech., 10:* 305–312.

9. Gasvik, K. J. (1987).*Optical Metrology,* John Wiley, Chichester.

10. Holloway, D. C. and Johnson, R. H., (1971). Advancements in holographic photoelasticity, *Exp. Mech., 11:* 57–63.

11. Stanford, R. J. (1980). Photoelastic holography– A modern tool for stress analysis, *Exp. Mech., 20:* 427–436.

12. Kobayashi, A. S., ed. (1987). *Handbook on Experimental Mechanics,* Prentice Hall, Englewood Cliffs, New Jersey.

13. Srinath, L. S. (1983). *Scattered Light Photoelasticity,* Tata McGraw-Hill, New Delhi.

CHAPTER 10

Fiber-Optic Sensors

10.1 INTRODUCTION

Sensing of a variable is essential for measurement and control of any process. The large number of variables encountered are measured by a variety of electrical, electronic, pneumatic, hydraulic, and other devices. Fiber-optic (FO) sensors are replacing some of these conventional sensors in some areas. FO sensors offer many advantages over conventional sensors. Some of these are the freedom from electromagnetic interference (EMI) due to the lack of electrical bias at the sensor; the capability of working in harsh environments contaminated with explosives, and corrosive gases (because FO sensors are chemically inert); compatibility with telemetry, and easy integrability with semiconductor devices; low weight and hence better frequency response over a wider range; low power consumption; capability of offering higher accuracy in measurements; the capability of multielement sensor systems of supporting a large number of high-bandwidth sensor elements; and the elimination of the requirement of transmitting electrical power remote from the monitoring site.

Fiber-optic sensors are used in areas such as aircraft engines and flight control, shipboard machinery and damage control, medical probing, and industrial process control.

10.2 A FIBER-OPTIC SENSOR CONFIGURATION

A configuration of a general fiber optic sensor is given in Fig. 10.1. It consists of a light source that may provide an input beam constant in intensity, frequency, phase, polarization, etc. The beam is launched into an optical fiber that conducts it to the region of measurement. The measurand modulates the light beam in amplitude/intensity, phase, frequency, polarization, or color.

The modulated light is returned from the measurement zone by another fiber to the detector. The output of the detector is then demodulated. It must be noted here that the detectors cannot follow optical frequencies and hence respond to intensity only. Hence any form of modulation is to be converted to intensity modulation before the light is detected.

Fiber-optic sensors can be classified into two groups. In the first group fall those sensors in which fibers serve as conduits to transmit light to the modulation zone and to return it to the detector. The modulation process takes place outside the fibers. The transmission properties are usually modulated by the measured variable. These sensors are called incoherent FO sensors. In the second group fall those sensors in which the measured variable interacts with the light in the fiber. The measurand may modulate the phase, polarization, and amplitude of the light wave in the fiber. These sensors are called the coherent FO sensors.

An FO sensor essentially consists of a light source, a detector, a modulation process, and a demodulation process. The details regarding sources and detectors are given in Chapter 1. There are a few parameters that should be considered while choosing a source. One of them is the radiance of the source. High-radiance sources are required for FO sensors. A typical launch power in a fiber is one mW. A criterion to determine the amount of power that can be launched into a fiber is area multiplied by the solid angle of the fiber and the source radiance. It may be noted that single mode fibers require the highest source radiance. The second parameter of interest is the noise in the source. The minimum noise level is determined by

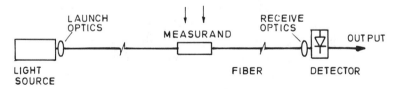

Figure 10.1   A schematic of a fiber-optic sensor.

the shot noise due to the photon nature of light. The noise in the source actually far exceeds the shot noise, and there are many reasons for this. The spectral characteristics of the source provide other parameters. The influence of environmental variables on the output and life of the source is also important for making a selection. Many detectors are available for use with FO sensors, for example PIN diodes, APDs, PMT, and their selection is again determined by performance requirements. An APD is used in situations where the shot noise, before multiplication, is well below the thermal noise. An optimum performance is obtained when the multiplication process brings the shot noise up to the thermal noise level.

In this chapter, we study various modulation and demodulation schemes and some FO sensors in greater detail.

10.3 MODULATION OF A LIGHT WAVE

There are a number of methods for modulating a light wave; of these, intensity modulation, phase modulation, frequency modulation, and polarization modulation are discussed here.

10.3.1 Intensity Modulation

Some kind of measurand-dependent attenuation of the light is obtained in intensity modulation. This can be achieved outside the fiber and within it. A few configurations for introducing attenuation externally are shown in Fig. 10.2. Attenuation can be introduced by the reflector movement, as shown in Fig. 10.2 (a). The movement of a mask perpendicular to the direction of the beam also introduces attenuation. The input fiber end is imaged on the entrance face of the output fiber, and the mask moves perpendicular to the direction of the beam in between the lenses, as shown in Fig. 10.2 (b). The mask movement changes the amount of light energy coupled to the output fiber. The sensitivity is expected to be better than 1 part in 10^6. A simpler arrangement, as shown in Fig. 10.2 (c) works without a lens, but the sensitivity is reduced because of loss over the gap.

Some simple sensors are based on direct coupling of energy between input and output fibers; Figs. 10.3 (a) and (b) show arrangements in which position dependent coupling between the two fibers, which move relative to each other, is used for sensing. Both lateral and longitudinal translations of the fiber can be used for intensity modulation. Intensity modulation internally in the fiber can be obtained by microbending, as shown in Fig. 10.4. Microbending induces losses by coupling the guided modes to the cladding.

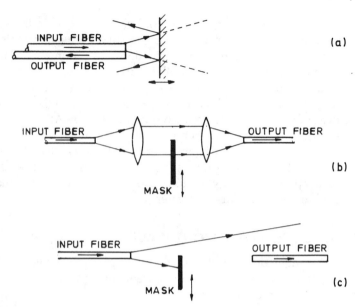

Figure 10.2   Intensity modulation by (a) motion of a reflector, (b) and (c) motion of a mask.

10.3.2 Phase Modulation

The phase of a light wave propagating along an optical fiber depends on its length, on the refractive index or index profile, and on the geometric transverse dimensions of the guide (the core of the fiber). Therefore phase modulation can be achieved by the variation of any of these parameters. For example, the total length of the fiber can be varied by an application of longitudinal strain, by thermal expansion, or by an application of hydrostatic pressure, which causes expansion via the Poisson ratio. The refractive index varies with temperature, pressure, and longitudinal strain via the photoelastic effect. The guide dimensions vary with radial strain in a pressure field, longitudinal strain through the Poisson ratio, and thermal expansion.

The detection of phase variations is accomplished by interferometry; the phase variations are converted into intensity variations to which the detectors respond in the optical region. Interferometers are used for a wide variety of applications in such diverse areas as aerodynamics, combustion, fluid flow, and optical testing. The advantages of using fibers in interferometric sensors lie both in easing the alignment difficulties inherent in

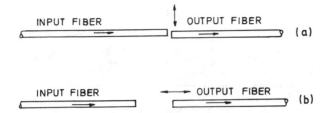

Figure 10.3 Direct intensity modulation by (a) lateral shift of a fiber and (b) longitudinal shift of a fiber.

setting up long-path interferometers and in increasing the sensitivity by exposing long lengths to the measurand. Compact and mechanically rugged interferometers can be constructed with fibers. Fiber-optical interferometric sensors are by far the most sensitive devices. Hydrophones, magnetometers, accelerometers, strain gauges, and thermometers have been fabricated around fiber interferometers, and all have achieved sensitivities far exceeding those available with other techniques. Fiber-optic interferometric sensors tend to exhibit sensitivity towards environmental parameters and hence their packaging is extremely difficult.

10.3.3 Frequency Modulation

The frequency of light waves scattered or reflected from moving targets is Doppler shifted. The Doppler shift is linearly related to velocity, so measurement of the Doppler shift provides a very sensitive means of measuring velocities. A Doppler probe is capable of measuring velocities over a wide range from a few μm/sec to 100 m/sec. It has an extremely large dynamic response, and its response is linear. As an example, a probe using a He-Ne laser will give a Doppler shift of 1.6 MHz per m/sec. Because measurement of a few Hz is possible, extremely small velocities can be monitored. Frequency modulation is frequently used for the measurement of fluid flow; the instrument is known as a laser Doppler anemometer (LDA). The fiber version LDA is also used for measuring fluid flow, In particular it is used to measure the flow of blood in the arteries.

10.3.4 Polarization Modulation

The state of polarization of a light wave is influenced by a variety of physical phenomena like birefringence, optical activity, the Faraday effect,

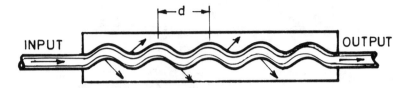

Figure 10.4   Intensity modulation by microbending.

and the electrooptic effect. Birefringence may be induced by the application of a stress or strain field. Variation in electrical current and consequently in magnetic field will modulate the polarization of the wave through the Faraday effect. Optical activity is influenced by temperature, so temperature may be used to change the polarization state through optical activity. The polarization of the wave can also be varied by the application of an electric field when the wave passes through an electrooptic medium.

10.4 DEMODULATION OF A LIGHT WAVE

Optical detectors respond to intensity, that is they are square-law detectors. The input optical power is converted to an electrical current in the detection process. Thus the electrical power is proportional to the square of the optical power.

We describe various schemes to detect phase modulation, frequency modulation, and polarization modulation. These forms of modulation are to be converted to intensity modulation prior to detection.

10.4.1 The Detection of Phase Modulation

The detection of phase modulation is usually performed by interferometry. The interferometric process converts the phase variations into intensity variations. Consider a Mach-Zehnder interferometer as shown in Fig. 10.5. One arm of the interferometer contains the phase modulation process. The detectors D_1 and D_2 receive the optical fields traversing each arm. The resultant field at detector D_1 is given by

$$E = E_0 \cos(kz - wt) + E_0 \cos[kz - wt + \delta(t)] \qquad (10.1)$$

where E_0 is the amplitude of the light wave in each arm at the detector

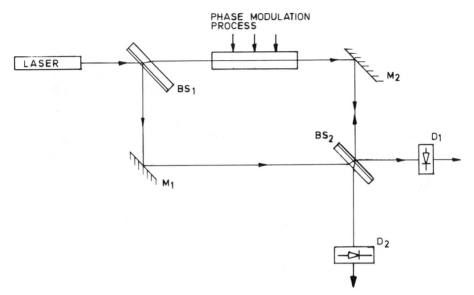

Figure 10.5   A schematic of a Mach-Zehnder interferometer used for phase modulation.

plane, w is the frequency, $k = 2\pi/\lambda$ is the wave vector, and $\delta(t)$ is the time-varying phase introduced by the phase modulation process. The photodetector responds to the intensity $|E|^2$ given by

$$|E|^2 = I_0[1 + \cos \delta(t)] \qquad I_0 = |E_0|^2 \tag{10.2}$$

Thus we find that the phase variation $\delta(t)$ has been converted to intensity variation by interferometry. The photocurrent in detector D_1 is proportional to $|E|^2$ and will be expressed as

$$i_1(t) \alpha 1 + \cos \delta(t) \tag{10.3a}$$

Similarly, the photocurrent in detector D_2 will be expressed as

$$i_2(t) \alpha 1- \cos \delta(t) \tag{10.3b}$$

The photocurrent depends on the phase variations. The output of detector D_1 is maximum when $\delta(t) = 2m\pi$, $m = 0, 1, 2$, and minimum (zero) when $\delta(t) = (2m+1)\pi$. Thus the output of D_1 varies with time. The response of the interferometer to small changes $\Delta\delta(t)$ between the two beams may be obtained by differentiating Eq. (10.3):

$\Delta i_1(t) \; \alpha \; \sin \delta \; \Delta\delta$ (10.4)

Thus the change in photocurrent depends both on the initial phase setting of the interferometer and on the phase change $\Delta\delta$. If we set $\sin \delta = 1$, that is, the phases of the two beams in the interferometer are in quadrature, the photocurrent is directly proportional to $\Delta\delta$. The relationship between the incremental phase change and incremental photocurrent change is linear within 1% over a range of $\pm 9°$ at the quadrature condition.

There are, however, two principal difficulties with this interferometric arrangement. Maintaining a stable quadrature condition is far from trivial: some kind of automatic feedback technique must be used. The other difficulty is that output of the detector is sensitive to both the source intensity and the phase change. It may be seen that a 3.5% drift in source intensity produces exactly the same effect as a 2° phase change. In addition to these difficulties, interferometers with nonzero path differences require sources of high temporal coherence.

The Heterodyne Technique

Many of these difficulties with interferometric detection may be overcome by the use of the heterodyne technique. In a heterodyne interferometer, the reference beam is frequency shifted by passing it through a suitable modulator. The total field at the detector plane is

$$E = E_0 \cos(kz - w't) + E_0 \cos[kz - wt + \delta(t)]$$ (10.5)

where $E_0 \cos(kz - w't)$ is the reference beam with frequency w', which is different from the source frequency w. The intensity incident on the detector is $|E|^2$, that is,

$$I = |E|^2 = I_0[1 + \cos\{(w' - w)t + \delta(t)\}]$$ (10.6)

Therefore the photocurrent $i(t)$ in a detector is given by

$$i(t) \; \alpha \; 1 + \cos[(w' - w)t + \delta(t)]$$ (10.7)

If $\delta(t)$ is periodic, then it will modulate the intermediate frequency $w' - w$, and standard phase modulation detection techniques may be used to extract the modulation. Estimates based on the reception of a 1mW local oscillator signal with a 100-kHz bandwidth indicate a theoretical signal-to-noise ratio of about 100 dB. Phase modulation of the order of 10^{-5} radians should be detectable. In practice, a signal-to-noise ratio of about 80–90 dB is achievable, and maximum detectable levels in the range of 10^{-4} radians of phase modulation may be obtained.

Figure 10.6 shows a heterodyne interferometer using a Bragg cell for imposing a frequency shift w_B. The phase modulation $\delta(t)$ is induced by

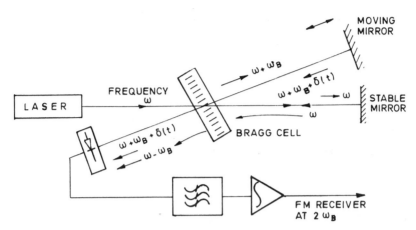

Figure 10.6   A heterodyne Michelson interferometer. The Bragg cell shifts the frequency of the light wave. (From Ref. 2, used with permission.)

vibrating one of the mirrors of the interferometer. The reference wave can be expressed as $E_0 \cos[kz - (w - w_B)t]$. The total field incident on the detector is

$$E = E_0 \cos[kz - (w - w_B)t] + E_0 \cos[kz - (w + w_B)t + \delta(t)] \qquad (10.8)$$

Therefore the photo current i(t) is given by

$$i(t) \propto 1 + \cos[2w_B t + \delta(t)] \qquad (10.9)$$

The phase variation $\delta(t)$ modulates the intermediate frequency $2w_B$.

10.4.2 Detection of Frequency Modulation

It has been mentioned that the frequency of light scattered or reflected from moving targets is Doppler shifted. The frequency shift can be measured using interferometric methods. These methods are homodyne or heterodyne. In the homodyne method a reference beam derived from the beam used for illuminating the moving target is mixed with the scattered beam. The output carries the Doppler signal. In the heterodyne method the reference beam is frequency shifted and then mixed with the scattered beam. The homodyne technique has the great advantage of simplicity, but it cannot yield the sign of the Doppler shift; whereas the heterodyne technique gives the sign of the Doppler shift at the expense of increased complexity.

10.4.3 The Detection of Polarization Modulation

The change in the orientation of a linearly polarized light can be converted into intensity variation by letting it through a polarizer (or analyzer). If the angle between the azimuth of the linearly polarized light and the transmission axis of the polarizer is θ, then the intensity $I(\theta)$ transmitted through the polarizer is given by the Malus law:

$$I(\theta) = I_0 \cos^2 \theta \tag{10.10}$$

The output of the detector, assuming linear response, will be

$$i(\theta) \; \alpha \; \cos^2 \theta \tag{10.11}$$

An incremental change $\Delta\theta$ in orientation results in a change $\Delta i(\theta)$ in the current output of the detector, where

$$\Delta i(\theta) \; \alpha \; \sin 2\theta \; \Delta\theta \tag{10.12}$$

We usually set $\sin 2\theta = 1$ so that the output is linearly related to the change in the azimuth of the linearly polarized light. It may be seen that the output varies with the change in the source intensity, thereby requiring a source whose intensity is constant.

To overcome the problem arizing out of source intensity variation, a Wollaston prism is used instead of a polarizer. The Wollaston prism produces two orthogonally polarized beams that are angularly shifted, as shown in Fig. 10.7 (a). The intensities of these two beams are equal when the azimuth of the incident linearly polarized beam is at an angle of 45° with the principal section of the prism. If the azimuth deviates by an angle θ' from 45°, then the amplitudes of two orthogonally polarized beams are $A/\sin(\pi/4+\theta')$ and $A/\cos(\pi/4+\theta')$, where A is the amplitude of the incident linearly polarized beam. This is shown in Fig. 10.7 (b). Thus the orientation error θ' is obtained as

$$\sin 2\theta' = \frac{I_1 - I_2}{I_1 + I_2} \tag{10.13}$$

where I_1 and I_2 are the intensities of the two beams. The outputs i_1 and i_2 of the two detectors are proportional to the corresponding intensities. The outputs of the detectors are subtracted and added, and then division is performed to obtained the value of the azimuth change θ' from the relation

$$\sin 2\theta' = \frac{i_1 - i_2}{i_1 + i_2} \tag{10.14}$$

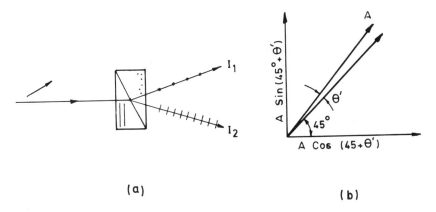

Figure 10.7 (a) A Wollaston prism as a beam splitter and (b) resolution of linearly polarized light into components. (From Ref. 2, used with permission.)

This measurement is convenient in that it is purely radiometric and is nearly independent of source intensity fluctuations and variations in the attenuation of the link between the source and the detector. The measurement of a general state of polarization is achieved by the measurement of the Stokes parameters of the beam.

10.5 SOME FIBER-OPTIC SENSORS

All fiber-optic sensors employ at least one light source for injecting light into the fiber, a light detector for receiving the signal after the light wave has been modulated by the measurand and demodulated, and electronics for converting the detected signal into useful output. It has been mentioned earlier that FO sensors can be classified into two groups, incoherent and coherent. Some of these sensors are described in the following sections.

10.6 INCOHERENT SENSORS

Multimode fibers with LED sources offer simple and reliable sources for many applications. Some of them are described here.

10.6.1 The Fiber-Optic Position Sensor

A number of variables can be measured using intensity modulation through a mirror as shown in Fig. 10.2(a)[1,2]. The axial motion of the mirror induces position dependent attenuation. The variable to be measured is transduced to displacement which is then monitored by the sensor. The obvious choice of variables is pressure, temperature, acceleration, etc.

A sensor known as Photonic was developed for a number of specialized applications. It consists of a fiber bundle in which half the fibers are used to illuminate the mirror and the remaining half are used to carry the returned signal to the detector. The output of the detector changes because of the motion of the mirror, and intensity variations can also change the output. These intensity variations may be caused by the aging of the source, a change in connector characteristics, a decrease in the reflectivity of the mirror, the breaking of some fibers in the bundle, etc. With the exception of last effect, these problems may be compensated for by a relatively simple arrangement that uses two sets of return fibers separated by a known distance, as shown in Fig. 10.8. Measurement of the ratio of the returned intensities from each of these two separate fiber bundles will give a unique value for the position of the mirror, provided the correct portion of the sensor characteristics curve is used, as the sensor exhibits a nonlinear response. In another position sensor, ten output fibers are illuminated by nine input fibers via reflection from the mirror. With appropriate spacing between input and output fibers, resolution of 0.05% of the fiber diameter is possible.

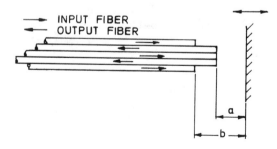

Figure 10.8   A fiber-optic position sensor.

10.6.2 The Object Proximity Sensor

This is a relatively simple sensor. It makes use of a multimode fiber bundle in which half the fibers are used to illuminate the object while the other half receive the reflected light and transmit it to the detector as shown in Fig. 10.9 (a). The detector output as a function of distance between the fiber end and the object is shown in Fig. 10.9(b). The presence of an object is ascertained if the signal is higher than a prespecified value. The proximity of the object to the fiber end can also be quantified if the objects are of consistent reflectivity.

Flow measurement using a similiar sensor in a turbine flow meter can be performed [3].

10.6.3 Optical Microphones

Numerous schemes have been suggested to realize an optical microphone. In one scheme, a fiber bundle similar to that used in a proximity sensor is used to illiuminate a diaphragm. The diaphragm modulates the amount of light reflected back. Performance acceptable for telephony has been achieved.

An alternative scheme uses a fixed fiber illuminated by a source and another fiber attached to the core of the speaker diaphragm, as shown in Fig. 10.10. The moving core displaces the fiber with respect to the fixed fiber, thereby introducing attenuation. This results in a varying signal from

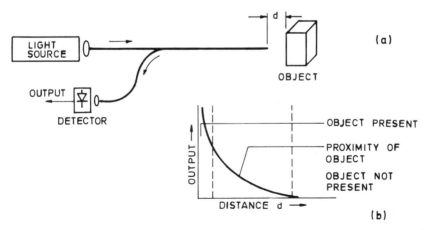

Figure 10.9   (a) A schematic of a proximity sensor and (b) variation of output with distance.

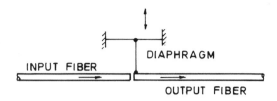

Figure 10.10   A schematic of an optical microphone.

the detector. A sensor can be designed that will have a fairly linear response over a lateral displacement equal to its diameter. For a fiber of 100
μm core diameter, the minimum detectable movement is found to be 10
nm and the dynamic range 110 dB.

A more sensitive approach is to place fine Ronchi gratings on the ends
of the two fibers. A transverse displacement equal to the grating pitch
causes 100% modulation, so that very small lateral displacement can be
detected.

Yet another scheme, that is based on frustrated total internal reflection
(FTIR), has been proposed. The schematic of an optical microphone is
shown in Fig. 10.11. The ends of the fibers are cut at the critical angle and
polished. The fibers are aligned and placed very close so that an appreciable amount of light can be coupled by an evanescent wave. Static
displacement as small as a fraction of an angstrom can be detected.

10.6.4 Temperature Sensors

The absorption edge of a semiconductor shifts with temperature. This
phenomenon has been employed in the design of a temperature sensor
that is commercially available. A semiconductor plate encapsuled in a
stainless steel body is the sensor head; the semiconductor plate is illuminated by a multimode fiber coupled to a suitably chosen diode. The transmitted light is received by another multimode fiber, which transmits it to
the detector as shown in Fig. 10.12(a). The temperature change causes the
shift of the absorption edge and hence the transmission of the semiconductor plate. Fig. 10.12 (b) shows the variation of absorption edge with
temperature along with the diode spectrum. The variation of energy gap
with temperature in the majority of semiconductors can be expressed [4]
by the empirical relation

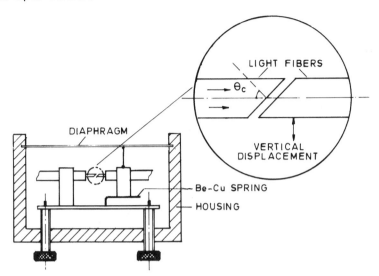

Figure 10.11   An optical microphone based on FTIR.

$$E_G(T) = E_G(0) - \frac{\alpha T^2}{T + \theta} \qquad (10.15)$$

where $E_G(0)$ is the energy gap at the absolute zero and α and θ are the constants.

As an example, a thin plate of GaAs with some Te impurity is used for the sensor head. The value of α is 5.405×10^{-4} eV/K, and $\delta = 204$ K. If the operation is limited in the linear range, the temperature dependence of the energy gap is expressed as

$$E_G(T) = E_G(0) - \beta\, \Delta T, \qquad (10.16)$$

where $\beta = \partial E_G/\partial T = 1.5 \times 10^{-4}$ eV/K. This corresponds to a variation of the absorption edge $\partial \lambda_G/\partial T = 0.092$ nm/K. Figure 10.12(c) shows the variation of the detector output as a function of temperature. The relationship is linear over the range 0–160°C.

In a commercial instrument, the shift of the absorption edge is measured by measuring the transmission of a GaAs sample at the two widely separated wavelengths of 0.88 and 1.3 μm. The longer wavelength is transmitted unattenuated and thus provides the reference. The shorter wavelength suffers a temperature-dependent attenuation. Using a tempe-

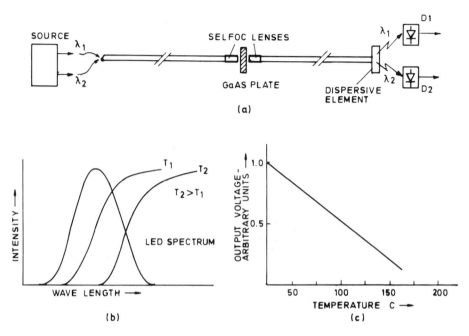

Figure 10.12 (a) Fiber-optic thermometer using a simiconductor plate as a sensor, (b)shift of absorption edge with temperature along with the LED emission spectrum, and (c) variation of output with temperature.

rature sensor based on the shift of the absorption edge, an accuracy of ± 2 K can be achieved over the whole range.

Temperature measurement is also done by monitoring the fluorescence emission of both diavalent and trivalent materials [5]. The decay time and the intensity of emission lines depend on temperature. A 15% linear change in decay time of a highly doped Nd^{3+} sample has been observed over a range of 0–200°C. The readability of the sensor is ± 2 K in the linear range. A piece of pink ruby is used as a sensor and the temperature is measured by measuring the intensity of R lines; the total fluorescence output is used to supply the reference. The loss of energy in the R lines due to the increase in temperature is redistributed to higher and lower wavelengths on either side of it, thus maintaining the total fluorescent output constant up to 500 K.

A temperature sensor can be realized from a position sensor when the mirror is replaced by a highly reflective end face of a bimetallic strip arranged so that the temperature changes the gap.

10.6.5 The Pressure Sensor

A modification of the position sensor is used for pressure measurement. Figure 10.13 shows the schematic of the sensor. Three multimode fibers are fed through a pressure conduit to the sensing head. One of the three fibers is the input fiber, which transmits light to a mirror coating on a membrane that deflects with pressure. The other two fibers receive light after reflection from the mirror. As the fibers are at different distances, they have different characteristic outputs as a function of the deflection of the membrane or pressure. The optical output from these fibers is equal when the membrane is undeflected. This null condition is realized by supplying pressure to the sensing head. In other words, the external pressure is equal to the pressure supplied to the sensing head when the outputs of the two fibers are equal.

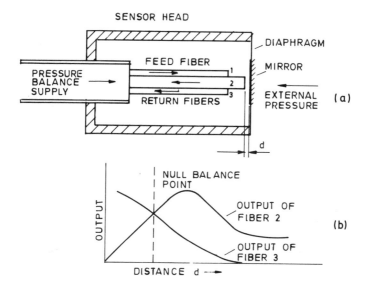

Figure 10.13 (a) A schematic of a pressure sensor and (b) variation of output with distance d.

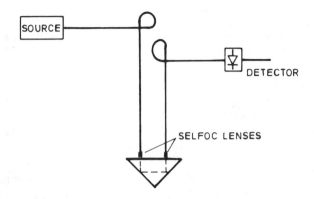

Figure 10.14   A liquid level fiber-optic sensor.

10.6.6 The Liquid Level Sensor

The liquid level sensor [6] acts as a switch. It works on the principle that when an unclad fiber is immersed in a liquid, the increased refractive index at the interface allows light to escape from the fiber. This enables a go/no-go type of detection. In another version, as shown in Fig. 10.14, the light from the input fiber, after total internal reflection at the prism surfaces, is received by the output fiber. As the prism surfaces come in contact with the liquid, the condition for total internal reflection is violated, and some of the light escapes to the liquid. This results in a sharp decrease in the output. Such level sensors are used for monitoring levels in petroleum tanks and cryogenic propellant tanks.

10.6.7 The Refractive Index Sensor

Let us consider a multimode fiber of diameter $2a_1$ that has been tapered uniformly over a small length and then to a fiber of diameter $2a_2$. This is shown in Fig. 10.15. The tapered portion forms the sensing head. If a Lambert source is used to inject optical power into the input fiber, then the power coupled into fiber 2 through the taper would vary linearly with the dielectric constant $\varepsilon = n^2$ of the medium surrounding the sensing head. The power P coupled to the second fiber [7] is given by

$$P = P_0 \left[\frac{n_1^2 - n^2}{R^2(n_1^2 - n_2^2)} \right] \tag{10.17}$$

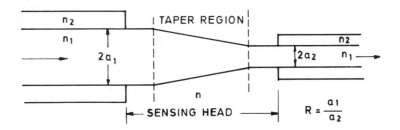

Figure 10.15   Geometry of a fiber-optic refractometer.

where n_1 is the refractive index of the core of fiber 1, taper, and fiber 2, n_2 is the refractive index of the cladding, and $R = a_1/a_2$. It is therefore seen that the power coupled decreases linearly with increasing n^2. The largest refractive index that can be measured is equal to that of the core. On the other hand, all the power launched into fiber 1 will be coupled to fiber 2 when $n = [n_1^2 - R^2 (n_1^2 - n_2^2)]^{1/2}$. This lower value can be changed by changing R.

10.6.8 Micro-bend Loss Sensors

When an optical fiber is bent in an alternating series of curves or microbends the light traveling in the fiber is attenuated. Possible microbending loss mechanisms for the light are radiation loss, because light traveling a steep bend may be radiated sideways out of the fiber core and into the cladding, and mode coupling loss, because bending of a fiber results in coupling between the modes, normally into higher-order modes or more leaky modes [8]. A number of sensors are designed around microbending. Figure 10.16 shows a microbending transducer. A multimode fiber is sandwiched between a comb structure with spatial period d. The difference in the propagation constants of adjacent modes in a fiber is given by

$$\Delta\beta = \beta_{m+1} - \beta_m = \sqrt{\frac{\alpha}{\alpha + 2}} \frac{2\sqrt{\Delta}}{a} \left(\frac{m}{M}\right)^{(2-\alpha)/(2+\alpha)} \tag{10.18}$$

where α is the grading constant of the fiber, a is its radius , M is the total number of modes, m is the mode label, and Δ is the refractive index difference between core and cladding. β_m is the propagation constant of the mth mode. For a parabolic index fiber, $\alpha = 2$ and hence

$$\beta = \frac{\sqrt{2\Delta}}{a} \tag{10.19}$$

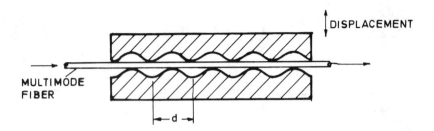

Figure 10.16   A microbend displacement transducer.

It should be noted that the values of microbending loss vary considerably for different fibers. In general, the losses increase for fibers of lower numerical aperture.

The two modes with propagation constants β and β' will couple when the microbend periodicity satisfies the equation

$$\Delta\beta = |\beta - \beta'| = \frac{2\pi}{d} \tag{10.20}$$

Therefore the spatial period of the comb structure is chosen so that it matches the difference of propagation constants between suitably chosen modes. Usually this period lies in the mm range.

The microbend transducer essentially responds to displacement, which may be caused by pressure, temperature, force, etc. Displacement sensitivity better than 0.01 mm is easily achieved. By increasing the input power and using an optimum detector, sensitivities in the range 10^{-4} mm can be achieved.

The relationship between applied force and displacement is found to be linear over a very large range, thereby making the microbend transducer a good candidate for a force transducer. The small size and very small movement required makes possible a rugged design of a microbend transducer.

10.6.9 The Fiber-Optic Doppler Probe

The frequency of light scattered from a moving target (solid or fluid) is Doppler shifted, i.e., the frequency of the received light is different from that of the incident light. Consider a particle moving with velocity v, illuminated by a wave with propagation vector k_i and observed in the direction of propagation vector k_0. The frequency of the observed wave

w_s is given by

$$w_s = w_i + n_0 \bar{v} \cdot (\bar{k}_0 - \bar{k}_i) \tag{10.21}$$

where n_0 is the refractive index of the medium. The Doppler shift $\Delta \upsilon_D = (w_s - w_0)/2\pi$ can be expressed as

$$\Delta \upsilon_D = \frac{2}{\lambda} \, n_0 v \sin \frac{\theta}{2} \sin \beta \tag{10.22}$$

where θ and β are as shown in Fig. 10.17(a). When $\beta = \pi/2$, the velocity of the moving target is given by

$$v = \frac{\lambda}{2 n_0 \sin(\theta/2)} \, \Delta \upsilon_D \tag{10.23}$$

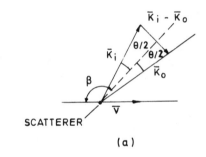

(a)

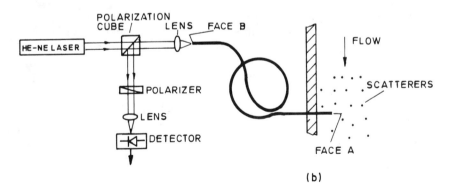

(b)

Figure 10.17   (a) Geometry of scattering from a moving particle and (b) schematic of a fiber-optic Doppler anemometer. (Modified from Ref. 2, used with permission.)

The velocity is linearly related to the Doppler shift. Instruments developed on this principle for measuring velocity are called laser Doppler anemometers (LDAs). Doppler anemometry is a powerful tool for noncontact measurement of fluid flow velocities over a very large range. It provides high spatial resolution. The basic advantage of a fiber-optic probe is that the position of the measuring region may be adjusted without recourse to the alignment of the system's launch and receive optics. The disadvantages are that the probe volume may be a few mm^3. Figure 10.17(b) shows a schematic diagram of a fiber-optic Doppler probe. A beam from a laser source is launched into a multimode fiber via a polarizing beam splitter and launch optics. The other end of the fiber is immersed into the moving seeded fluid. The light is scattered within the fluid by the scatterers, and some of the back scattered light is collected by the fiber. The scattered light is randomly polarized so that half of the returned light is reflected to the detector by the polarizing beam splitter. The reference beam is derived by the Fresnel reflection from the end A of the fiber. The multimode fiber rapidly depolarizes the input beam within a short distance. The returned signal from the end face A is depolarized, and hence the polarizing beam splitter reflects half of this to the detector. The two light beams, one scattered from the moving fluid and the other reflected from the end face A, are photo-mixed in the detector, whose output is the Doppler signal. The Doppler shift is obtained either by autocorrelation of the signal and then Fourier transforming it or directly from the spectrum analyzer. It may be noted that the back reflection from the end face B of the fiber has the same state of polarization as the incident beam and hence is transmitted by the polarizing beam splitter. The fiber-optic Doppler probe has been used to measure the blood flow in the veins as well.

The back scattered signal is extremely weak and presents a problem in adequate detection and measurement. A solution to this problem in some cases has been found by coating a retroreflective paint on the moving target, which enhances the back scattered radiation by a factor of several hundred over that of a white diffuse surface.

10.6.10 Fiber-Optic Faraday Monitor

When linearly polarized light passes through a transparent material of length L that is in the magnetic field H, its plane of polarization is rotated [9,10,11,23]. The rotation θ in orientation is given by

$$\theta = V \int_L H\ell\, d\ell \tag{10.24}$$

where Hℓ denotes the component of the magnetic field in the direction of the light, dℓ is a small element of the material, and V is the Verdet constant of the material. The measurement of the angle of rotation gives the magnitude of the magnetic field. The most suitable materials for this application are diamagentic materials, since in this case the Verdet constant is temperature independent.

The current measurement is done by measuring the magnetic field it produces. An ideal nonbirefringent optical fiber of length L is wound in N turns of radius r around a conductor carrying a current I. The Faraday rotation is then given by

$$\theta = VHL = VIN \tag{10.25}$$

since $H = I/2\pi r$. The Verdet constant for fused silica is 4.68×10^{-6} rad/A at 633 nm. Therefore a kilo amperes current in a conductor will produce a rotation of 21 min per loop of the fiber. The rotation depends only on the enclosed current. The technique is immume to vibrations of the current conductor or busbar.

Figure 10.18 shows a schematic of a current monitor. The beam from a He-Ne laser is launched into a single mode fiber. The light is polarized before launch, and it maintains the polarization until it reaches the busbar. The polarization is rotated by the magnetic field produced by the current.

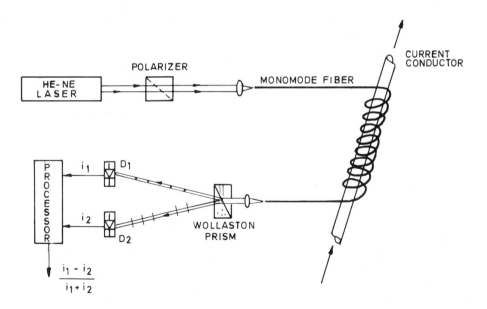

Figure 10.18 A schematic of a fiber-optic current monitor.

The beam is then transmitted to an analyzer for measuring the angle of rotation θ. The analyzer is a Wollaston prism oriented so that its principal section makes an angle of 45° with the incident polarization. The intensities of the two beams are

$$I_1(\theta) = I_0(1 + \sin 2\theta)$$

$$I_2(\theta) = I_0(1 - \sin 2\theta)$$

Assuming the photocurrent to be proportional to the intensities, the output T is

$$T = \frac{i_1 - i_2}{i_1 + i_2} = \sin 2\theta \qquad (10.26)$$

The output is a nonlinear function of θ; the linear output can be obtained over a range of ±10° using a simple linearization procedure. Although the method appears appropriate for the measurement of currents in high-tension lines, there are a large number of problem areas. For example, intrinsic birefringence and bend-induced birefrigence can introduce large errors. In principle it is possible to separate the influence of Faraday rotation from that of birefringence. The vibration of the fiber induced by an acoustic field can result in the rotation of the plane of polarization by a comparable magnitude. The fiber coil is kept in a jelly-filled tube to mitigate this problem.

The voltage can be measured by placing two quartz crystals on opposite sides of a conductor and passing separate linearly polarized beams perpendicular to the conductor through the crystals. In this configuration, the electric field is in the same direction while the magnetic field is in opposite directions. As a result the rotations caused by electric and magnetic fields will be added in one beam and subtracted in the other beam. The contributions of both fields then can be separated and the fields measured.

10.6.11 The Fiber-Optic Temperature Sensor

The temperature sensor is based on the variation of the optical activity of a quartz crystal with temperature. Figure 10.19 shows the schematic of a temperature probe. A beam from a He-Ne laser is launched into a multimode input fiber, which is at the center of a bundle of fibers. The light is transmitted to the detector D_2 by the remaining fibers in the bundle.

The multimode input fiber completely depolarizes the input polarization of the laser light. Therefore the polarization is completely defined by the polarizer P placed in the sensor head. The linearly polarized light

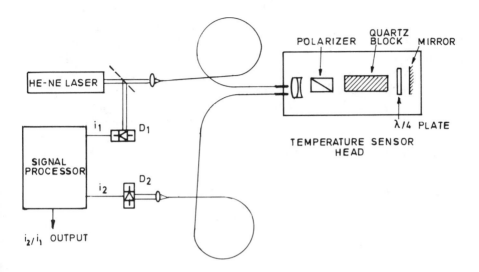

Figure 10.19   A schematic of a fiber-optic temperature sensor.

passes through the quartz crystal, which rotates its orientation. Faraday rotation may also be present, but its contribution is cancelled by double passage through the quartz block. The function of the $\lambda/4$ plate is to rotate the input polarization by double passage through it. The polarization prism also acts as an analyzer. The relative orientations of the polarizer and the $\lambda/4$ plate can be set so that the probe has the optimum sensitivity. The temperature is sensed by measuring the rotation of the plane of polarization that arizes from the temperature dependence of optical activity. A device having a quartz block 9 mm in diameter and 65 mm in length provides a resolution of 2°C over a range of 20–180 C. Such temperature sensors are well suited for applications in the electrical power supply industry. It may be noted that the fibers used in temperature probes perform the task of light conduction only.

10.7 COHERENT SENSORS

Coherent sensors measure the optical phase shift introduced by the action of the measurand. Thus they use single mode fibers in their design. The phase modulation is converted into intensity modulation by interferometry. There are four basic configurations for an interferometer, Michelson,

Mach-Zehnder, Fabry-Perot, and Sagnac, that have been used for sensing. Figure 10.20(a) shows the layout of a Mach-Zehnder interferometer for the measurement of variables that may induce phase variations. The reference beam can be frequency shifted for heterodyning. An all-fiber interferometer can be realized as shown in Fig. 10.20(b). The quadrature conditin is maintained by a PZT cylinder energized via a feedback loop. An all-fiber interferometer offers the advantages of mechanical stability and interface possibility with integrated optics. The heterodyne scheme cannot be incorporated without breaking the fiber to introduce a frequency shifter, etc., thus introducing mechanically sensitive interfaces.

10.7.1 The Fiber-Optic Temperature Sensor

There are two properties of a fiber that change with the temperature, its dimensions and its refractive index [12,13]. The relationship [12] between the phase change $\Delta\delta$ and the temperature change ΔT is given by

$$\Delta\delta = -\frac{2\pi L}{\lambda_0}\left[(n + \frac{a\lambda_0}{2\pi}\frac{\partial\beta}{\partial a})\alpha + \frac{\partial n}{\partial T}\right]\Delta T \qquad (10.27)$$

where L is the interaction length, a is the radius of the core of the fiber, $\partial\beta/\partial a$ is the rate of change of the propagation constant with respect to the radius of the core, α is the linear expansion coefficient, $\partial n/\partial T$ in the refractive index temperature coefficient, and n is the mean refractive index of the core. If a glass fiber is used for a sensor, one obtains a phase change of 100 radians per degree Celcius per meter of interaction length. This corresponds to roughly 30 fringes per degree Celcius per meter. of interaction length. Therefore an FO thermometer is an extremely sensitive device.

The sensing fiber lies in one arm of the Mach-Zehnder interferometer. The fringes are counted electronically. In order to determine the direction of the fringe movement, and hence whether the temperature is above or below the ambient, two detectors in phase quadrature are used. Use of four detectors compensates for source intensity variations.

The extremely high sensitivity of the interferometer demands temperature stability of the reference arm and also of the mechanical and optical components of the interferometer.

10.7.2 The Fiber-Optic Pressure Sensor

When the hydrostatic pressure applied to a length L of a fiber is changed by ΔP, a phase change $\Delta\delta$ is introduced [12] which is given by

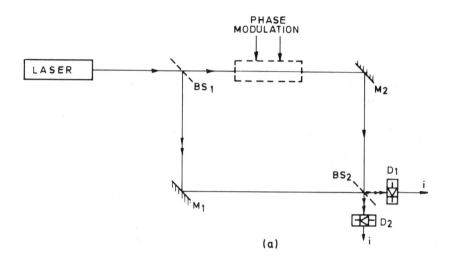

(a)

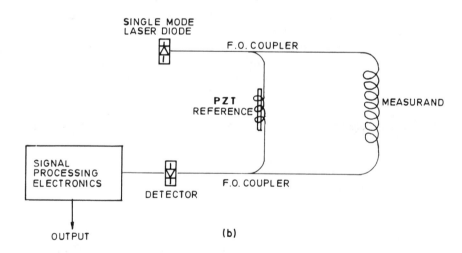

(b)

Figure 10.20   (a) A Mach-Zehnder interferometer and (b) a fiber-optic Mach-Zehnder interferometer. (Modified from Ref. 2, used with permission.)

$$\Delta\delta = \Delta P \frac{\pi L}{\lambda_0} \left[\frac{\lambda_0 a}{\pi} \frac{\partial \beta}{\partial a} - n^3(P_{11} + P_{12}) \right] \frac{1 - \mu - 2\mu^2}{E} \tag{10.28}$$

where P_{11} and P_{12} are the photoelastic constants, E is Young's modulus, and μ is the Poisson ratio of the fiber core material.

The sensing fiber is a silica fiber that is usually coated with a medium that transforms a pressure field into a longitudinal strain. The coating normally is of a compliant plastic drawn onto the fiber during manufacture. This enhances acoustic sensitivity [14,15]. The sensing fiber lies in one arm of the interferometer. Like any inteferometric sensor, the pressure sensor is extremely sensitive. As an example, a fiber 10 m long will detect an acoustic field at the threshold of hearing. Further, it has a good response over a wide frequency range from dc to 20 MHz, although at frequencies higher than 100 kHz the efficiency of conversion of pressure field into longitudinal strain decreases; hence the use of coated fibers is restricted to frequencies less than 100 kHz. Silica fiber itself may be used in the frequency range 100 kHz to 20 MHz.

The FO pressure sensor, apart from its very high sensitivity, offers the advantage of geometrical flexibility. A fiber sensor formed in a coil whose diameter is much less than the acoustic wavelength will act as an omnidirectional sensor. Two such coils can be used in a pressure gradient sensor; one is put in the reference arm and the other in the sensor arm of the interferometer. The output of the interferometer will be proportional to the pressure difference between the locations of these two coils.

A highly directional sensor can be realized by winding the fiber coil on a thin long cylinder; the diameter of the cylinder is kept much smaller than the wavelength and its length many times the acoustic wavelength.

10.7.3 Fiber-Optic Accelerometers

FO accelerometers [16] employ the property that acceleration-induced stress produces a strain in a fiber. The strain causes a phase change in the wave propagating in the fiber, that is measured using a Mach-Zehnder interferometer. In one configuration the accelerometer is in the form of a simple harmonic oscillator consisting of a mass suspended between two fibers or from a single fiber. When the device is accelerated along the length of the fiber, a strain is induced in the supporting fiber. The force acting on mass m is ma, where a is the acceleration. Therefore, from Hooke's law,

$$\frac{\delta L}{L} = \frac{ma}{EA} \tag{10.29}$$

where $\delta L/L$ is the strain, A the cross-sectional area, and E the Young's modulus of the fiber material. The phase shift $d\delta$ introduced due to the change in length δL is

$$d\delta = \frac{2\pi}{\lambda_0} n\delta L = \frac{2\pi}{\lambda_0} n \frac{maL}{EA} = \frac{8nmaL}{\lambda_0 Ed^2} \tag{10.30}$$

where n is the refractive index and d the diameter of the fiber. We define the sensitivity S of the accelerometer as the ratio of phase change to acceleration, so that

$$S = \frac{d\delta}{a} = \frac{8nmL}{\lambda_0 Ed^2} \tag{10.31}$$

For a typical fiber used in the sensor, n=1.5, E = 7.3×10^{10} N/m², d = 80 μm, and λ_0 = 633 nm, and hence the sensitivity S is given by

$$S = 4 \times 10^4 \text{ mL} \quad \text{rad/g} \tag{10.32}$$

For a mass of 10 g suspended on a fiber 10 cm long, this value works out ot be 40 rad|g. It is easy to measure a phase of a few degrees, and hence the accelerometer is highly sensitive.

Fiber-optic accelerometers may also be realized using the principle employed in hydrophones. A freely mounted mass may exert a pressure proportional to the acceleration of the fiber coil. The phase change thus induced may be detected using an all-fiber Mach-Zehnder interferometer.

10.7.4 The Fiber-Optic Rotation Rate Sensor: The Fiber-Optic Gyroscope

Of all FO sensors, the rotation rate sensor is the most important; it has a theoretical sensitivity of about 10°/hr. For the sake of comparison, the earth's rotation rate is about 14°/hr. Apart from sensitivity, the FO rotation rate sensor possesses some important features: it is a passive device: it has no moving parts; it consumes less power than other gyroscopes; and it does not suffer from errors due to null shift and lock-in.

The FO gyroscope is based on the Sagnac effect [17,18]. It consists of a long single mode fiber in the form of a coil capable of rotation about an axis. Coherent light is launched through both ends of the fiber as shown in Fig. 10.21. The light beams travel in opposite directions (clockwise and counter-clockwise in the coil) and are collected on emergence and super-posed for interference. The circular fringe pattern is then observed. The fringe pattern remains stationary if the coil is stationary. When the coil rotates about its axis, a phase difference is introduced between the beams traveling clockwise and that counter-clockwise. The phase difference $\Delta\delta$ is

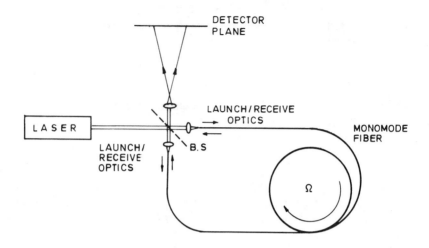

Figure 10.21   A schematic of a laser gyroscope.

given by

$$\Delta\delta = \frac{4\pi LR}{\lambda c}\Omega \qquad\qquad (10.33)$$

$$= \frac{8\pi AN}{\lambda c}\Omega \qquad \text{for a circular coil}$$

where Ω is the angular frequency with which the fiber coil is rotating, L is the fiber length, R is the radius of the coil, λ is the wavelength of light used, and c is the velocity of light. The area A of the circular coil is πR^2 and the total length of the fiber L is $2\pi RN$, where N is the number of turns in the coil.

The phase difference $\Delta\delta$ causes the fringes to expand or collapse depending on the direction of rotation. For slow rotation rates only a small change in the intensity of the central fringe is observed. The intensity variation in the inference pattern is expressed as

$$I = I_o(1+ \cos \Delta\delta) \qquad\qquad (10.34)$$

where I_o is the maximum intensity, and we assume that both the interfering beams are coherent and of equal amplitude. The sensitivity of the sensor will depend on the change in intensity with the incremental change in rotation rate. Differentiating Eq. (10.34) we obtain

$$\left| \frac{dI}{\delta\Omega} \right| = I_0 \left[\frac{4\pi LR}{\lambda c} \right] \sin \Delta\delta \qquad (10.35)$$

In Eq. (10.35) the bracketed factor can be called the sensitivity of the interferometer in a particular setup. Hence in principle one should use long fibers, in large-diameter loops for detecting very small rotation rates. It should not be inferred from this that one can measure any minute rotation rate by using longer lengths of fiber in larger-diameter loops. In practice the packaging criterion limits the size of the loop; optical fiber loss sets an upper limit on the length L; nonlinear damage effects forbid the use of high-power laser sources; and signal strength, scattered light, and the quantum efficiency of the detector utilimately limit the system's sensitivity. The sensitivity will increase by the use of radiation of shorter wavelengths. It may be noted that the phase shift $\Delta\delta$ does not depend on the material of the fiber.

There are various ways of measuring the phase shift. A pair of detectors in quadrature may be placed at the plane of the interference pattern. Both the direction and magnitude of fringe movement can then be obtained. Heterodyning and phase nulling schemes may also be used. It can be shown that for the dc detection scheme, the minimum rotation rate $\Omega_{\min}$ detectable for a shot noise limited detector (assuming a signal-to-noise ratio of 1) is given by

$$\Omega_{\min} = \left(\frac{4\, hvB_0}{\eta P_0} \right)^{1/2} \frac{\lambda c}{4\pi RL} \exp \frac{\alpha_T L}{2} \qquad (10.36)$$

where h is the Planck's constant, v the laser frequency, B_0 the detector bandwidth, η the quantum efficiency of the detector, P_0 the power launched, and α the attentuation in the fiber. If we take as some typical values $\alpha_T = 2 dB$, L = 4.3 Km, $P_0 = 3$ mW, R = 0.15 m, $B_0 = 10$ cm, $\eta = 0.5$, and, $\lambda = 633$ nm, we obtain

$$\Omega_{\min} = 10^{-4} \text{ deg/hr}$$

For the heterodyne scheme we would expect it to be lower by a factor of $1/2\sqrt{2}$. Very large changes in the absolute phase of the fiber path are introduced by variations in the environmental parameters temperature and pressure. The essential feature of the FO gyroscope is that the reference and signal beams in the interferometer travel exactly the same path in the fiber and any air gap, so that in principle the phase changes due to environmental variables are fully compensated and the only difference in phase is due to the Sagnac effect. The Sagnac interferometer is a true zero-path-difference interferometer, so that detection of very small phase differences is fundamentally impossible unless the operating point of the interferometer is shifted to satisfy the quadrature condition. Operation at the quadrature point is achieved by the introduction of a nonreciprocal phase bias of $\pi/2$ between counterpropagating beams. The $\pi/2$ bias is achieved by slightly defocusing the coupling lenses or by using a lossy beam splitter. In a dual input FO gyroscope the operating point automatically shifts to the quadrature condition when it is rotated [19].

The FO gyroscope is a low cost, all solid-state device that has a very wide application range. It does not exhibit the lock-in effect that prevents the ring laser gyroscope from performing below a certain rotation rate.

10.7.5 The Magnetic Field Sensor

Magnetic field sensors [20,23] based on the Faraday effect may require very large lengths of the fiber to measure feeble fields. Magnetic field measurement can also be accomplished by using the magnetostriction effect. Magnetometers using fibers coated with magnetostrictive materials or wound on magnetostrictive mandrels have the potential of far greater sensitivity than is achieved by methods using Faraday rotation. Under the action of a magnetic field, the length of the fiber wound on the magnetostrictive mandrel changes, and consequently the phase of the light wave propagating through the fiber changes, which is detected using a Mach-Zehnder interferometer. In one arrangement the fiber is wound on a nickel cylinder that is kept in a magnetic field. The commercially available metallic glasses are good candidates for a magnetostrictive material. The fiber is either wrapped around the tube of a metallic glass or bonded onto a strip of metallic glass [21]. When properly biased, they exhibit a linear response over a fairly large range. Their frequency response is satisfactory up to about 1 kHz. The sensitivities of these sensors are extremely good; the minimum detectable magnetic fields are in the range of 5×10^{-9} Oe/m of fiber. This corresponds to a measurable phase change of 10^{-6} radians.

It may be mentioned that fiber bonded to a metallic glass may serve as a phase modulator.

10.7.6 The Strain Sensors

The strain sensor utilizes the change in optical path length caused by a strain in a fiber. Let a monomode fiber of length L be subjected to a longitudinal strain ε. This causes [22] a phase change $\Delta\delta$ of the wave propagating in the fiber

$$\Delta\delta = \Delta(\beta L) = \beta\ \Delta L + L\ \Delta\beta \tag{10.37}$$

where β is the propagation constant in the fiber. The first term represents the phase change due to change in the length, that is, $\beta\Delta L = \beta\epsilon L$. The second term contains the phase change due to the change in β that can arise from the stress-optic effect, in which the strain changes the refractive index of the fiber, or the waveguide mode dispersion effect, due to the change in the fiber diameter (2a) produced by the longitudinal strain. Thus

$$L\ \Delta\beta = L\frac{d\beta}{dn}\ \Delta n + L\ \frac{d\beta}{da}\ \Delta a \tag{10.38}$$

Usually $d\beta/dn = \beta/n$, and Δn for a homogeneous isotropic medium is given by

$$\Delta n = -\tfrac{1}{2}\ n^3\ [\epsilon\ (1 - \mu)\ P_{12} - \mu\epsilon\ P_{11}] \tag{10.39}$$

where P_{11} and P_{12} are the photoelastic coefficients and $\Delta a = \mu\epsilon a$. Substituting for ΔL and $\Delta\beta$, the phase change $\Delta\delta$ is given [12] as

$$\Delta\delta = \beta\epsilon L - L\ \frac{\beta}{2}\ n^2[\epsilon(1 - \mu)P_{12} - \mu\epsilon P_{11}] - L\mu\epsilon\ \frac{d\beta}{da}\ a$$

or

$$\frac{\Delta\delta}{\epsilon L} = \beta[1 - \frac{n^2}{2}\ \{(1 - \mu)P_{12} - \mu P_{11}\} - \frac{\mu}{\beta}\ a\ \frac{d\beta}{da}] \tag{10.40}$$

The phase change can be measured for static strain by counting fringes in a Mach-Zehnder interferometer. The fibers in both arms of the interferometer may be mounted in such a way that they experience compressive and tensile strains and thereby give enhanced sensitivity.

The sensitivity of an interferometric sensor depends on the characteristics of the fiber. Table 10.1 compares temperature, pressure, and strain sensors realized from a single mode silica fiber.

In this chapter we have described various kinds of fiber sensors. They are classified as incoherent and coherent FO sensors. Coherent FO sensors are by far the most sensitive ones. FO sensors offer many advantages over conventional sensors, the most important being freedom from EMI, adaptability to a hostile environment, and compatibility with integrated optical devices and telemetry. An all-fiber sensor is compact and has no interfaces. The possibility of multipoint sensing using time multiplexing, and also the capability of measuring more than one parameter simultaneously, are additional features of FO sensors.

Table 10.1 Sensitivity and Characteristics of Sensors Employing Single Mode Silica Fiber

Variable	Interferometric configuration	Sensitivity	Charactersitics
Temperature	Mach-Zehnder interferometer with bidirectional counter	30 fringes/C/m	Dynamic range depends on interferometric configuration and stability. Linearity depends on fiber itself
Pressure		2 fringes/bar/m	
Strain		2 fringes/ μ strain/m	

REFERENCES

1. Oliver, M. R., Spooncer, R. C., and Ghezelayagh, M. H. (1986). An auto-referenced two state optical fiber reflective sensor, *SPIE 630*; 233–238.
2. Pinnock, R. A., Extance P, Hazelden, R. J., Pacaud, S. J., and Cockshott, C. P. (1987). Color-modulation optical displacement sensors, *SPIE, 734*; 238–244.
3. Place,J. D., and Maurer R. (1986). A non-invasive fiber optic pick-up for a turbine flowmeter, *SPIE, 630*: 226–232.
4. Domanski, A. and Kille, A. (1986). Investigation of parameters of Fibre-optic temperature sensor, *SPIE, 670*: 123–126.
5. Grattan, K.T.V., and Palmer, A. W. (1968). Fluorescence monitoring for optical temperature sensing,*SPIE, 630*: 256–265
6. Harmer, A. L. (1987). Testing of practical fibre-optic sensors for process control systems, *SPIE, 734*: 231–236.
7. Kumar, A. Subrahmanyam, T. V. B., Sharma, A. D., Thyagarajan, K, Pal, B. P., and Goyal, I. C. (1984). Novel refractometer using a tapered optical fibre, *Elect. Letts, 20*: 534–535.
8. Oscroft, G. (1987). Intrinsic fibre-optic sensors, *SPIE, 734*: 207–213.
9. Papp, A. and Harms, H. (1980). Magneto-optical current transformer 1: Principles, *Appl. Opt., 19*, 3729–3734.
10. Aulich, H., Beck, W., Dauklias, N., Harms, H., Papp, A., and Schneider, H. Magneto-optical current transformer 2: Components, *Appl. Opt., 19*: 3735–3740.
11. Harms, H. and Papp, A. (1980). Magneto-optical current transformer 3: Measdurements, *Appl. Opt., 19*: 3741–3745.
12. Culshaw, B., Davies, D. E. N., and Kingsley, S. A., (1978). Fibre optic strain, pressure and temperature sensors, *Proc. Fourth European Conference on Optical Communication*, Geneva, pp. 115–126.

13. Hocker, G. B. (1979). Fiber-optic sensing of pressure and temperature, *Appl. Opt. 18*, 1445–1448.
14. Giallorenzi, T. G. (1981). Fiber optic sensors, *Optics and Laser Technology*, April, 73–78.
15. Bucaro, J. A., and Hickman, T. R. (1979). Measurement of sensitivity of optical fibers for acoustic detection, *Appl. Opt., 18:* 938–940.
16. Tveten, A. B., Dandridge, A., Davis, C. M. and Giallorenzi, T. G. (1980). *Elect. Letts, 16:* 854–856.
17. Lin, S. C., and Giallorenzi, T. G. (1979). Sensitivity analysis of the Sagnac-effect optical-fiber ring interferometer, *Appl. Opt., 18:* 915–930.
18. Bohm, M. (1986). Achievements and perspectives of fiber gyroscopes; *SPIE, 630:*205–212.
19. Rashleigh, S. C. and Burns, W. K. (1980). Dual-input fiber-optic gyroscope, *Opt. Letts, 5:* 482–484.
20. Buchholtz, F., Koo, K. P., Kersey, A. D., and Dandridge, A. (1986). Fiber optic magnetic sensor development, *SPIE, 718:* 56–65.
21. Trowbridge, F. R., and Phillips, R. L. (1981). Metallic-glass fiber-optic phase modulators, *Opt. Letts, 6:* 636-638.
22. Butter, C. D. and Hocker, G. B. (1978). Fiber optics strain gauge, *Appl. Opt., 17:* 2867–2869.
23. Yoshino, T. (1987). Optical fiber sensors for electric industry, *SPIE, 798:* 258–265.

ADDITIONAL READINGS

Culshaw, B. (1984). *Optical fiber sensing and Signal Processing,* Peter Peregrines, London.

du Chastel, Marie-Hélène, (1987). Fiber optic sensors begin moving from laboratory to market place, *Laser Focus| Electro-optics* May, pp. 110–120.

Jones, J. D. C., and Jackson, D. A. (1985). Research advances in fiber optic interferometric and polarimetric sensors, *Laser Focus|* Electro-optics, October, pp. 142–146.

Penny, C. M., and Caulfield, H. J. eds. (1986). Optical techniques for industrial measurement and control, *Proceedings of the Fifth International Congress on Application of Lasers and Electro-optics, IFS Publications.*

CHAPTER 11

Miscellaneous Techniques

This chapter contains discussion of a few more useful optical techniques. These include noncontact surface sensors, surface evaluation methods, scanning methods, laser Doppler anemometry, and ellipsometry.

11.1 THE OPTICAL PROBE

An optical probe is a noncontacting surface sensor that measures at short distance small changes in the axial position of a body. The advantage of a noncontact sensor is in its application to soft materials and the possibility of high scanning rates. Objects with large height variations (very rough surfaces) can be measured faster than with contacting devices. Such sensors are obviously useful in coordinate measuring machines, surface profiling, optical disc technology, etc. A variety of schemes have been proposed [1-10].

Figure 11.1 shows a design of an optical probe [1]. Light from a source L illuminates a prism P through an aperture A, which has two holes symmetrically placed about the optical axis by a condenser lens C. The prism P is a truncated pyramid with a clear nose but reflecting surfaces. The nose of the prism is focused by a lens O on the surface of a body W. The hatched fans of rays show the rays passing through the two apertures of A. Both the fans converge to form a single image of the nose on the surface when it is at the correct focus of the lens O.

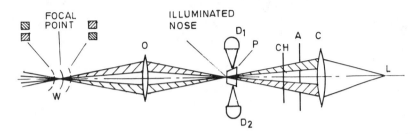

Figure 11.1 Optical arrangement of an optical probe. (From Ref. 1, used with permission.)

The light reflected or scattered from the surface will be collected by the lens O, and the nose image will again be reimaged (autoreflected) on the nose of the prism. Neglecting diffraction effects, no light reaches the reflecting sides of the prism or the two detectors D_1 and D_2. When the surface is outside the focus of the probe, however, the spot splits and the two move laterally in opposite directions. The autoreflected images therefore fall on the sides of the prism P, and the detectors D_1 and D_2 start receiving light. The direction of movement of the spots is reversed as the body goes from inside the focus to outside. In practice, a chopper CH placed in front of the aperture A allows only one hole to pass light at a time as the rotating chopper disc blocks one or the other hole.

Whenever there is a displacement of the surface from the focus, light from one holes falls on one detector and light from the other hole on the other detector. This relation reverses as the surface moves closer to or farther away from the probe. In the presence of a distance error, each detector receives light whose magnitude is proportional to the error magnitude. The relation between a particular hole in the aperture A transmitting and a detector receiving light determines the error direction. The signals received from the detectors are compared, the error magnitude and direction are determined and appropriate adjustment is done to arrive at the focus.

It can be shown that the focus error Δ,L the spot separation $\Delta\sigma$, and the beam convergence angle ψ are related as

$$\Delta\sigma = 2\psi/\Delta L \tag{11.1}$$

It has been demonstrated that distance error resolution of 0.1 μm down to 1 nm can be achieved depending on whether the surface is diffuse or specular. The probe is insensitive to tilt of the surface for small tilt angles. For larger tilt angles it is possible to set the probe for normality by obtaining error information from the detector signals.

Several optical probes have been designed using lasers as light sources because they give high-intensity focused spots. Figure 11.2 shows a laser probe that employs a linear diode array or linear position detector [2,3]. A low power He-Ne or diode laser projects a spot of light on a diffuse surface. Part of the light scattered from the surfaced is collected by a lens to image the spot on an array or position detector. If the body is displaced from the focus position by a small amount Δz, a slightly defocused spot will be formed on it. This spot will be imaged by the lens on the detector so that the center of the spot is displaced by an amount Δd from the original position. The measurement of this displacement can be used to determine the displacement of the body and is given by

$$\Delta z = \frac{\Delta d}{m \sin \theta} \qquad (11.2)$$

where m is the magnification of the imaging geometry. This is only an approximate relation. Once the probe is constructed, it can be calibrated by giving known Δz inputs.

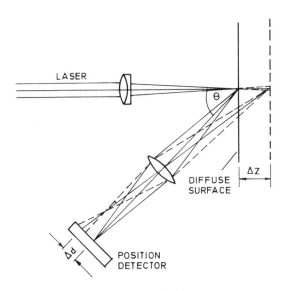

Figure 11.2   Laser probe with a position detector.

The system shown in Fig. 11.2 does not keep the image in focus as it is displaced on the detector. To keep the image in focus, the detector must be oriented to an angle to the lens axis as shown as Fig. 11.3. This system gives

$$\Delta z = \frac{\Delta d \, \sin \phi}{m \, \sin \theta} \tag{11.3}$$

This equation again is not exact and a calibration is necessary. The range and resolution of the probe depends on the magnification of the lens systems used. A resolution of a few micrometers can be achieved. The determination of the image location on the array is an important consideration. Figure 11.4 shows the record of a profile obtained with a three-dimensional profiling device based on this technique. Scanning can be in the x–y direction or around a circle.

Figure 11.5 shows another probe based on the above principle, which makes use of a projected slit spot on the surface [4]. A slit is projected with the help of a lens L_1 whose axis is inclined at an angle θ to the normal. The reflected image of the slit is seen with another lens L_2 from the other side of the normal with its axis again inclined at angle θ. The location of the

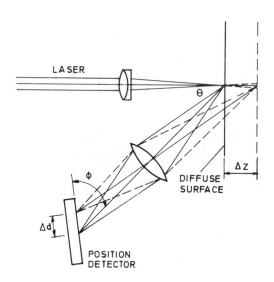

Figure 11.3   Laser probe with position detector at an angle relative to the lens axis.

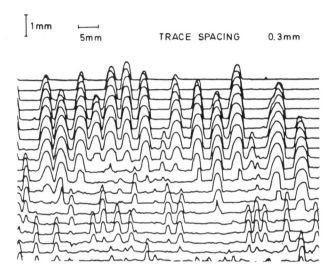

$\updownarrow$ 1mm

$\longmapsto$
5mm

TRACE SPACING   0.3mm

Figure 11.4   Profiles of weld deposits obtained with a laser probe. (From Ref. 3, used with permission.)

reflected spot (and hence of the surface) is determined with the help of a differential-position-sensitive photodiode pair. The detector produces a zero signal when the spot falls symmetrically on the diode pair, indicating that the projected spot is imaged exactly on the work surface. Any displacement of the surface linearly displaces the slit image on the detector, upsetting the balance, and a signal proportional to the displacement is produced. A linear range of ±75 µm and a resolution of 0.01 µm for polished surfaces has been reported.

A two-beam optical probe similar in principle to one discussed in Ref. [1] using a position-sensitive detector pair has been proposed [5]. Figure 11.6(a) shows two light beams B_1 and B_2 intersecting at a spot S_1 on the work surface W_1. A lens images the spot at the position I_1; see Fig. 11.6(b). When the object is in position W_2 or W_3, the spot splits into two spots S_2 and S_2' or S_3 and S_3'. The corresponding images on the detector will be I_2 and I_2' or I_3 and I_3'. The positional relationship between the images (and hence the signals) is related to the displacement and inclination of the surface. In practice the two beams are switched on alternately to discriminate between the images of B_1 and B_2 and decide whether surface is at 2 or 3.

When the surface is normal to the optical axis, the spots S_2 and S_2' or

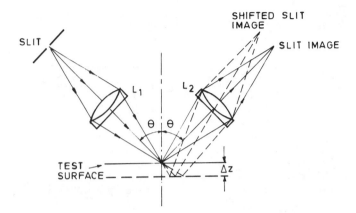

Figure 11.5   Optical probe using a projected slit. (From Ref. 4, used with permission.)

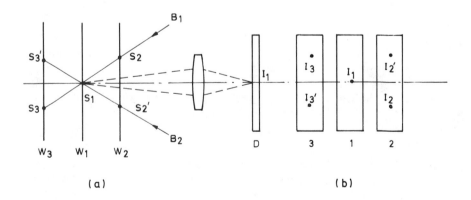

Figure 11.6   Principle of a two-beam laser probe. (From Ref. 5, used with permission.)

the images I_2 and I_2' are symmetrically separated. If the surface is inclined, the images are not symmetrically separated, and the asymmetry is related to the angle of inclination (Fig. 11.7). The detector signals can be processed to measure inclination. The optical arrangement of an actual probe is shown in Fig. 11.8. Two laser diodes are used as light sources. The lenses L_3 and L_4 collimate the light from the diode lasers, and the lens L_1 focuses them on the surface. Images of the spot are formed on a position-sensitive photodetector by lenses L_1 and L_2. Measurements have shown linearity over ±1 mm. Resolution in the submicrometer range has been achieved.

Surface sensors have also been constructed using the total internal reflection principle [6]. Figure 11.9 explains the principle of such a device. A point source S_1 placed at the focus of a lens gives a collimated beam. This beam is deviated through 90° by a prism acting as a total internal reflector. The light intensity is uniform across the beam and an equal amount of light is received on two detectors D_1 and D_2. If the source moves to the S_2 or S_3 position, the light leaving the lens will be divergent (rays 2) or convergent (rays 3) upsetting the critical angle condition at the upper or the lower edge of the light beam, respectively. Consequently the detectors will receive different amounts of light that can be used to sense the focus error. The principle is also used in the pickup for an optical disc. A practical high-precision optical surface sensors has been developed based on this principle with a resolution of 1 nm for measurement of surface roughness [6]. The point source is obtained by projecting a light spot on the surface from

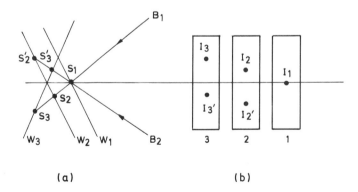

(a)   (b)

Figure 11.7 Measurement of inclination of a surface with a two beam-laser probe. (From Ref. 5, used with permission.)

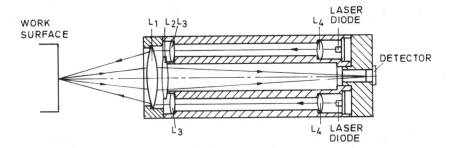

Figure 11.8 Optical arrangement of a two-beam laser probe. (From Ref. 5, used with permission.)

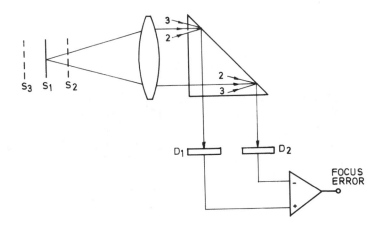

Figure 11.9 Principle of an optical probe using total internal reflection. (From Ref. 6, used with permission.)

a laser diode with a microsope objective, as shown in Fig. 11.10. The reflected or scattered light is used for position sensing using the above mentioned principle. To avoid the influence of surface inclination and spatial nonuniformity of the light beam, two identical paths consisting of parallel prisms and split photodetectors are used. The error component is eliminated from the signal by the operation $\{(A-B) + (C-D)\}/(A+B+C+D)$.

Focusing properties of a lens are made use of in another type of surface sensor [7,8]. The principle is illustrated in Fig. 11.11. A collimated light beam is focused by a lens L_1 on the surface. The back scattered light is collected by L_1 and focused at the two focal planes F_2 and F_2' by lens L_2 and received by detectors D_1 and D_2. B_1 and B_2 are beam splitters. If two identical matched spatial filters SF and SF' are placed at F_2 and F_2' respectively, they allow all the light to pass through. If displaced from the focus, the filters allow less light to pass through. It is possible to locate the filters on either side of the foci F_2 and F_2' so that the transmitted powers are equal. The effect of a small longitudinal displacement of the surface O will be an effective shift of the focal planes F_2 and F_2' as they come closer to one spatial filter and move away from the other. As a result, the power transmitted to the two detectors will be proportional to the magnitude and

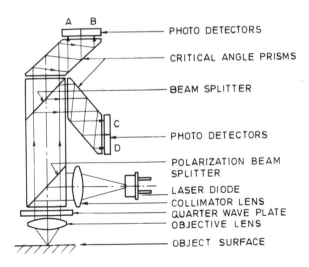

Figure 11.10   Optical arrangement of an optical probe using total internal reflection. (From Ref. 6, used with permission.)

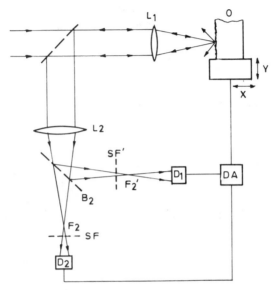

Figure 11.11   Principle of an optical probe using the focusing property of a lens. (From Ref. 7, used with permission.)

direction of displacement. The outputs of the detectors D_1 and D_2 are fed to a differential amplifier DA. A submicron resolution has been achieved.

11.2 SURFACE EVALUATION

A variety of optical techniques have been developed for surface evaluation. Some of the optical techniques are likely to play a major role in nondestructive on-line evaluation of surfaces produced by manufacturing processes [11]. Surface errors can be classified as form errors, waviness, and roughness. Basically this classification is based on the wavelength of the irregularities present on the surface.

Engineering surfaces are produced by a wide variety of processes. Depending on the manufacturing technique, the structure of the surfaces vary and they may be highly complex. Some surfaces, such as milled surfaces, may be regular and periodic. Bead-blasted surfaces, on the other hand, are highly random. A number of statistical parameters and functions have been developed to describe surface topography. The basic aspects of the topography are the heights of the irregularities (amplitude) and the longitudinal spacings between them (wavelength). Some of the common parameters to describe surface topography are R_t, which is the peak-to-valley height; R_a or roughness average, which represents the average deviation of the surface profile about its mean line; and R_q, which is the rms roughness defined as one standard deviation of the profile about the mean line. We discuss here some of the optical techniques useful in surface evaluation.

11.2.1 The Light Section Microscope

One of the earliest instruments used for measurement of a surface is the light section microscope introduced by Schmaltz in 1932. A fine slit image is projected onto the surface at an angle to the surface normal, with the help of a microscope, and it is seen by another microscope, again at an angle but on the other side of the surface normal, an arrangement similar to that in Fig. 11.5 with θ usually equal to 45° and the split detector replaced by a micrometer eyepiece. The slit image seen in the second microscope has the profile of the surface as shown in Fig. 11.12. The

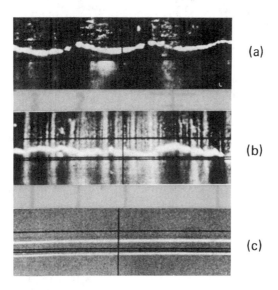

(a)

(b)

(c)

Figure 11.12   Slit image in a light section microscope from a rough surface and (c) from an anodized aluminum specimen (Courtesy of Gaertner Scientific Corporation, USA).

method is equivalent to sectioning the object with a light fan. This technique can measure R_t. Since the magnification of the microscope is limited to about 500X, the light section microscope is used to examine fairly rough surfaces with R_t in the range 0.5 to 50 µm. The light section technique is also useful for measuring the thickness of transparent thin films on surfaces in the above range. Figure 11.13 shows the principle of thickness measurement. The slit focused on the transparent surface is seen split into two in the image due to partial reflections both at the top A and bottom B of the layer as shown in Fig. 11.12. Because of refraction B appears at B' in the transparent film. The separation AB' is measured by the micrometer eyepiece and converted into a thickness value using the refractive index data of the film.

11.2.2 Interference Surface Profilers

An interference fringe connects the points of the same height of a test surface. A fringe therefore represents the section of a surface and shows the details of the surface profile. To examine the micro surface profile, inteference microscopes are used that may have the Twyman-Green, Fizeau, Mireau etc. configuration. Figure 11.14 shows an interference microscope based on the Twyman-Green configuration. A typical interference pattern of a rough surface is also shown. Visual magnification of the

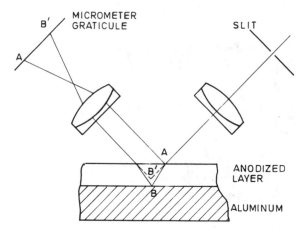

Figure 11.13   Measurement of the thickness of a transparent layer with a light section microscope.

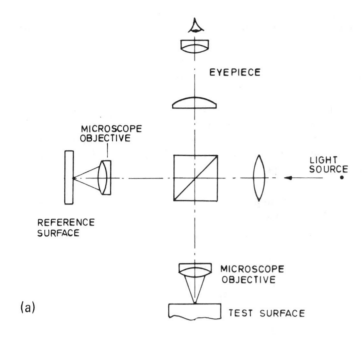

(a)

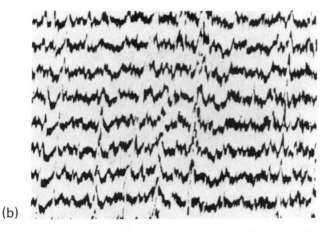

(b)

Figure 11.14 (a) An interference microscope and (b) a typical interference pattern of a rough surface.

microscope is 500 to 700X. Visual interpolation can give height differences to better than 0.05 μm. The test surface should be reflective to get interference. Hence this method is applicable where the surface finish is high.

In the two-beam interference pattern, the fringes are broad and the intensity varies sinusoidally. Consequently, the clarity of the profile is poor. This can be avoided by photographing the fringes and using a microdensitometer. This procedure is, however, time consuming. A fringe profile can also be obtained by analyzing the fringes from a video monitor. The fringes can be made sharp using the multiple beam interference technique. One interferometer based on multiple beam interference is the FECO scanning interferometer [12]. It uses a television camera to scan the image of a single interference fringe, and processing is done by a microcomputer. The instrument has a vertical resolution of about 1 nm, RMS and a horizontal resolution of 2 μm.

Recent developments in the area of interferometry, namely heterodyne and phase shifting interferometry (Chap. 6) have resulted in high performance inteference surrace profilers with very good height resolution. The lateral resolution, however, is inferior to that of mechanical profilers because of the size of the diffraction spot on the surface.

A common path dual spot interferometric profiler based on the heterodyne technique was discussed in Sec. 6.1.3. A polarimetric interferometer profiler has been proposed based on the same common path principle [13]. The two spots of different diameters are produced here using a birefringent lens in conjunction with a microscope objective, as shown in Fig. 11.15(a), giving two foci corresponding to orthogonally linearly polarized light beams with o and e polarizations. It is clear that on a sample surface one beam will produce a sharp focus (a small spot) and the other a larger defocused spot as shown in Fig. 11.15(b). As the sample is scanned across the light beams, the phase difference between the two polarized components varies according to the surface profile, as explained in Sec. 6.1.3. This phase difference is measured by a polarimetric system. Figure 11.16 shows the optical scheme of the profiler. An expanded beam from a He-Ne laser passes through a polarizer P with its transmission axis at 45° to the optic axis of thre quartz lens QL. The $\lambda/2$ plate is used to control the intensity of light in the system. The quartz lens splits the light into two orthogonally polarized beams of equal intensity, producing longitudinally shifted foci. The $\lambda/4$ plate Q_1 reverses the polarization states of the two components after reflection from the surface, their phase difference being governed by the surface profile. The reflector R, consisting of an aluminium film overcoated with a 130-nm layer of magnesium fluoride, does not introduce any phase or amplitude difference between the beams. These beams produce a linear vibration on passing through the $\lambda/4$ plate Q_2 oriented at 45° to the

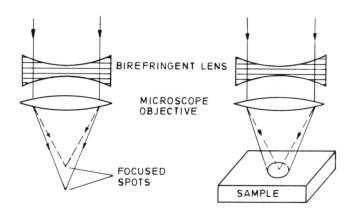

Figure 11.15   Principle of a dual-spot polarimetric interference profiler. (From Ref. 13, used with permission.)

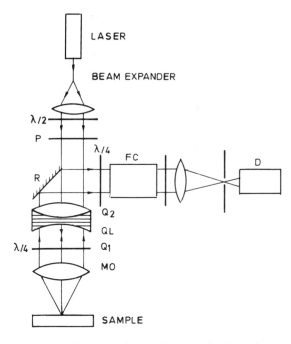

Figure 11.16   Optical scheme of a polarimetric interference profiler. (From Ref. 13, used with permission.)

orthogonally polarized components of the beam. The azimuth of this linear polarization depends on the phase difference between the components and hence the surface topography. The variations in the azimuth of this beam as the sample is scanned across the light beam is measured using a Faraday effect polarimeter consisting of a Faraday cell FC, an analyzer A and a detector D. The sensitivity of the instrument is better than 0.1 nm.

11.2.3 Optical Profilers

In the previous section, several optical probes have been discussed. Some of these have been developed as high-resolution surface profilers [6–8]. Here again the lateral resolution is limited by diffraction.

11.2.4 Scattering Techniques

The profilers result in surface profiles that must be analyzed to derive the roughness parameter. Some of the optical techniques directly produce a measurable quantity related to roughness. These are based on the light reflected or scattered from the surface [11,14–19].

A collimated monochromatic light beam incident on a rough surface is scattered into an angular distribution; see Fig. 11.17. For a perfectly smooth surface ($R_q <<\lambda$), all the light is specularly reflected (direction 1). The distribution pattern of the scattered radiation depends on the roughness height, the spatial wavelengths, and the wavelength of light. As in the case of a grating, the diffracted light from a small spatial component goes into large angles (direction 3, 3'). For a practical surface the spatial components are distributed over a broad spectrum; consequently the light is diffracted over a range of angles. The intensity of the specularly reflected light

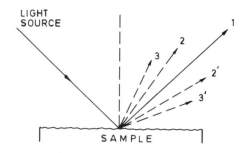

Figure 11.17   Reflection from a rough surface.

decreases with increasing R_q, and that of the diffracted light increases and becomes more diffuse. The light reflected or scattered from a rough surface therefore serves as the basis for several surface measurement techniques.

Specular Reflection

Assuming $R_q \ll \lambda$ and Gaussian height distribution for surface irregularities, the specular reflectance ρ of a rough surface is given by

$$\rho = \rho_0 \exp -\left(\frac{4\pi R_q \cos \theta_i}{\lambda} \right)^2 \tag{11.4}$$

where θ_i is the angle of incidence and ρ_0 is the total reflectance (including all scattered light) of the rough surface. It is seen that the specular reflectance is inversely correlated to roughness. The specular reflectance can be measured by placing a detector with a small aperture in direction 1 (Fig. 11.17). The instruments based on this technique are known as glossmeters. A properly calibrated instrument gives the result immediately, which is the main advantage of this method. It is ideal for the comparison of similiar surfaces.

One of the difficulties in measurement is that the detector should have a small aperture for measurement of reflectance to exclude any diffuse reflection in that direction. This makes the alignment of the optical system critical and sensitive to surface vibrations. The vibration problems may be minimized by using an optical fiber bundle to transmit and receive light.

From theoretical considerations R_q must be much less than the optical wavelength λ. For $\lambda = 0.6328$ μm (the He-Ne laser), R_q should be less than 0.1 μm. Hence for engineering surfaces, where R_q in the range 0.1 to 2 μm is expected, infrared radiation would be more practical. A resolution of 1 nm can be obtained by the use of this technique.

The Total Integrated Scatter Technique

The total integrated scatter (TIS) technique is complementary to the specular reflectance technique. One measures here the total intensity of the diffusely scattered light. This is defined as

$$TIS = \frac{\rho_0 - \rho}{\rho_0} = 1 - \exp -\left(\frac{4\pi R_q \cos \theta_i}{\lambda} \right)^2$$

$$\approx \left(\frac{4\pi R_q \cos \theta_i}{\lambda} \right)^2 \tag{11.5}$$

It is clear that the TIS is directly related to R_q. There are difficulties in implementing this techniques for on-line measurement. Figure 11.18 shows a scheme for TIS measurement. Light from a He-Ne laser strikes the test surface. The scattered radiation is collected by a hemispherical mirror and focused on a detector D_1. The specularly reflected beam is received on the detector D_2. A detector D_3 is arranaged to measure the incident beam with the help of a chopper mirror CH.

There are two additional ways to relate the light scattered from a rough surface to roughness. One way is to the measure the ratio of the specular intensity to the intensity at one off-specular angle. This ratio generally decreases with increasing surface roughness and thus could provide a measure of the roughness itself.

The angular distribution of the scattered radiation contains a great deal of informtion about the surface topography. Therefore in another method this distrbution is measured with the help of an array of detectors or a single scanning detector and related to surface parameters. This technique has been suitable for obtaining details about the amplitude and wavelength composition of surface topography. When the amplitude distribution is measured by a scanning detector the method is slow, but commmercial instruments have used diode arrays to allow for high-speed data collection.

Speckle Techniques

Speckles are produced when a rough surface is illuminated by coherent or partially coherent light (Chap. 7.) The spatial pattern and contrast of the speckle depend on the optical system used for observation, the coherence condition of the illumination, and the roughness of the surface. The speckle pattern therefore has been utilized for roughness measurement and could be applied to on-line measurement. The methods used to evaluate the surface roughness include the measurement of speckle contrast (laser and polychromatic) and speckle pattern decorrelation [11,18].

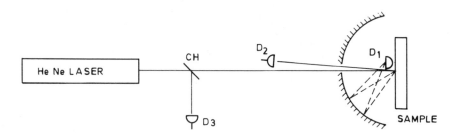

Figure 11.18   Measurement of total integrated scatter (TIS).

11.3 SCANNING TECHNIQUES

In several applications we must sample an object point by point in succession over the whole field. Alternately, an object field may be created point by point. This involves a scanning process. The devices that achieve this are called scanners. Basically scanners deflect light and may be classified according to the physical principle they use for light deflection: reflection, refraction, or diffraction [2,20–25]. Scanners have been employed in laser printers, as scanning analyzers for flaw detection in materials, in metrological devices, in infrared cameras, as line and page scanners, in character generators in laser systems for computer output on films, in point-of-sale terminals for bar code reading, and in facsimile transmission, reprography, etc.

11.3.1 Reflective Scanners

Mirrors and prisms are used as reflective scanners. Figure 11.19 shows a mirror scanner for a laser beam. As the mirror oscillates, the laser beam scans a line. Using a lens or a mirror system as shown in Fig. 11.20(a), the scanning beam can be made to scan parallel to a fixed axis. The laser beam itself converges at the focal plane of the lens. If necessary, this can be avoided by using the scheme in Fig. 11.20(b). In a lens system designed to meet the Abbe sine condition, the following relation holds good:

$$y = f \sin \theta \tag{11.6}$$

where y is the linear displacement of the beam, f is the focal length of the

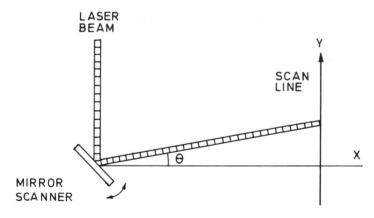

Figure 11.19   A mirror scanner.

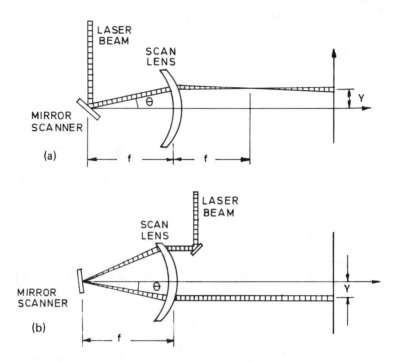

Figure 11.20   (a) A scanner with the scanning mirror at the focus of a lens and (b) a scanner system without a focused scan beam.

lens, and θ is the scan angle. θ may be written as

$$\theta = 2wt \tag{11.7}$$

where w is the angular speed of the mirror. The transverse speed (the scan speed) of the scan beam will be

$$v = \frac{dy}{dt} = 2wf \cos 2wt \tag{11.8}$$

For small θ this reduces to

$$v = 2wf \tag{11.9}$$

indicating linearity between the angular speed w and the scan speed.

   The lenses of the scanner system can be replaced by mirrors for large apertures. Parabolic mirrors are used to eliminate spherical aberration, as shown in Fig. 11.21. The linear displacement of the beam is given by

$$y = 2f \tan\frac{\theta}{2} \qquad (11.10)$$

Here f is the focal length of the parabolic mirror.

The scan mirror is given a limited rotation for scanning. Galvanometers are used for this purpose. The same effect but with higher speed can be achieved by continuously rotating a polygon with its surfaces coated so that each face acts as a mirror (Fig. 11.22). Such polygons are accurately fabricated to avoid face-to-face error. Typical values of the scan angle for the mirror or polygon scanner are in the range of 30 to 40° at a scan rate of several thousands.

One of the applications of a laser scanner system using a polygon sscanner is for gauging, as shown in Fig. 11.23. The object whose dimension is being measured is back lighted with the scan beam, which will be obstructed by the object. The duration of obstruction is a measure of the dimension and can be determined with the help of a detector. Fig. 11.24 shows the principle of the laser flaw detector, which forms the basis of a laser scanning analyzer. There is a sudden drop in the detector signal due to scattering of the laser beam at the site of the defect.

11.3.2 Refractive Scanners

Refractive components that can be used for scanning include wedge plates, lens, parallel plates, and polygons.

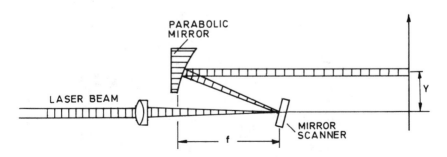

Figure 11.21   A scanner system with a parabolic mirror.

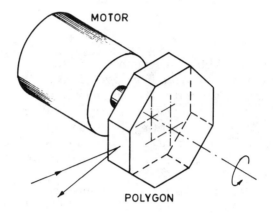

Figure 11.22 Rotating polygon scanner.

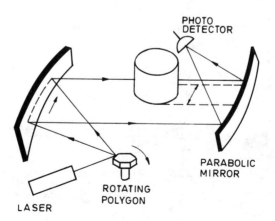

Figure 11.23 Scanner system for gauging.

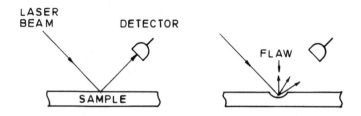

Figure 11.24   Principle of flaw detection using laser scanning.

The Single Wedge Plate

A light beam incident normally on a wedge plate is deviated through an angle $\theta = (n-1)A$, where A is the wedge angle and n the refractive index of the wedge material. The plane of deviation for normal incidence is normal to the refracting edge (the edge of the wedge). It is clear that when the wedge is rotated about the incident beam, the emergent beam sweeps out a cone whose half angle is θ.

Two Wedge Plates

When two wedge plates are combined, the total deviation θ of the beam passing through the assembly is the vector sum $\theta_1 + \theta_2$ of the individual deviations. If the two wedges are rotated in opposite directions, then the resultant beam sweeps out an elliptical cone with major and minor diameters proportional to $2(\theta_1+\theta_2)$ and $2(\theta_1-\theta_2)$, respectively. If $\theta_1 = \theta_2$ the beam effectively sweeps a straight line.

Lens

Figure 11.25 shows how a lens can be used as a scanner. Moving a lens transverse to the optical axis across a fixed incident beam at a speed v results in a refracted beam that sweeps at an angular frequency $w = v/f$, where f is the focal length of the lens.

Parallel Plate

A light beam passing through a parallel plate is displaced by an amount y given by

$$y = \frac{(n-1)tI}{n} \tag{11.11}$$

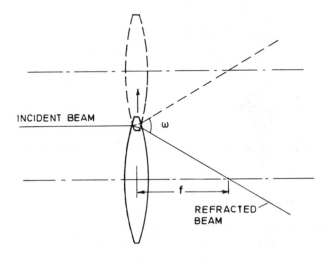

Figure 11.25   Lens scanner.

where I is the angle of incidence on the plate and n and t represent the refractive index and thickness, respectively, of the plate. By periodically oscillating the plate, a scan line can be obtained. The same effect can be achieved in a light beam transmitted by a rotational multifaceted polygon with parallel opposite faces that are parallel to the axis of rotation.

11.3.3 Diffractive Scanners

Diffractive scanners make use of the diffraction of light and are useful with monochromatic light. One of the scanner types makes use of the acoustooptic effect; see Chap. 2. When an acoustic wave propagates through a photoelastic material, a phase grating is set up in the medium as a result of refractive index variations due to the propagation of the wave. A monochromatic light beam is diffracted by such a grating into several orders. The relative amplitudes of the diffracted orders depend essentially on the thickness and the intensity of the sound field. In the Bragg regime, which concerns thick sound fields, only the zero and first orders exist. The first-order diffraction angle is given by

$$\theta = \frac{\lambda}{\Lambda} \tag{11.12}$$

where Λ is the acoustic wavelength. If the frequency of the wave is v then $\Lambda = v/v$ where v is the acoustic velocity. Then

$$\theta = \frac{\lambda}{v} v \tag{11.13}$$

If v varies over a band of frequencies Δv, the diffraction angle will vary over a range $\Delta\theta$ where

$$\Delta\theta = \frac{\lambda}{v} \Delta v \tag{11.14}$$

Thus by sweeping the frequency, the diffracted beam can be made to scan over a range of angles. These scanners work at high speed but have limitations of aperture $\approx$ 6mm and scan angle $\cong 3°$.

Another type of diffractive scanner uses a rotating holographic optical element for scanning in the same way as a mirror polygon. The hologram can be made into circular, cylindrical, or spherical sectors and can be either reflective or transmissive. One design [22] uses a flat disc with linear grating sectors as shown in Fig. 11.26. It can generate straight scans over nearly 40°. When $\phi_1 = \phi_2$, it is relatively free from cross-scan error due to its increase tolerance to wobble and centration error. This scanner has found application in printers.

Figure 11.27 shows another holographic scanner design, which uses

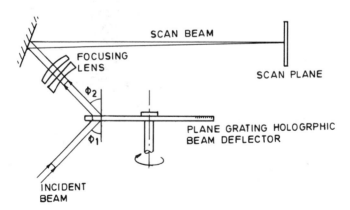

Figure 11.26 Holographic scanner using a linear grating. (From Ref. 22, used with permission.)

generalized zone plate structures for facets, does not require a scan lens for focusing, and gives focused flat-field straight line scanning [22].

11.4 FLOW VELOCITY MEASUREMENT

Measurement of flow velocity [26-28] is very important in the areas of fluid mechanics, aerodynamics, etc. Fluctuating flows are often encountered in practice, and they necessitate good spatial resolution and large frequency response. The hot-wire anemometer was developed to meet these demands. It has its shortcoming, however; the probe interferes with flow, its output is a nonlinear function of velocity, it requires calibration, and the interpretation of data may be difficult when turbulence is being measured. The flow visualization method is one of the earliest and still one of the most powerful flow measurement techniques. The problem lies in the quantitative analysis of data, since its involves tracking the motions of thousands of particles through the flow.

With the advent of the laser, a number of flow measuring techniques were developed. All of these techniques are appplicable only to seeded flows; they rely for their operation on the detection of scattered light from these seed particles. They fall in the domain of laser anemometry, which offers certain advantages. It is a form of noncontact measurement so that proble interference to the flow is avoided; it has excellent spatial resolu-

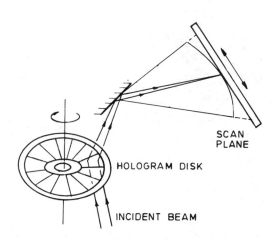

SCAN PLANE

HOLOGRAM DISK

INCIDENT BEAM

Figure 11.27   Holographic scanner using a zone plate. (From Ref. 22, used with permission.)

tion, it has very fast response fluctuating velocities; no transfer function is involved and the output voltage is linearly related to the velocity; it has a very large measurement range; and there are measurement possibilities in both liquid and gas flows. These features establish the superiority of the laser anemometer over the hot-wire anemometer. The laser anemometer does not measure the fluid flow but the motion of scatterers. The particles are, therefore, expected to follow the fluid flow faithfully. The particle density of the scatterers should not be less than 10^{10} particles per mm^3. The particle size in gases ranges from 1 to 5 μm and in water from 2 to 10 μm.

Laser anemometry may be performed in reference beam mode, the fringe mode, or the dual focus mode in the time-of-flight anemometer.

To measure both the velocity components simultaneously, modified versions of the anemometer are used. The measurements can be made in both the forward and the backward scattering directions. The intensity of backward scattering is many orders of magnitude lower than that of foward scattering, but some situations, where through ports are not possible, demand the use of backward scattering geometry.

11.4.1 Scattering from a Moving Participle

Consider a particle moving with velocity $\bar{V}$, illuminated by a wave with propagation vector $\bar{k}_i$. Let us observe the scattered wave with propagation vector $\bar{k}_0$. The frequency w_o of the observed wave is given by

$$w_o = w_i + n_0 \, \bar{V} \cdot (\bar{k}_0 - \bar{k}_i) \tag{11.15}$$

where w_i is the frequency of the illuminating wave and n_0 is the refractive index of the medium. The scattered light is Doppler Shifted. The Doppler shift $\Delta\upsilon_D = (w_o - w_i)/2\pi$ can be expressed as

$$\Delta\upsilon_D = \frac{2}{\lambda} \, n_0 V \sin \frac{\theta}{2} \sin \beta \tag{11.16}$$

where angles θ and β are shown in Fig. 11.28. When $\beta = \pi/2$, the Doppler frequency $\Delta\upsilon_D$ is

$$\Delta\upsilon_D = \frac{2}{\lambda} \, n_0 V \sin \frac{\theta}{2} \tag{11.17}$$

The velocity V of the flow is obtained from the relation

$$V = \frac{\lambda}{2n_0 \sin(\theta/2)} \Delta\upsilon_D \tag{11.18}$$

The Doppler frequency is measured by heterodyning the scattered signal with a reference signal.

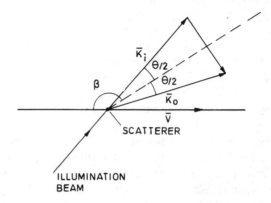

Figure 11.28   Scattering geometry.

Reference Beam Mode

Figure 11.29 shows the schematic of a reference beam mode laser ane-momter. The laser beam is divided into two beams, a reference beam and an illumination beam. The illumination beam is focused in the flow in the region of interest. The scattered beam is Doppler shifter. The Doppler shift is measured by adding a reference beam at the detector's surface. The reference beam need not traverse the flow region; it should , however, be collinear with the scattered beam. Since the scattered light is very weak, the illumination beam is made very strong compared to the reference beam. Let the reference beam be expressed as

$$E_r(t) = E_{ro} \exp{(iw_it)} \tag{11.19}$$

and the scattered beam be expressed as

$$E_s(t) = E_{so} \exp[i \, (w_i + (\bar{k}_0 - \bar{k}_i) \cdot \bar{V})t]$$

$$= E_{so} \exp[i(w_i + w_D)t] \tag{11.20}$$

We assume that the laser beam has sufficiently large coherence length so that the influence of any path difference between the two beams can be ignored. These two beams, when added on a photocathode, yield a photo-current i(t) given by

$$i(t) = B(E_{ro}^2 + E_{so}^2 + 2E_{ro}E_{so} \cos{w_Dt}) \tag{11.21}$$

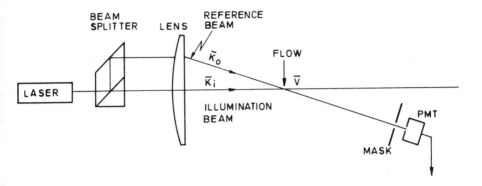

Figure 11.29   A schematic of a reference beam anemometer.

where B is a constant. Usually $E_{ro} \gg E_{so}$, and hence

$$i(t) = B(E_{ro}^2 + 2E_{ro}E_{so} \cos w_D t) \qquad (11.22)$$

This expression represents a photocurrent due to a single particle moving through an illuminating beam of an infinite size. In practice, the illuminating beam is of finite size, typically 0.1 mm, and hence the scattering particle has a finite residence time. Further, the laser beam will have a Gaussian intensity distribution. The photocurrent i(t) due to a single particle moving through the probe volume [24] will be

$$i(t) = B[E_{ro}^2 + 2E_{ro}E_{so}p(t) \cos w_D t] \qquad (11.23)$$

where p(t) is a blocking function such that $p(t) \neq 0$ when the particle is in the beam. In practice there will be N scattering centers distributed in the sample volume, and the photocurrent will then depend on the relative phases of all scattered waves.

The photocurrent signal is processed by a number of methods that will be discussed later.

Fringe Mode

Figure 11.30 (a) shows the schematic of fringe mode arrangement. Both beams are of equal intensity. The beams are superposed in the region where the flow velocity is to be measured. In the region of superposition, high-contrast intereference fringes called Young's fringes are formed. It is assumed that beam waists coincide. The fringe planes in the interfer-

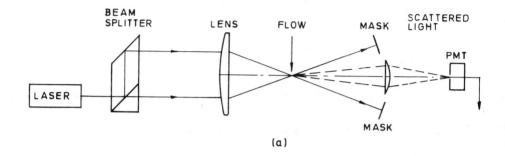

(a)

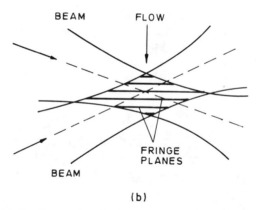

(b)

Figure 11.30   (a) A schematic of a fringe-mode laser anemometer and (b) fringe pattern in the overlap region.

erence pattern are perpendicular to the direction of flow, as shown in Fig. 11.30(b). The fringe spacing p is

$$p = \frac{\lambda}{2n_0 \sin(\theta/2)} = \frac{\lambda_m}{2 \sin(\theta/2)} \qquad (11.24)$$

where λ_m is the wavelength in the medium. The probe volume is obtained by the figure of revolution of ABCD. As the scatterer follows the flow, it encounters the fringe planes, i.e., the regions of maximum brightness. Whenever it is on a bright fringe it scatters, while there is no scattered field when it is on a dark fringe. Therefore the observed signal will have a frequency Δv_D where

$$\Delta v_D = \frac{V}{p} = \frac{2n_0 V}{\lambda} \sin \frac{\theta}{2} \qquad (11.25)$$

This equation may be compared with that obtained from Doppler shift considerations. In fact the configurations are equivalent. The fringe mode configuration is often called the differential Doppler technique. The scattered light due to each beam from the focused region is Doppler shifted, and the two reach the detector simultaneously. The frequencies of the scattered beams due to each illumination beam are different, and they beat to form the Doppler signal. An arrangement where the probe volume is illuminated by a single beam, and the scattered light is collected in two different directions, performs in the same way [28]. Figure 11.31 shows the photosignal. A particle tranversing the fringe volume formed due to two beams of equal intensity is an ideal signal shown in Fig. 11.31(a). An imperfectly modulated signal as shown in Fig. 11.31(b) results when the interfering beams are of unequal intensity or when a particle whose size is not smaller than the fringe spacing is moving through the fringe volume. The portion of the signal associated with the particle motion through the

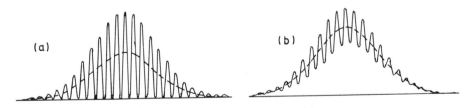

Figure 11.31 (a) A perfectly modulated and (b) an imperfectly modulated photosignal.

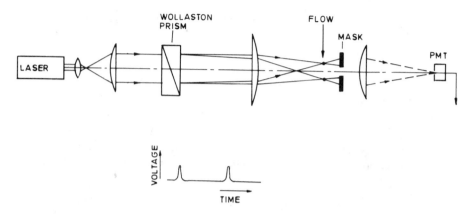

Figure 11.32   A schematic of a dual focus laser anemometer.

fringe volume is called the pedestal signal. It is normally removed by high pass filtering.

The analysis of the Doppler signal gives the velocity but not the sign. There are numerous situations where velocity reverses directions, so that it is necessary to obtain both magnitude and direction. This can be easily achieved by shifting the frequency of one of the interfering beams; the fringe planes thereby acquire a velocity that is either added to or subtracted from the particle velocity depending on the direction of motion [27,29]

Signal Processing

The Doppler signal usually takes two forms depending on the concentration of scatterers in the flow. For low particle concentration, the Doppler signal consists of a sequence of randomly occuring nonoverlapping pulses of random amplitude. These pulses appear whenever a scatterer crosses the probe volume. When the particle concentration is high, so that many particles cross the probe volume, continuous random amplitude signals result. The low-frequency components or pedestals are suppressed by a high-pass filter, which constitutes the first stage of a signal processing system.

There are a number of signal processing methods employed with a laser Doppler anemometer. The choice of a particular method is dictated by the signal-to-noise ratio of the input signal, the type of information required, for example, average frequency, instantaneous value of the frequency, etc, the accuracy of measurement required, and other factors. It is often difficult to recommend decisively a particular processing method. The most often used methods include frequency analysis, tracking , and counting. When the signal is extremely weak, the photon correlation method may be used [27,29].

The Time-of-Flight Anemometer

Another approach for velocity measurement [31,32] uses two small adjacent illuminated spots. These spots, in a simple setup, may be realized by passing plane wave through a small-angle Wollaston prism that angularly shears the wave. The two angularly sheared waves are focused to two tiny spots by a lens as shown in Fig. 11.32. The diameters of the spots at their waists are about 20 μm and their separation typically 200 μm. The detector receives light scattered from both spots. Thus a particle passing through both spots gives rise to a pair of pulses whose separation is measured. The velocity is then deduced from the measured transit time. Obviously the particle must cross both spots to yield data for velocity measurement. Since there are two spots in the anemometer, it is often called a two-spot or a dual-focus anemometer. The method works well for sparsely seeded flows. The relative merits of the laser Doppler anemometer and the dual-focus anemometers have been discussed by Lading [33]. It is however, pointed out, that in an environment where the plane wave on propagation gets distorted, and hence the spot size becomes very large, the dual-focus anemometer may yield erroneous results.

11.5 ELLIPSOMETRY

The technique of ellipsometry [34,35] involves the measurement of changes in the state of polarization of light upon reflection from a surface. For a clean reflecting surface, the optical constants of the surface may be calculated from these changes. But polarization state is influenced by the presence of thin films on the surface. The thickness of the thin film and its refractive index can therefore be measured. Physical and chemical absorption, oxidation, and corrosion also modify the state of polarization, and ellipsometry has contributed to their investigations [34]. Other areas where ellipsometry has been used are surface metrology [36,37], biology, and medicine [34].

Figure 11.33 shows the components of an ellipsometer. A parallel beams of monochromatic light is obtained from a laser or a monochrometer. The polarization state of the light before reflection from the sample surface is determined by the combination of polarizer and compensator.

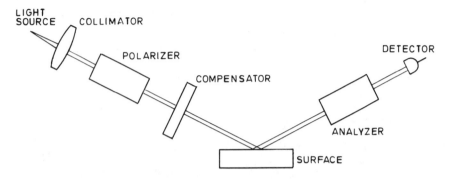

Figure 11.33   Schematic of an ellipsometer.

After reflection from the surface, the light beam passes through a second polarizer (an analyzer) and the transmitted light is sensed by a detector. The position of the compensator is arbitrary; it can as well be placed after reflection. The polarizers and the compensator are mounted on precision divided circles (resolution $\approx 0.01°$). The polarizer azimuth reads $0°$ when its transmission axis is in the plane of incidence of the sample. Similarly, the compensator azimuth reads $0°$ when its fast axis is in the plane of incidence.

Two widely used methods in ellipsometry are: null ellipsometry and photometric ellipsometry. Ellipsometric data in null ellipsometry are obtained by adjusting the polarization components for zero transmitted light intensity. In photometric ellipsometry, the time dependence of the transmitted intensity resulting from a periodic rotation of one of the components (e.g., the analyzer) is studied. In both cases the general expression for complex amplitude of the electric field vector of the light transmitted by the analyzer is the same. This can be obtained using the Jones calculus [34] and is given by

$$E_A = K[r_p \cos C \cos P^* \cos A - r_p \rho_c \sin C \sin P^* \cos A$$

$$+ r_s \sin C \cos P^* \sin A + r_s \rho_c \cos C \sin P^* \sin A] \qquad (11.26)$$

where K is a constant, r_p and r_s are the Fresnel coefficients, P and A are azimuths of the transmission axes of the polarizer and analyzer, C is the azimuth of the fast axis of the compensator, P^* $P- C$, $\rho_c = \rho_o \exp{(-i\delta)}$ is the transmittance ratio of the compensator, and

$$\rho = \frac{r_p}{r_s} = \left| \frac{r_p}{r_s} \right| \exp\left[i(\delta_p - \delta_s)\right] = \tan \psi \exp(i\Delta) \tag{11.27}$$

Here ψ and Δ are known as the ellipsometric angles of the reflector (surface).

11.5.1 Null Ellipsometry

In null ellipsometry the components are adjusted so that no light is transmitted through the analyzer. Setting $E_A = 0$, Eq. (11.26) gives

$$\rho = \frac{(\rho_c \tan P^* + \tan C) \tan A}{\rho_c \tan P^* \tan C - 1} \tag{11.28}$$

This equation can be used to determine ρ using ρ_c, and the nulling angles A, P, and C. In practice, however, a fixed compensator nulling scheme is most commonly employed. This makes the data reduction simple. In this scheme the compensator azimuth is set at $C = \pm\pi/4$ and the polarizer and analyzer are adjusted for null. The compensator is assumed to be an ideal-quarter-wave plate, i.e., $\rho_o = 1$ and $\delta = 90°$. For each setting of the compensator, two sets of polarizer and analyzer angle settings are obtained. These are (P_1/A_1) and (P_3/A_3) for $C = -\pi/4$ and (P_2/A_2) and (P_4/A_4) for $C = +\pi/4$. It can be shown from Eq. 11.28 that

$$\Delta = \frac{\pi}{2} + 2P_1, \qquad \psi = A_1$$

$$\Delta = \frac{-\pi}{2} + 2P_3, \qquad \psi = -A_3$$

$$\Delta = \frac{-\pi}{2} - 2P_2, \qquad \psi = A_2$$

$$\Delta = \frac{\pi}{2} - 2P_4, \qquad \psi = -A_4 \tag{11.29}$$

Thus the parameters of interest ψ and Δ can be detemined directly from the ellipsometer angle settings.

11.5.2 Rotating Analyzer Ellipsometry

In rotating analyzer ellipsometry the analyzer is rotated at a constant speed while other components are kept stationary. This results in a periodically varying detector signal, which is Fourier analyzed. The transmitted intensity $I(t) = E_A E_A^*$ is obtained from Eq. 11.26 and is given by

$$I(t) = \tfrac{1}{2}I_0(|r_s|^2 \sin^2 P + |r_p|^2 \cos^2 P)(1 + \alpha \cos 2A + \beta \sin 2A) \quad (11.30)$$

when $C = 0$. I_0 is the transmitted intensity in the straight-through position without the sample, and A is the instantaneous analyzer azimuth. α and β are normalized Fourier coefficients that are determined experimentally from the values of the dc, cosine, and sine components of the transmitted intensity. α and β are related as

$$\alpha + i\beta = \frac{\tan^2 \psi - \tan^2 P + i2 \tan \psi \tan P \cos (\Delta - \delta)}{\tan^2 \psi + \tan^2 P} \quad (11.31)$$

The desired parameters ψ and Δ are given by

$$\tan \psi = \sqrt{\frac{1 + \alpha}{1 - \alpha}} \tan P \quad (11.32)$$

$$\cos (\Delta - \delta) = \frac{\beta}{\sqrt{1 - \alpha^2 - \beta^2}} \quad (11.33)$$

Thus ψ and Δ can be determined from the knowledge of α, β, P, and δ. P and δ are known from the instrument setting while α and β are obtained by analyzing the detector signal.

11.5.3 Measurement of Optical Constants of a Bare Surface

The Fresnel coefficients for reflection from an interface between two isotropic media are given by

$$r_p = \frac{n_1 \cos \phi_2 - n_2 \cos \phi_1}{n_1 \cos \phi_2 + n_2 \cos \phi_1} \quad (11.34)$$

$$r_s = \frac{n_1 \cos \phi_1 - n_2 \cos \phi_2}{n_1 \cos \phi_1 + n_2 \cos \phi_2} \quad (11.35)$$

where n_1 is the refractive index of the first medium (usually air), n_2 is the refractive index of the second medium (the sample), and ϕ_1 and ϕ_2 are angles of incidence and refraction, respectively, at the interface. When the second medium is absorbing, n_2 is complex ($n_2 = n^* = n - ik$); n and k are called the optical constants of the sample. Substituting in Eq. (11.34) from Eq. (11.35) and rearranging, we obtain

$$\frac{1 - \tan \psi \exp (i\Delta)}{1 + \tan \psi \exp(i\Delta)} = \frac{\cos \phi_1 \sqrt{n_2 - n_1 \sin^2 \phi_1}}{n_1 \sin^2 \phi_1} \quad (11.36)$$

Separation of real and imaginery parts gives

$$n^2 - k^2 = n_1 \sin^2 \phi_1 \left[1 + \tan^2 \phi_1 \frac{(\cos 2\psi - i \sin 2\psi \sin \Delta)^2}{(1 + \sin 2\psi \cos \Delta)^2} \right] \quad (11.37)$$

$$2nk = \frac{n_1^2 \sin^2 \phi_1 \tan^2 \phi_1 \sin 4\psi \sin \Delta}{(1 + \sin 2\psi \cos \Delta)^2} \quad (11.38)$$

From Eqs. (11.37) and (11.38), n and k can be determined if ψ and Δ are measured at the angle of incidence ϕ_1 and n_1 is known.

11.5.4 Measurement of Optical Constants of a Thin Film

The case reflection of polarized light from surface covered by a single thin film is of considerable importance in ellipsometry. A film of thickness $d_1$is sandwiches between ambient and substrate (sample) media. The refractive indices of the ambient, film, and substrate are n_0, n_1 and n_2, respectively. The amplitudes of the light reflected from the film are given by

$$r_p = \frac{r_{01p} + r_{12p}\exp(-i2\beta)}{1 + r_{01p}r_{12p}\exp(-i2\beta)} \quad (11.39)$$

$$r_s = \frac{r_{01s} + r_{12s}\exp(-i2\beta)}{1 + r_{01s}r_{12s}\exp(-i2\beta)} \quad (11.40)$$

where the film phase thickness β is given by

$$\beta = \frac{2\pi}{\lambda} dn_1 \cos \phi_1 = \frac{2\pi}{\lambda} d(n_1^2 - n_0^2 \sin^2 \phi_0)^{\frac{1}{2}} \quad (11.41)$$

Here ϕ_0 and ϕ_1 are angles of incidence in the ambients and film media. In Eqs. (11.39) and (11.40), r_{01p}, r_{12p}, r_{01s}, and r_{12s} are the Fresnel coefficients at the ambient-film and film-substrate interfaces, respectively. We may therefore express ρ for the film as

$$\rho = \tan \psi \ wxp(-i\Delta) = \frac{r_p}{r_s} \quad (11.42)$$

where r_p and r_s are given by Eqs. (11.39) and (11.40). Eq. (11.42) may be functionally written as

$$\tan \psi \exp (i\Delta) = \rho(n_0, n_1, n_2, d, \phi_0, \lambda) \quad (11.43)$$

Eq. (11.43) can be separated into real and imaginary parts as

$$\psi = \tan^{-1}| \ \rho(n_0, n_1, n_2, d, \phi_1, \lambda)| \quad (11.44)$$

$$\Delta = \arg \rho(n_0, n_1, n_2, d, \phi_0, \lambda) \quad (11.45)$$

Eqs. (11.44) and (11.45) are quite complicated and require electronic computation. In general, ρ depends on nine real parameters including real and imaginary parts of n_0, n_1, and n_2, d, ϕ_0, and λ. In special cases simpler solutions are possible. For example, if a transparent (non-absorbing) film is deposited on a known substrate, a graphical solution is feasible. Values

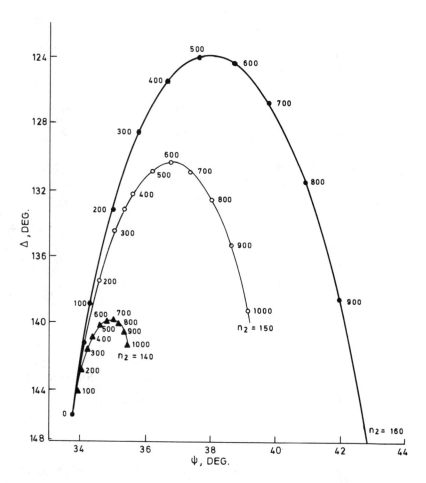

Figure 11.34   A (4, Δ) graph for different thicknesses of three materials (n_2= 1.40, 1.50, and 1.60) on a chrome substrate in an ambient medium of refractive index 1.359. (From Ref. 38, used with permission.)

of ψ and Δ can be calculated as a function of thicknesses of the film for various refractive indices of the film. Equal refractive index contours for various thicknesses can thus be obtained. Typical results are shown in Fig. 11.34, which shows ψ and Δ for films of refractive indices 1.4, 1.5, and 1.6 and thicknesses of 0 to 1000Å on a chromium substrate. The refractive index of the ambient medium is 1.359.

11.5.5 Surface Roughness

The surface roughness of a sample influences its ellipsometric angles ψ and Δ. Hence studies have been made in the use of ellipsometry for the measurement of surface roughness, and useful results have been obtained [11,36,37].

REFERENCES

1. Vyce, J. R. (1969), Non-contacting surface sensor (optical probe), *Appl. Opt. 8:* 2301–2310.
2. Luxon, J. J., and Parker, D. E. (1985). *Industrial Lasers and their Applications,* Prentice Hall, Englewood Cliffs, NJ, Chap. 9.
3. Thwaite, E. G. (1984). "Advances in optical methods of measuring form and Roundness," All India Machine Tool Design and Research Conference, December 20–22. madras, India.
4. Tanwar, L. S., and Kunzmann, H. (1984). An Electrooptical sensor for micro-displacement and control. *J. Phys. E: Sci. Instrum.,* 17: 364–366.
5. Schimokohbe, A., Osada, H., and Gotoh, K. (1985), An optical non-contact probe, *Precision Engineering,* 7: 195–200.
6. Kohno, T., Ozawa, N., Miyamoto, K., and Musha T. (1985). Practical non-contact surface measuring instrument with one nanometer resolution, *Precision Engineering,* 7: 231–232.
7. Fainman, Y., Lenz, E., and Shamir, J. (1982). Optical profilometer: A new method for high sensitivity and wide dynamic range, *Appl. Opt.,* 21: 3200–3208.
8. Dobosz, M. (1983). Optical profilometer: a practical approximate method of analysis, *Appl. Opt.,* 22: 3983-3987.
9. Lou D. Y., Martinez, A., and Stanton, D. (1984). Surface profile measurement with a dual beam optical system, *Appl. Opt,* 23: 746–751.
10. Whitefield, R. J. (1975). Non-contact optical profilometer, *Appl. Opt.,* 14: 2480–2485.
11. Vorburger, T. V., and Teague, E. C. (1981). Optical techniques for on-line measurement of surface topography, *Precision Engineering,* 3: 61–83. Most of the discussion in Sec. 11.2 is based on this review article.
12. Bennett, J. M. (1976). Measurement of the RMS roughness, autocovariance function and other statistical properties of optical surfaces using a FECO scanning interferometer, *Appl. Opt.,* 15: 2705–2721.

13. Downs, M . J. McGivern, W. H., and Ferguson, H. J. (1985). Optical system for measuring the profiles of super smooth surfaces, *precision Engineering, 7:* 211–215.

14. Bennett, J. M. (1985). Comparison of techniques for measuring the roughness of optical surfaces,*Opt. Eng. 24:* 380–387.

15. Church, E. L. and Takais, P. Z. (1985). Survey of the finish characteristics of machined optical surfaces, *Opt. Eng. 24:* 396–403.

16. Stover, J. C., Serreti, S. A., and Gillespie, C. H. (1984). Calculation of surface statistics from light scatter, *Opt. Eng. 23:* 406–412.

17. Detrio, J. A., and Miner, S. M. (1985). Standardized total integrated scatter measurements of optical surfaces, *Opt. Eng. 24:* 419–422.

18. Ahlers, R. J. (1985). White light method: A new sensor for optical evaluation of rough surfaces, *Opt. Eng. 24:* 423-427.

19. Bennett, J. M. (1985). Scattering and surface evaluation techniques for the optics of future, *Optics News, 11:* 17–27.

20. Kingslake, R., and Thompson, B. J., eds. (1980). *Applied Optics and Optical Engineering,* Academic Press, New York. Chap. 6.

21. Gottliebs, M., Ireland, C. L. M. and Ley, J. M. (1983). *Electro-Optic and Acousto-Optic Scanning and Deflection,* Marcel Dekker, New York.

22. Rallison, R. (1984). Applications of holographic optical elements, *Lasers and Applications, 3:* 61 68.

23. Beiser, L. (1985). Laser scanning and recording: Development and trends, *Laser Focus, 21:* 88–96.

24. Marshal, G. F. (1985). *Laser Beam Scanning,* Marcel Dekker, New York.

25. Beiser, L. (1988). *Holographic Scanning,* John Wiley, New York.

26. Watrasiewicz, B. M., and Rudd, M. J. (1976).*Laser Doppler Measurements,* Butterworth, London.

27. Durst, F., Melling, A. and Whitelaw, J. H. (1976) *Principles and practice of Laser-Doppler Anemometry,* Academic press, New York.

28. Durrani, T. S., and Greated, C. A. (1977). *Laser Systems in Flow Measurement, Plenum Press,* New York.

29. Drain, L. E. (1980), *The Laser Doppler Technique,* John Wiley, Chichester.

30. Durst, F. and Whitelaw, J. H. (1971). Optimization of optical anemometers, *Proc. Royal Soc A. 324:* 157–181.

31. Tanner, L. H. (1973). A particle timing laser velocity meter, *Optics and Laser Technology, 5:* 108–110.

32. Lading, L., Jensen, A. S., Fog, C., and Andersen, H. (1978). Time-of-flight laser anemometer for velocity measurements in the atmosphere, *Appl. Opt. 17:*1486–1488.

33. Lading, L. (1979). *Laser Velocimetry and Particle Sizing* (H.D. Thompson and W. K. Stevenson, eds.), Hemisphere, Washington.

34. Azzam, R. M. A., and Bashara, N. M. (1977). *Ellipsometry and Polarized Light,* North Holland.

35. Aspnes, D. E. (1976). *Optical Properties of Solids* (B.O. Seraphin, ed.), American Elsevier, New York.

36. Blanco, J. R., McMarr, P. J. and Vedam, J. (1985). Roughness measurement by

spectroscopic ellipsometry, *Appl. Opt. 24:* 3773–3779
37. Nee, S-M.F. (1988). Ellipsometric analysis of surface roughness and texture, *Appl. Opt. 27:* 2819–2831.
38. McCrackin, F. L., Passaglia, E., Stromberg, R. R., and Steinberg, H. L. (1963). Measurement of thickness and refractive index of very thin films and the optical properties of surfaces by ellipsometry, *Journal of Research, National Bureau of Standards A: Physics and Chemistry, 67A:* 363–377.

Index